Springer Tracts in Electrical and Electronics Engineering

Series Editors

Springer Tracts in Electrical and Electronics Engineering (STEEE) publishes the latest developments in Electrical and Electronics Engineering - quickly, informally and with high quality. The intent is to cover all the main branches of electrical and electronics engineering, both theoretical and applied, including:

- Signal, Speech and Image Processing
- Speech and Audio Processing
- Image Processing
- Human-Machine Interfaces
- Digital and Analog Signal Processing
- Microwaves, RF Engineering and Optical Communications
- Electronics and Microelectronics, Instrumentation
- Electronic Circuits and Systems
- Embedded Systems
- Electronics Design and Verification
- Cyber-Physical Systems
- Electrical Power Engineering
- Power Electronics
- Photovoltaics
- Energy Grids and Networks
- Electrical Machines
- Control, Robotics, Automation
- Robotic Engineering
- Mechatronics
- Control and Systems Theory
- Automation
- Communications Engineering, Networks
- Wireless and Mobile Communication
- Internet of Things
- Computer Networks

Within the scope of the series are monographs, professional books or graduate textbooks, edited volumes as well as outstanding PhD theses and books purposely devoted to support education in electrical and electronics engineering at graduate and post-graduate levels.

Review Process

The proposal for each volume is reviewed by the main editor and/or the advisory board. The books of this series are reviewed in a single blind peer review process.

Ethics Statement for this series can be found in the Springer standard guidelines here https://www.springer.com/us/authors-editors/journal-author/journal-author-helpdesk/before-you-start/before-you-start/1330#c14214

Koichi Mori • Yasuhisa Oda •
Takahashi Masayuki • Kohei Shimamura
Editors

Beamed-mobility Engineering

Wireless-power Beaming to Aircrafts, Spacecrafts and Rockets

Springer

Editors
Koichi Mori
Department of Aerospace Engineering
Osaka Prefecture University
Sakai, Osaka, Japan

Yasuhisa Oda
Department of Mechanical Engineering
Setsunan University
Neyagawa, Japan

Takahashi Masayuki
Department of Aerospace Engineering
Tohoku University
Sendai, Japan

Kohei Shimamura
Department of Aeronautics and Astronautics
Tokyo Metropolitan University
Tokyo, Japan

This work contains media enhancements, which are displayed with a "play" icon. Material in the print book can be viewed on a mobile device by downloading the Springer Nature "More Media" app available in the major app stores. The media enhancements in the online version of the work can be accessed directly by authorized users.

ISSN 2731-4200 ISSN 2731-4219 (electronic)
Springer Tracts in Electrical and Electronics Engineering
ISBN 978-981-99-4617-4 ISBN 978-981-99-4618-1 (eBook)
https://doi.org/10.1007/978-981-99-4618-1

This Springer imprint is published by the registered company Springer Nature Singapore Pte Ltd.
The registered company address is: 152 Beach Road, #21-01/04 Gateway East, Singapore 189721, Singapore

Springer Nature More Media App

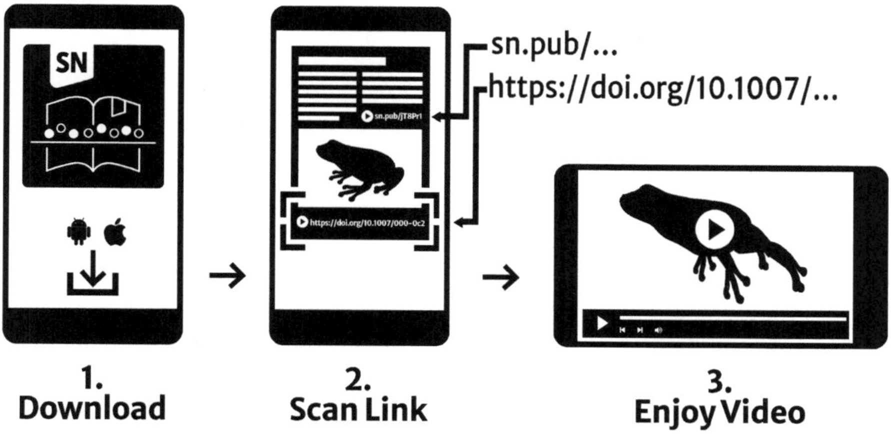

Support: customerservice@springernature.com

Foreword

Despite the trend away from fossil fuels and the international efforts to stop global warming, aerospace vehicles have no choice but to carry large amounts of fuel for takeoff and cruse, and this constraint is most severe for rocket launchers, which currently achieve a payload ratio of only about 1% to transport a satellite into geostationary orbit. As a means of freeing them from this constraint, attempts to supply propulsion power to the aerospace vehicles wirelessly using electromagnetic beams have shown promise, and various forms of beamed energy propulsions and laser/microwave power beaming technologies have been proposed and investigated.

This book is a collection of articles written by the professors active in electric propulsion, hypersonic flight vehicles, and power beaming technologies, with the goal of contributing to efficient transportation and environmental sustainability through wireless power transfer technology.

I hope that in the near future, these technologies will work together to create a new world of mobility.

Tokyo, Japan
August 2022

Kimiya Komurasaki

Preface

This book describes the technologies of wireless power beaming to aerospace crafts, such as the drone, flying car, aircraft, spacecraft, and rocket. Beamed mobility is the ability to move aerospace craft using a highly directional electromagnetic beam whose energy is efficiently converted to driving force. This technology will be indispensable for the full electrification of mobility in the future.

This book specializes in aerospace mobility, where weight and other constraints are strict, and was written by researchers in different disciplines such as rocket engineering, plasma engineering, laser engineering, and communications and control engineering. Beamed mobility forms a new area of integrated engineering. It combines optics and mechanical engineering to create a world where aerospace craft are mobilized by freely available energy for propulsion, whenever and wherever needed. It is expected to become the core technology of mobility, energy, infrastructure, and services of the future society that extends to outer space.

Beamed mobility has a long history. Rocket scientist A. Kantrowitz already proposed in 1972 a launch method that supplies propulsion energy from the ground using a laser beam. In recent years, the use of microwaves and millimeter waves has spread to wireless energy transmission to electric and electronic devices, cars, and aircraft. In Chap. 1 of this book, the latest research trends on this technology are explained. The tracking technology using a laser is introduced in Chap. 9.

Laser light has high directivity and is suitable for application to space propulsion. Advantages such as energy efficiency, cleanliness, and safety are attractive. However, the construction cost of the laser transmission base is enormous, and it cannot be said that the launch system will be completed at a low cost. On the other hand, although the directivity of millimeter waves is inferior to that of lasers, the gyrotron is based on vacuum tube technology similar to that of magnetrons, and its most attractive feature is that it can be manufactured several orders of magnitude cheaper than laser equipment. It is introduced in Chap. 8, and propulsion equipment is introduced in Chaps. 2 and 3.

The most important physical mechanism in this propulsion system is a phenomenon called detonation, in which laser or millimeter wave energy is directly converted into fluid pressure. Research on millimeter-wave discharge propagation was reported for the first time around 1988, and the history of research is still short. The cutting-edge research was introduced in Chaps. 4–7 of this book. Also, in Chap. 10, an energy conversion system using a non-detonation type high-power diode, which has been developed for industrial use in recent years, is introduced.

This book is good reference for graduate students, researcher, and engineer in the field of aerospace, electrical, and mechanical engineering.

Osaka, Japan — Koichi Mori
Osaka, Japan — Yasuhisa Oda
Tokyo, Japan — Kohei Shimamura
Miyagi, Japan — Masayuki Takahashi
September 2022

This work contains media enhancements, which are displayed with a "play" icon. Material in the print book can be viewed on a mobile device by downloading the Springer Nature "More Media" app available in the major app stores. The media enhancements in the online version of the work can be accessed directly by authorized users.

Contents

Chapter 1
Wireless Power Transfer Via Microwave for Mobilities

Kohei Shimamura and Maho Matsukura

Abstract This chapter provides a comprehensive review of microwave wireless power transmission, focusing on its application to the aircraft including the drone and unmanned aerial vehicles (UAVs). Focusing on the key element of these systems, the rectifier circuit, this work outlines the principles of rectification, application of high-frequency rectifiers, and specific band rectifier circuits. Furthermore, the chapter discusses the comparative advantages of wireless power transfer over battery systems for UAVs. It also delves into the feasibility of microwave-powered flight, examining factors such as beam collecting and capture efficiency, and RF-DC conversion efficiency. The dependency of operation frequency and the design of mobile devices for flight demonstration are analyzed as well. The research herein contributes significantly to the advancement of UAV technology and energy harvesting.

Keywords Microwave · Millimeter wave · Wireless power transmission · UAVs · Drone · Rectifier circuits · RF-to-DC conversion · High-frequency rectifiers · Flight demonstration, Beam tracking technology

Introduction

This chapter introduces the current status and issues of wireless power transfer technology using electromagnetic waves such as microwaves and lasers for airplane, drone, and other atmospheric mobility. This chapter will focus on mobility that requires high energy density, such as drone, flying car, airplane, and rocket. In

K. Shimamura (✉)
Department of Aeronautics and Astronautics, Tokyo Metropolitan University, 6-6 Asahigaoka, Hino, Tokyo, Japan
e-mail: shimamura@tmu.ac.jp

M. Matsukura
Department of Aerospace Engineering, Tohoku University, 6-6-01 Aramaki-Aza, Aoba, Sendai, Miyagi, Japan
e-mail: maho.matsukura.d8@tohoku.ac.jp

K. Mori et al. (eds.), *Beamed-mobility Engineering*, Springer Tracts in Electrical and Electronics Engineering, https://doi.org/10.1007/978-981-99-4618-1_1

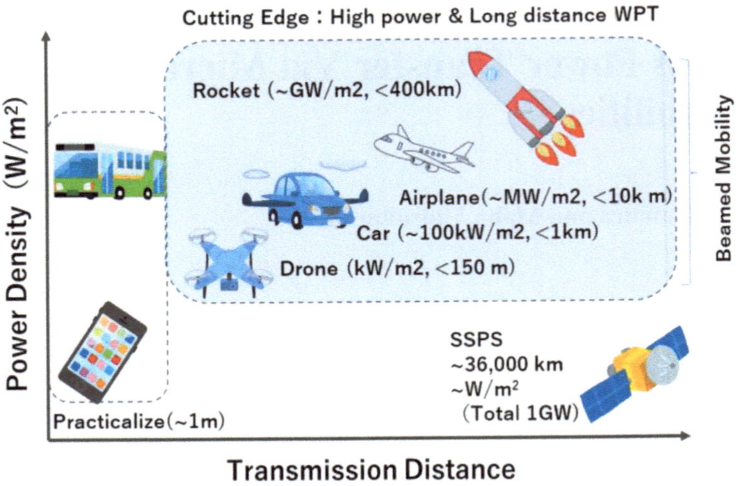

Fig. 1.1 Wireless power transfer for mobility: power density vs. transmission distance

particular, drone and flying mobility are now attracting attention due to the technological innovation of gyro sensor and Li-ion battery. Beamed mobility is an ideal technology because it can reduce the onboard fuel to zero and maximize the payload in principle. Figure 1.1 summarizes the power density and ground to object distance required for various mobilities.

The most historical and traditional wireless power transfer application via microwave is the space solar power satellite (SSPS). The power density is not so high because the beam radius is a few kilometers, while the transmission distance is 36,000 km from the Earth's surface to geostationary orbit. Wireless power transfer, which is relatively close to the ground and mobility such as cars and buses, has been studied in recent years by magnetic resonance coupling technology (Suh et al. 2015; Cabrera et al. 2020; Rim and Mi 2017). In this chapter, we focus on drone and micro aerial vehicles (MAV), which have relatively small energy density. This chapter would like to focus on the mobility of drone, micro aerial vehicles (MAV), flying drone, etc., which have relatively small energy density and can extract energy electrically at receiver side. For higher density rockets, plasmas, shock waves, and heat exchange technologies will be used, which will be introduced in the later chapters.

Table 1.1 summarizes the main methods of wireless power transfer and their applications and performance. Electromagnetic induction and magnetic resonance methods are classified as non-radiative wireless power transfer methods that transmit power over relatively short distances, while microwave wireless power transfer is a method that transmits power in the radio frequency (RF) domain and is classified as radiated wireless power transfer.

Microwave wireless power transfer is a wireless power transfer method that uses a rectenna, which is a device that integrates an antenna and a rectifier circuit, to extract microwave power irradiated from the outside directly as DC power. One of the problems for microwave wireless power transfer is that the power transfer efficiency

Table 1.1 Summary of main methods of wireless power transmission and its applications and performance

	Trans. distance	Trans. power	WPT efficiency (%)
Electric field coupling	Very short (~ cm)	<1 kW	<90
Magnetic field coupling	Short (<1 m)	<1 MW	<90
Microwave	Far (<100 m)	<100 kW	<20
Laser	Far (<10 m)	<1 kW	<20–30

is lower than that of the non-radiative type. The efficiency of microwave and laser wireless power transfer is less than 10%, while non-radiative type is expected to be around 90%. This is because there is a large loss in the process of converting DC to electromagnetic waves or electromagnetic waves to DC and also because the beam leaks outside the receiving antenna if the directivity is low. In addition, in microwave systems, the loss due to reradiation on the receiving circuit tends to be high in proportion to the operating frequency. The other problem is radio interference with communication systems. The higher the transmission frequency, such as millimeter wave or terahertz wave, the higher the directivity, and the easier it is to solve the radio interference problem.

Review of Microwave Power Transmission to UAVs

Microwave or millimeter-wave power transmission (MPT) is useful in this regard as it can provide charging in flight without reducing the payload. In particular with UAV logistics, MPT is expected to be a power source that does not compress the payload. Considering that typical flight times depend on the battery capacity, which is generally proportional to the weight, there is a trade-off between the extension of the flight time and the payload. For applications that require a certain payload, increasing the battery capacity for long flights is not effective. By combining UAVs with MPT, long-distance and quick transportation is possible. Various MPT experiments have been demonstrated in the frequency range of 2.45–28 GHz, as summarized in Table 1.2. This chapter reviews it from the viewpoint of application to flight mobility. Note that the review of microwave power transmission by Rodenbeck et al. is well summarized (Rodenbeck et al. 2021). In 1963, Brown et al. conducted an experiment of microwave power transmission to a small helicopter. In this experiment, 5 kW of 3.0 GHz microwaves were transmitted and received by a rectenna with an area of about 0.4 m^2 installed underneath the small helicopter at a height of 5.5 m. A parabolic antenna with a diameter of about 3 m was used as the transmission antenna. As a result, the transmission efficiency was 15%, and the received power was 750 W (Brown 1969). In 1992, Matsumoto et al. at Kyoto University conducted a power transfer experiment using a non-fueled airplane called MILAX. In this experiment, 1.25 kW of 2.45 GHz microwaves were transmitted and received by a 120-element rectenna installed underneath a fixed-wing

Table 1.2 Summary of rectenna based beamed mobility with flight demonstration (Nex and Remondino 2014; Han et al. 2016; Zhang and Kovacs 2012; Torresan et al. 2017; Song et al. 2018; Gur and Rosen 2009; Traub 2011; Avanzini and Giulietti 2013; Pines and Bohorquez 2006; Mukherjee et al. 2010; Brown 1969; Shinohara 2013; Schlesak et al. 1988; Fujino and Fujita 1998; Shimamura et al. 2017; Rodenbeck et al. 2021; Nguyen et al. 2020; Suganuma et al. 2021)

	Operation frequency GHz	Tx power W	WPT efficiency %	WPT distance m	Beam control system	Mobility control	Type of mobility	Gross weight kg
Brown (1963)	3.0	5 k	15	5.5	No	Fixed	Helicopter	2.3
MILAX (1992)	2.45	1 k	10	10	Yes	Manual	Airplane	4.0
SHARP (1987)	2.45	1 k	1.5	1.5	No	Manual	Airplane	4.1
ETHER (1995)	2.45	10 k	–	50	Yes	Manual	Airship	22.8
K. Shimamura (2017)	5.8	5	0.6	1.5	Yes	Fixed	MAV	0.1
Song (2019)	10	200	5.1	6	No	Manual	Airship	1.0
METI (2019)	5.8	1.6 k	2.6	30	Yes	Auto	Drone	
D. Hung (2020)	28	1	0.2	0.8	No	IFT	Drone	0.42
S. Suganuma (2021)	28	5	0.4	0.8	Yes	IFT	Drone	0.42

UAV flying 10 m above the ground. As a result of the experiment, the maximum power received during the flight was measured to be 88 W. Furthermore, the motor was driven by the power obtained from the rectenna, and the UAV successfully flew over the car loaded with the power transmission device while following it by image processing using a CCD camera (Shinohara 2013). In 1987, J.J. Schlesak et al. conducted an experiment in which an airplane with a wingspan of 4.5 m and a weight of 4.1 kg was irradiated with a microwave beam and flown. They used a 4.5-m-diameter parabolic antenna for the transmitter and an array of dual polarization rectennas for the receiver. As a result of the experiment, a total power transmission efficiency of 1.5% was measured, and enough power was supplied to fly a 4.1 kg airplane (Schlesak et al. 1988). In 1995, Fujino et al. built an airship called ETHER and conducted microwave power transmission experiments. A parabolic antenna was used on the transmission side, and a 3 m × 3 m rectenna with 1200 elements was used on the receiving side. A parabolic antenna was used on the transmitter side, and a 3 m × 3 m rectenna with 1200 elements was used on the receiver side (Fujino and Fujita 1998). In 2017, Shimamura conducted an experiment on feeding a lightweight felt rectenna using the frequency of 5.8 GHz band (Shimamura et al. 2017). They used an arrayed horn antenna on the transmission side and an arrayed felt rectenna on the receiving side for the experiment. In this experiment, it was shown that a fixed-wing UAV flying at an altitude of 50 m can be flown with microwave power below 5.8 GHz. However, it was concluded that it was difficult for a rotary-wing UAV to fly with microwave power below 5.8 GHz. This is due to the high power consumption of the rotary-wing UAV and the small footprint of the power receiving rectenna. Therefore, a microwave power supply of 5.8 GHz or higher is necessary to realize the flight of the rotary-wing UAV. In 2019, MEXT and JAXA conducted a microwave power transmission experiment using drones. In this experiment, a frequency of 5.8 GHz and a power of 1600 W were transmitted, and the drone hovering in the sky was fully controlled automatically. A phased array antenna with controllable beam direction was used for the transmitter, and a 200 mm × 186 mm arrayed circular microstrip antenna was used for the receiver. The experimental results show that the DC power is about 60 W at 10 m altitude and about 42 W at 30 m altitude (Rodenbeck et al. 2021). In 2020, Hung et al. conducted a microwave power transmission experiment at a frequency of 28 GHz for an autonomously controlled small flying vehicle. In this experiment, 28 GHz frequency and 2.04 W power are transmitted vertically upward, and a rotor-wing UAV is controlled autonomously using the IFT method to hover over the microwave beam and feed power efficiently. A single horn antenna is used for the transmitter, and a four-element microstrip patch antenna of 18 mm × 15 mm is used for the receiver. The experimental results show that the total transmission efficiency is 0.12% at a height of 80 cm (Nguyen et al. 2020). In 2021, Suganuma et al. conducted an experiment of microwave power transmission to a drone at 28 GHz (Suganuma et al. 2021). In this experimental system, 28 GHz frequency and 5.3 W power are transmitted and received by a rectenna installed under the drone. The antennas are arranged so that the beam radiates in the z-direction and is set to track the drone only in the x-direction by controlling the phase of the beam. The phase control is performed so that the beam tracks the drone. After taking off from the ground, the drone is controlled to hover for 30 s at 80 cm above the transmission antenna, where the ground effect is negligible. Two arrays of horn antennas are used for the transmission side, and four arrays of four-element patch antennas are used for the receiver side. The results show that the drone succeeded in providing almost uninterrupted microwave power for about 30 s of flight time. The maximum power output during the flight was at 15.4 s, and the DC power of 47.5 mW was measured. At 15.4 s, the maximum power transmission efficiency of 0.88% was measured. This result shows that the DC power and the efficiency of power transmission are low at present. In addition, although no flight demonstration has been confirmed, we will introduce the research on the system with fixed power transmission.

Ye et al. (2016) developed a rectenna that consisted of a 16-array patch antenna at 2.33 GHz and a rectifier circuit with a maximum rectification efficiency of 70% and conducted a power transmission experiment on a fixed-rotary UAV at 0.4 m between the transmitter and receiver. As a result, 14.3 W was received with a 110 W power transmission, giving a 13% conversion efficiency. Li et al. (2017) fabricated rectennas that consisted of a patch antenna at 2.45 GHz and a rectifier circuit with a maximum rectification efficiency of 46.5%, connected a power control unit to a 12-array rectenna, and performed power transmission experiments at distances between antennas of 1–4 m. As a result, an efficiency of 0.71% was recorded at 1 m and 0.055% at 4 m. They compared the results with analytic values using the Friis transmission equation, but those results deviated from their experimental values. Hashimoto et al. (2018) performed a power transmission experiment at 4 m using a 3-array rectenna consisting of a patch antenna and a rectifier circuit with a maximum RF-DC efficiency of 58% at 2.45 GHz. As a result, 0.8 W was output as DC at a 53 W power transmission, which successfully rotated the UAV rotor. Tornerro et al. (2016) fabricated an integrated LW antenna, multiplexer, and rectifier circuits to power small UAVs at 16.15–16.35 GHz. The power transmission horn antenna swiveled, and the power receiving the LW antenna achieved a scanning angle of 120° via electronic scanning.

Finally, some laser-powered UAVs are introduced here. Laser energy conversion mainly uses a solar panel as the power receiver. It is characterized by its high directivity compared to microwave. The energy level is a few watts and the receiver DC conversion efficiency is approximately 40–50%. Major examples include NASA's demonstrations and computation and the kite plane by Kawashima et al. (Blackwell 2005; Nugent and Kare 2010; Kawashima and Takeda 2008).

Receiver Systems

Principle of Rectifier

An overview of the microwave wireless transmission system is shown in Fig. 1.2. In designing this microwave transmission system, the characteristics of the rectifier circuit that converts the RF from the receiving antenna to DC are an important factor. Most rectifier circuits use semiconductor diodes and effectively use higher harmonics to achieve high efficiency. In addition, since the rectifier must have a high efficiency of about 80–90%, single-shunt rectifier circuits are often used. Figure 1.3 shows the most general diagram of a single-shunt rectifier circuit and the equivalent circuit of the diode. As shown in Fig. 1.3, the single-shunt rectifier circuit is a simple full-wave rectifier circuit with a single diode, a λ/4 transmission line, and a capacitor component on the output side (Ohira 2013; Gutmann and Borrego 1979).

Since the diode itself has resistance and parasitic capacitance, leading to losses, many circuits have one or two diodes per circuit. Conversion efficiency is expressed as the ratio of input RF power to output DC power. The conversion efficiency varies with input power and is highest near the diode breakdown voltage. Theoretical efficiency based on the equivalent circuit of a rectifier circuit says that 100% rectification is possible, but the efficiency tends to decrease, especially at higher frequencies, because of frequency-dependent diode switching losses, losses in resistors, and dielectric and conductor losses on the substrate (Yoo and Chang 1992). Figure 1.4 shows the operating frequencies and converting efficiencies of the circuits of major interest among the previous studies. In this section, high-

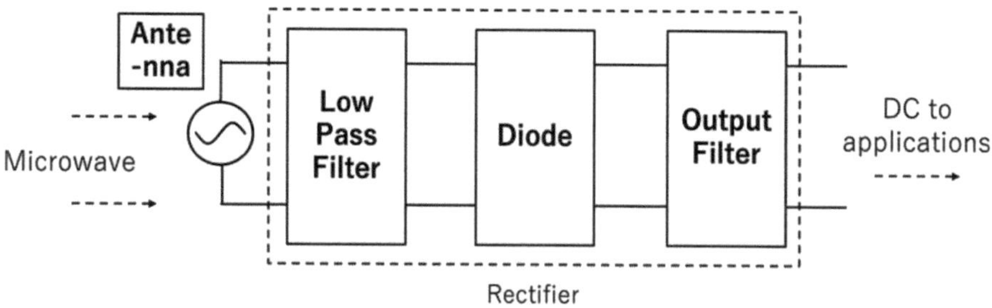

Fig. 1.2 Role of the rectifiers in wireless transmission systems

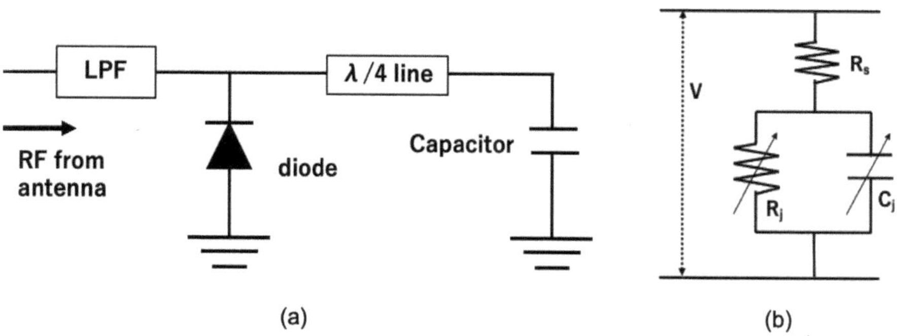

Fig. 1.3 Basic configuration of a single-shunt rectifier circuit (**a**) and typical equivalent circuit of a diode (**b**). Rs is parasitic series resistance, Cj is parasitic junction capacitance, and Rj is junction resistance

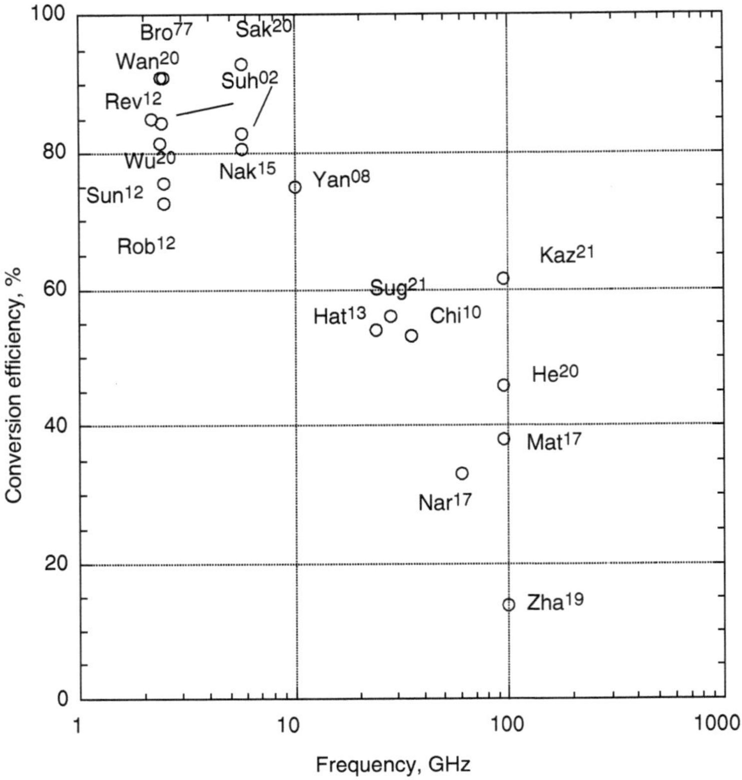

Fig. 1.4 Maps of conversion efficiency and frequency of previous studies (Suganuma et al. 2021; Brown 1977; Wang et al. 2021a; Reveyrand et al. 2012; Wu et al. 2020; Sun et al. 2012; Roberg et al. 2012; Sakai et al. 2020; Suh and Chang 2002; Nakamura et al. 2016; Yang et al. 2008; Hatano et al. 2013; Chiou and Chen 2010; Nariman et al. 2017; Kazemi 2022; He et al. 2020; Matsui et al. 2017; Zhao et al. 2019)

frequency circuits, high-efficiency operating circuits, and high-power operating circuits are presented.

High-Frequency Rectifiers (Millimeter Wave)

To achieve high conversion efficiency at high frequencies, it is important to use diodes with high switching operation (with high cutoff frequency) and low capacitance. At the same time, to achieve high-power operation, it is necessary to have a high breakdown voltage. Table 1.3 summarizes previous studies of high-frequency circuits with respect to diode type and power. Diodes using Si-CMOS technology have been used in many millimeter-wave frequency multipliers and mixers because of their high switching speeds (Chiou and Chen 2010) but are not suitable for high-power operation due to the breakdown voltage of their material properties. GaAs diodes are suitable for milli and sub-milli frequency operation due to their fast-switching frequency and moderate breakdown voltage (Kazemi 2022).

Methods of Operating Diodes at High Frequencies

First select a diode size that ensures sufficient voltage amplitude at the diode input and maintains a relatively low input impedance for concise input matching. Optimize diode finger width and number to improve cutoff frequency.

Matching circuit between antenna and diodes. A DC pass filter is provided on the output side to supply clean DC. The Schottky diode has an almost infinite impedance at the fundamental frequency (Zhao et al. 2019).

In 2015, Hemour et al. of Canada fabricated a 94 GHz rectenna for use in an I-SWARM micro-robot. Although the matching frequency was lower than the simulated design due to over-edging, resulting in a resonant frequency of 90 GHz, the conversion efficiency of the rectifier circuit was 37.7% for an input power of 3 dBm and 32.3% at 94 GHz. The rectifier size and the rectenna size are 1.17 mm^2 and 5.62 mm^2, respectively (Hemour et al. 2015). A 94 GHz rectifier circuit was developed with the conversion efficiency of 38% at 105 mW input power and a power density of 2.38 kW/m^2, and the GaAs diode was used (Matsui et al. 2017). At 99 GHz, Zhao et al. achieved the rectifier with the conversion efficiency of 13.8% at 22 dBm input. The rectifier circuit was fabricated in 100-nm InGaAs technology

Table 1.3 Summary of previous research on high-frequency rectifiers with respect to diode types and power (Nariman et al. 2017; Kazemi 2022; He et al. 2020; Matsui et al. 2017; Zhao et al. 2019; Hemour et al. 2015; Weissman et al. 2014; Mizojiri et al. 2018; Shaulov et al. 2017)

	Frequency GHz	Conversion efficiency %	Input power mW	Output power mW	Diode
Nariman (2017)	60	32.8	3.7	1.2	CMOS
Kazemi (2021)	95	61.5	9.0	5.7	GaN nano-SBD
P. He (2020)	94	45.8	10	4.58	40 nm bulk CMOS
Matsui (2017)	94	38	105	39	GaAs SBD
Zhao (2019)	99	13.8	158	21.8	SBD
Hemour (2015)	94	32.3	2.0	0.65	GaAs SBD
Weissman (2014)	94	10	2.8	0.28	65 nm CMOS
Mizojiri (2018)	303	2.17	342	7.4	GaAs SBD
Shaulov (2017)	94	20.5	73	15	Mott diode

(Zhao et al. 2019). In 2021, at a frequency of 95 GHz, H. Kazemi et al. in the USA presented a rectifier circuit with the conversion efficiency of 61.5% and an input power density of 3.61 kW/m^2. According to the paper, the power density was expected to be higher because the input power of the circuit could still be increased. The forward current was 1.8 A/mm and the breakdown voltage was 30 V. The diode was a GaN-based nano-Schottky diode. The cutoff frequency of this diode was 928 GHz (Kazemi 2022). A rectenna with an antipodal linearly tapered slot antenna attached to a rectifier circuit with the conversion efficiency of 45.8% (rectifier circuit area of 0.0756 mm^2) at a frequency of 94 GHz and an input power of 10 dBm has been developed. This rectenna achieved a power density of 0.9 mW/cm^2 at a distance of 30 cm for an input of 13.5 dBm (He et al. 2020). A rectenna for energy harvesting at 95 GHz was developed, and a 65 nm CMOS rectifier circuit with peak conversion efficiency of 21.3% at 89 GHz and 20% at 95 GHz was fabricated for an input power of 18.6 mW. A rectenna was developed by integrating the rectifier circuit with a ring antenna, and transmission experiments were conducted at 95 GHz, obtaining an acquired power of 0.2 mW with a transmission efficiency of 6.2% at 20 cm (Shaulov et al. 2017).

S and B Band Rectifier Circuits

Table 1.4 summarizes previous studies that achieved high efficiency and significant power for frequency operation below 30 GHz. At 2.45 and 5.8 GHz, there has been much research on rectifier circuits, including energy harvesting, high conversion efficiency, and circuits for high output power, since the remarkable work by Brown et al. Rectifier circuits for energy harvesting need to extract DC to the application from incoming microwave at low power levels. Therefore, in order to input as much power as possible to the rectifier circuit, it is important to achieve a high-gain yet compact antenna. In addition, the antenna and rectifier circuit are codesigned to optimize the low power matching. One of the most important achievements was a rectenna with efficiency of 75.5% at a frequency of 2.45 GHz and an input power of −7.7 dBm. A coupled coplanar and MSL circuit was used, with a designed antenna and Schottky barrier diode (SBD) (Sun et al. 2012).

The high-efficiency rectenna by Brown et al. had a rectification efficiency of 91% for 3 W input power, and no rectifier circuit has been developed that exceeds that efficiency for many years (Brown 1977). In 2020, C. Wang et al. designed and fabricated a rectifier circuit that operates at a frequency of 2.45 GHz with a conversion efficiency of 91% for an input power of 5 W, which is comparable to Brown's rectifier circuit. To achieve a high-efficiency high-power rectifier circuit, a GaAs SBD and SOD-523 package were used, which maintain high breakdown voltage and low diode series resistance. In addition to the SPICE parameters, the package model characteristics Ls (parasitic inductance of the diode pins) and Cp (parasitic capacitance between the diode pins and the GND plane) were built to enable an accurate design. The substrate has a low dielectric constant of 2.55 and a

Table 1.4 Summary of previous rectifiers with respect to power and the diode types (under 30 GHz) (Suganuma et al. 2021; Wang et al. 2021a,b; Reveyrand et al. 2012; Wu et al. 2020; Sun et al. 2012; Roberg et al. 2012; Nakamura et al. 2016; Yang et al. 2008; Schafer et al. 2015; Yoshida et al. 2019)

	Frequency GHz	Conversion efficiency %	Input power mW	Output power mW	Diode
Reveyrand (2012)	2.14	85	12,600	10,700	GaN HEMT
P. Wu (2020)	2.4	81.2	50	41	SBD (HSMS-282, 286)
Roberg (2012)	2.45	72.8	6.3	4.5	SBD
H. Sun (2012)	2.45	75.5	0.17	0.13	SBD (HSMS-2852)
C. Wang (2021)	2.45	91	5000	4550	GaAs SBD
Yoshida (2019)	5.78	44.8	13,180	5830	GaN HEMT
Nakamura (2015)	5.8	80.3	631	507	GaAs SBD
Y. Wang (2021)	5.8	68	12,000	8160	GaN HEMT
S. Schafer (2015)	9.9	52.2	8000	4176	MMIC
X. Yang (2008)	9.98	75	99	74.5	SBD
Hatano (2013)	24	54	130	70	MMIC
Suganuma (2021)	28	56	245	137	GaAs SBD

loss tangent of 0.0018 to reduce transmission loss. Furthermore, the circuit is single-shunt and the output filter is F-class (Wang et al. 2021a). In the same year, Sakai et al. in Japan presented a rectifier circuit with a frequency of 5.8 GHz, an input power of 1 W, and a conversion efficiency of 92.8% (Sakai et al. 2020).

Dual-Band Rectifiers

In 2010, Chiou et al. fabricated the rectifier operating at dual-band frequencies of 35 GHz and 94 GHz using CMOS 0.13 μm technology. They achieved conversion efficiencies of 53% and 37% at the respective frequencies. The circuit has two SBDs, with a finite-ground-coplanar-waveguide on the load side and a slot line on the antenna side. A bandpass filter (BPF) is inserted between the antenna and the diodes to suppress the harmonics. The linear tapered slot antenna (LTSA), integrated with the rectifier circuit, has a wide bandwidth at 35 GHz and 94 GHz with high directivity of 7.4 dBi and 6.5 dBi at each frequency (Chiou and Chen 2010).

The 2.45 GHz and 5.8 GHz dual-band rectenna was designed by Suh et al. in 2002. It consists of a dual-frequency dipole antenna, a coplanar stripline (CPS) input low-pass filter, 7.35 GHz and 4.9 GHz bandstop filters, and a rectifying diode. The antenna gains were 5 dBi and 5.4 dBi at the respective frequencies (Suh and Chang 2002).

High-Power Operation Rectifiers

Figure 1.5 shows a rectifier circuit capable of output power close to or exceeding 1 W plotted in frequency. As shown in Fig. 1.5, rectifier circuits capable of outputting W-class DC power have been increasing since the mid-2010s. In particular, an increasing number of studies are developing harmonic terminated power amplifiers that can also function as rectifiers using the time-reversal duality principle of power amplifiers for bidirectional high-power wireless transmission. A conceptual diagram of a power amplifier rectifier circuit is shown in Fig. 1.6.

In the amplifier mode, the RF input signal is input through the gate terminal. In the rectifier mode, the drain DC power supply is replaced by a DC load resistor R_{DC}. The RF is input from the drain port and the output is obtained from the drain DC port. In rectifier mode measurements, a conversion efficiency of 85% was obtained for 13 W (Reveyrand et al. 2012). In 2015, Schafer et al. designed a two-stage class AB X-band MMIC and achieved a conversion efficiency of 52.2% for an input RF power of 8 W. In a WPT system, the drain supply power can be changed to a DC load, and the gate bias voltage can be changed for use on the transmit and receive sides (Schafer et al. 2015).

Tube-Type Rectifier

Some devices use electron acceleration techniques to rectify high-power microwaves. These devices convert microwaves to DC by imparting microwave energy to the electron beam and finally extracting DC power as the product of current and voltage at the load. Cyclotron wave converter (CWC) was proposed by Watson et al. as a device capable of high-power rectification as shown in Fig. 1.7. The CWC consists of a DC high-voltage power supply, an electron gun, a body part, a permanent magnet, and a load resistor.

The power supply is used for the heater power supply to draw thermal electrons from the electron gun cathode and for the electron acceleration power supply. The body part can be divided into three sections based on the way electron energy is transferred, namely, the resonator, the motion direction conversion region, and the collector. Inside the cavity, there is an axial static magnetic field due to an external

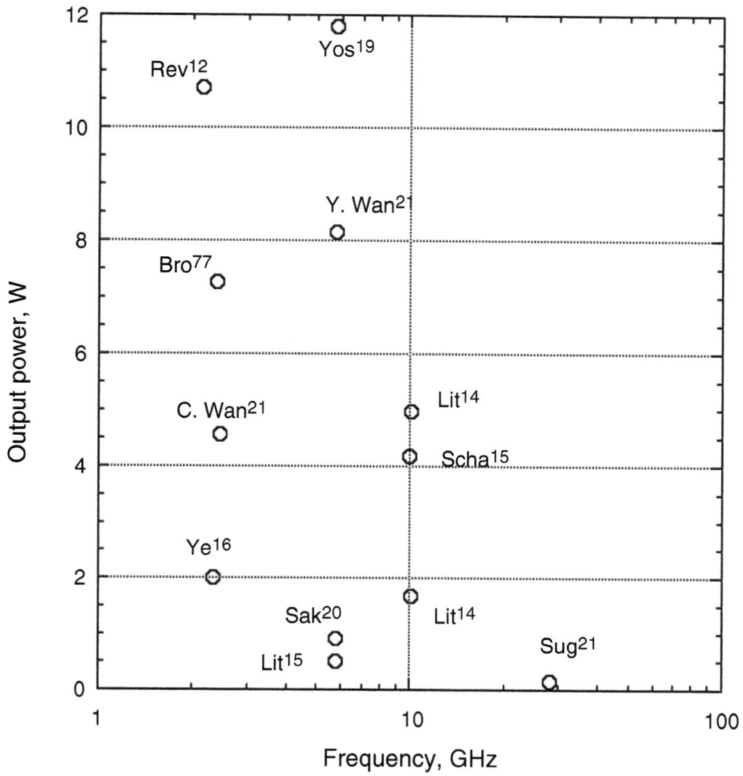

Fig. 1.5 Operating frequency and output power of the previous studies (Suganuma et al. 2021; Ye et al. 2016; Brown 1977; Wang et al. 2021a,b; Reveyrand et al. 2012; Sakai et al. 2020; Nakamura et al. 2016; Schafer et al. 2015; Yoshida et al. 2019; Litchfield et al. 2014)

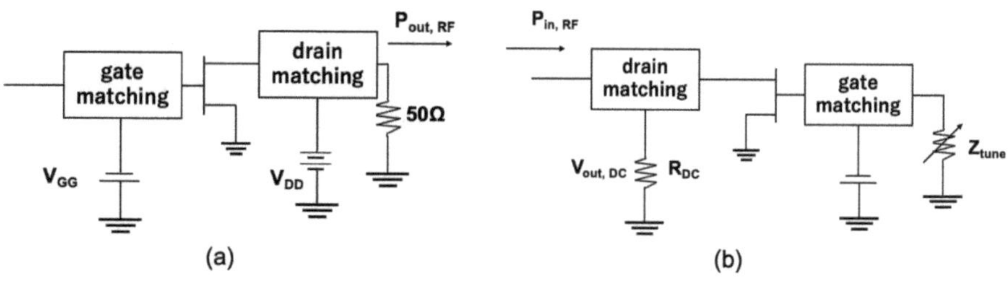

Fig. 1.6 Diagram of power amplifier circuit in PA mode (**a**) and rectification mode (**b**)

permanent magnet and a vertical microwave excited by the entering microwaves. When an electron beam with a strong axial velocity from the electron gun is inserted, cyclotron resonance occurs between the electrons and the electric field, and the energy of the electric field is absorbed by the electrons as rotational energy. The resonance occurs when the cyclotron frequency $f_c = eB_0/m$, which depends on the applied magnetic field B_0, coincides with the frequency f of the incident

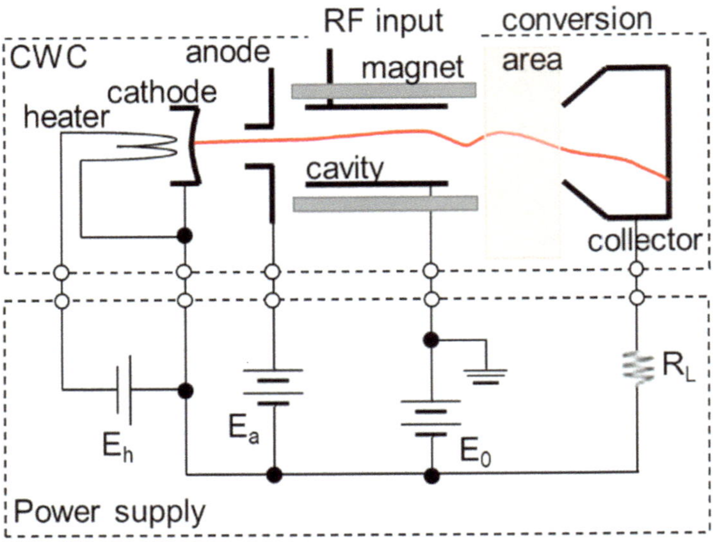

Fig. 1.7 Overview of the CWC and the power supply system

microwave electromagnetic field. Here, e and m are charge and electron mass, respectively. Therefore, as the operating frequency of the CWC increases, the required magnetic field also increases.

In the motion direction conversion region, a gradient is given to the axial magnetic flux density by a magnetic circuit using a yoke, and the Lorentz force due to the radial magnetic flux density is used to convert the electron rotation kinetic energy into the linear kinetic one. The electrons are then collected by a potential depression collector, and DC power is extracted from the load resistance based on the difference between the cathode and collector potentials.

In 1971, Watson et al. obtained real-world production measurements of 56% efficiency for 1–1.5 W microwave input (Watson et al. 1971). Vanke et al. achieved efficiencies of 60–80% for 10–20 kW input power in their evaluation of an actual device (Vanke et al. 1995). In recent years, they have again attracted attention as high-power devices with low losses and on a scale that is unexplored by semiconductor diodes alone. Research has mainly focused on the particle-in-cell method, which can calculate the internal electromagnetic field-electron interaction, and reported the result that 95% conversion is possible with resonator partial efficiency at a frequency of 2.45 GHz input power of 875 kW (Zhao et al. 2017). An experimental design for wireless power transmission over 10 km using magnetrons, CWCs, and parabolas was also reported, with an assumed DC output power of 8.5 kW and a total DC-DC efficiency of 1% (Hu et al. 2020).

In addition, a traveling wave tube (TWT), one of the microwave electron tubes, was used to compare and verify the results of simulation calculations and demonstrations of RF-DC operation instead of the usual DC-RF conversion operation. Sugimori et al. reported an overall efficiency of 21%, including beam power and heater power. They said that the efficiency could be improved by designing the traveling wave tube for the amplifier tube for RF-DC and by using high-power rectification (Sugimori et al. 1999).

Wireless Power Transfer vs. Battery

Wireless power transmission via microwave has been attempted for unmanned aerial vehicles such as drone, toy airplane, and airship. This section will discuss the technical possibilities of this technology in comparison with conventional battery technology. Unmanned aerial vehicles (UAVs) have been widely used in engineering and academic applications, such as 3D mapping and monitoring of terrain and buildings (Rim and Mi 2017), infrastructure (Nex and Remondino 2014), agriculture (Han et al. 2016), forestry (Zhang and Kovacs 2012), and logistics (Torresan et al. 2017). Governments and commercial entities throughout the world are actively conducting demonstration tests of delivery services using UAVs. The UAV logistics not only has the advantage of delivering goods quickly to remote islands and mountainous areas where transportation by vehicles is difficult but also eliminates the labor shortage in the logistics industry. For the aviation studies, the performance of UAVs in terms of the flight range and time has been evaluated. Gur et al. show the ratio between the battery capacity, and its weight is 180 Wh/kg to suggest the optimal UAV design (Song et al. 2018). Traub shows the relationship of the endurance and the range depending on the battery capacity (Gur and Rosen 2009). This research proves increasing battery weight is the reduction of the endurance and the maximum range and increasing the ratio of the battery weight makes a small increase of them. Avanzini et al. estimate accurately the airspeed in the battery flight (Traub 2011). They prove the maximum cruising speed is 11.3 m/s, the maximum range is 20 km, and the endurance is 30 min with 1 Ah lithium polymer battery. Therefore, extending the endurance and the range is necessary, although increasing the battery capacity is not the effective solution.

Next, to examine the specifications required for WPT, the flight endurance and minimum lift power were compared and analyzed for battery-powered UAVs with a gross weight of less than 1 kg. Figure 1.8 summarizes the flight endurance and gross weight required for UAV (less than 1 kg) for each propulsion system (Avanzini and Giulietti 2013; Pines and Bohorquez 2006). Total gross weight and flight endurance have a nonlinear relationship, $T_F \sim m^{0.64}$ for fixed wing type and $T_R \sim m^{0.24}$ for rotary wing type. As shown in Fig. 1.8, the flight endurance of a battery-based UAV weighing less than 1 kg is limited to 20–30 min for a fixed wing and more than 10 min for a rotary wing. Figure 1.8 also shows the battery performance. The energy density of the Li-ion battery has a roughly linear relationship, and this value has improved over the years in performance.

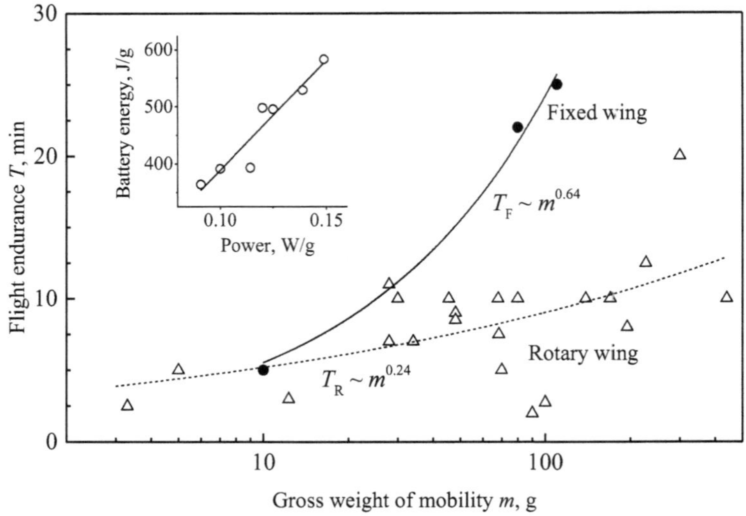

Fig. 1.8 Summary of the flight endurance and gross weight required for UAV (less than 1 kg) for each propulsion system (Avanzini and Giulietti 2013; Pines and Bohorquez 2006)

Feasibility of Microwave Powered Flight of UAVs

This section investigates the required power transmission to operate a UAV using 28 GHz millimeter waves, which is currently optimal, and considers possible transmission distances based on numeric values obtained from experiments (Brown 1969; Shinohara 2013). Besides, the present paper updated our 28 GHz MPT experiment in terms of the beam tracking system. The previous system has the problem that the time-series data of the analysis does not agree with the experimental values in the overall flight time and is overestimated by approximately ten times. These problems were caused by flights out of the beam range due to the errors of the indoor GPS system, which can be improved with beamforming toward the UAV. In the present study, a retrodirective tracking system is introduced into the previously reported system to compensate for the reduced efficiency from drone displacement. We also showed the theoretical analysis for all of power transmission efficacy of UAVs. This analysis helps the design of the effective aircraft with the high endurance and long range. Finally, we showed one example of feasibility study that the estimation assuming a 50 m transmission was performed and the superiority of the performance relative to batteries is shown.

In this study, the ratio of the DC power output from the rectenna attached to the flying object to the transmission power radiated to the flying object is defined as the efficiency between transmitter and receiver. The transmission efficiency in MPT experiments, η_{exp}, can be easily obtained by measuring the transmission power and the received DC power. To obtain this η_{exp} theoretically, we subdivide the theoretical transmission-to-reception efficiency η_{theory} into four parts. First, the ratio of

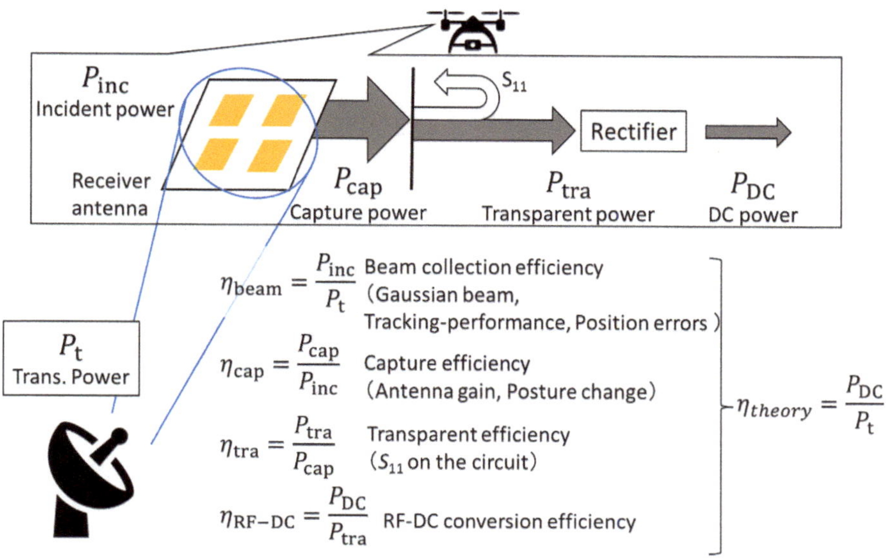

Fig. 1.9 Breakdown of microwave power transmission system with η_{beam}, η_{cap}, η_{tra}, η_{RF-DC}

the power irradiated to the receiving antenna to the power transmitted is defined as the beam collection efficiency η_{beam}. Second, the ratio of the power collected in the circuit to the power irradiated to the antenna is denoted as the collection efficiency η_{cap}. Third, the ratio of the power transmitted to the rectifier circuit without reflection to the collected power is defined as the transmission efficiency η_{tra}. The breakdown of η_{beam}, η_{cap}, η_{tra}, η_{RF-DC} is shown in Fig. 1.9. The theoretical transmission efficiency η_{theory} is the product of these four efficiencies and is given by the following equation:

$$\eta_{theory} = \eta_{beam}\eta_{cap}\eta_{tra}\eta_{RF-DC} \tag{1.1}$$

In the following sections, we present the detailed derivation of each efficiency.

Beam Collecting Efficiency η_{beam}

When an electromagnetic wave radiated from an antenna propagates in free space, the plane shape of the radiated beam can be regarded as a Gaussian distribution in the far field. A beam with such a shape is called a Gaussian beam. In the cylindrical coordinate system where the antenna radiation direction is the z-axis, the power density $S(r)$ of the Gaussian beam is given by the following equation:

$$S(r) = S(0)\exp\left(-\frac{2r^2}{\omega_z^2}\right), \tag{1.2}$$

$$S(0) = \frac{2}{\pi\omega_z^2}P_t. \tag{1.3}$$

Here r, ω_z, P_t, and $S(0)$ are, respectively, the radial distance from the center of the beam, the beam radius, the transmission power, and the central power density. The beam radius is the radius that is $1/e^2$ of the central power density at the position where $r = \omega_z$. The central power density $S(0)$ is twice the average value of the transmission power in the area of a circle of radius ω_z. Due to diffraction of electromagnetic waves, the beam wavefront widens as the distance z increases. The beam radius ω_z as a function of wavelength and distance z is given by the following equation:

$$\omega_z = \omega_0\sqrt{1 + \left(\frac{\lambda z}{\pi\omega_0^2}\right)^2} \tag{1.4}$$

where ω_0 is the beam waist. The beam waist ω_0 is the beam radius at which the Gaussian beam is most focused. With respect to Eq. 1.4, if the beam waist is small, the beam radius at a certain distance will be large. Therefore, if we want to suppress the diffusion, we need to use an antenna with a large beam waist.

When the beam shape can be assumed to be a Gaussian beam at the altitude of the flying object, the power irradiated on the surface of the rectenna can be calculated by determining the beam waist of the transmission antenna. In Eq. 1.2, we consider a flying object that is perfectly stationary at an altitude z just above the transmission antenna. If the transmission antenna has different apertures in the x- and y-directions in the Cartesian coordinate system, there are beam waists ω_{0x}, ω_{0y} in the x- and y-directions, respectively. The beam radii in the x- and y-directions at altitude z are ω_{zx}, ω_{zy}. Assume that the rectenna is a rectangle with the length l_x in the x-direction and l_y in the y-direction. The irradiation power P_{inc} to the receiving antenna is expressed by dividing both sides of Eqs. 1.11.2 by the area:

$$\mathrm{P_{inc}} = P_t \mathrm{erf}\left(\frac{\sqrt{2}l_x}{2\omega_{zx}}\right) erf\left(\frac{\sqrt{2}l_y}{2\omega_{yz}}\right) \tag{1.5}$$

where *erf* denotes the error function. Equation 1.8 is valid when the flying object is perfectly stationary at the altitude z just above the transmission antenna, and the beam collection efficiency η_{beam} takes the maximum value. The maximum value of the beam collection efficiency is $\eta_{\mathrm{beam,\ max}}$, and $\eta_{\mathrm{beam,\ max}}$ is given by the following equation from Eq. 1.5:

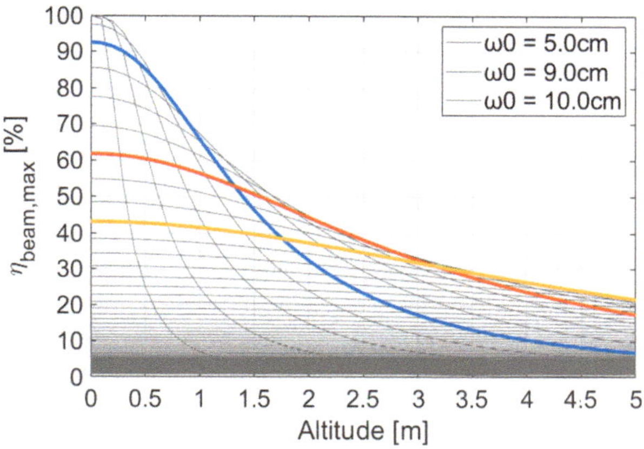

Fig. 1.10 $\eta_{\text{beam, max}}$ vs. vehicle altitude for 9×15 cm receiving area

Table 1.5 $\eta_{\text{beam, max}}$ and beam waist for the different altitude

Altitude (m)	$\eta_{\text{beam, max}}$ (%)	ω_0 (cm)
0.8	74.3	5
2.5	38.1	9
5.0	21.6	10

$$\frac{P_{\text{inc}}}{P_t} = \eta_{\text{beam, max}}(z) = \text{erf}\left(\frac{\sqrt{2}l_x}{2\omega_{zx}(z)}\right)\text{erf}\left(\frac{\sqrt{2}l_y}{2\omega_{yz}(z)}\right) \tag{1.6}$$

Equation 1.6 is a monotonically decreasing function that takes a maximum value at $z = 0$. The beam radius at a given distance is inversely proportional to the beam waist diameter from Eq. 1.4. Since the area where the rectenna can be mounted depends on the flying object, the transmission distance and the flying object determine a single beam waist diameter that maximizes the beam collection efficiency $\eta_{\text{beam, max}}$. As an example, in the case of MPT using a horn antenna with a square aperture on a flying object with a rectenna loading area of 9×15 cm, Fig. 1.10 shows $\eta_{\text{beam, max}}$ versus altitude. The beam radius in 1 cm increments of the beam waist is plotted in Fig. 1.10, and the maximum value of $\eta_{\text{beam, max}}$ and the beam waist at each altitude are shown in Table 1.5.

However, it is difficult for a general flying object to stay perfectly stationary in the sky. In the MPT of a flying vehicle, the loss of efficiency due to position fluctuation during hovering can be suppressed by tracking. The retrodirective system is one of the main tracking methods, but the gain of the transmission antenna decreases as the beam angle increases. It is necessary to take into account this decrease in gain due to tracking. If the tracking is only in one axis direction, or if no tracking is used, the position variation in the direction without tracking must be considered separately.

The beam acquisition efficiency η_{beam}, which takes into account the gain reduction due to tracking and the position variation in the direction without tracking, is given by the following equation:

$$\eta_{\text{beam}}(z) = \eta_{\text{beam, max}}(z) \times f_{\text{track}}(\alpha, \beta) \tag{1.7}$$

$$\eta_{\text{beam}} = \eta_{\text{beam, max}}(z) \times f_{\text{track}}(\alpha) \times \exp\left(-\frac{2y^2}{\omega_{zy}{}^2}\right) \tag{1.8}$$

$$\eta_{\text{beam}} = \eta_{\text{beam, max}}(z) \times exp\left(-\frac{2x^2}{\omega_{zx}{}^2}\right) \times \exp\left(-\frac{2y^2}{\omega_{zy}{}^2}\right) \tag{1.9}$$

where Eqs. 1.7, 1.8, and 1.9 show η_{beam} with xy-biaxial tracking, x-axis tracking only, and no tracking, respectively, and α. β, (x, y) are beam angle in x-direction, beam angle in y-direction, and position of the flying object. The function f_{track} in Eqs. 1.7 and 1.8 indicates the gain reduction due to tracking. The exp. terms in Eqs. 1.7 and 1.8 show the effect of position variation in the direction without tracking due to the characteristics of Gaussian beams.

Beam Capture Efficiency on Rectenna η_{cap}

The effective area A_e, which represents the area where the receiving antenna can take power into the circuit, is given by the following equation:

$$A_{\text{e}} = \frac{\lambda^2 G_{\text{r}}}{4\pi} \tag{1.10}$$

Here, G_{r} is the gain of the receiving antenna. Although the irradiated power is distributed over the entire area of the rectenna, only the component of the effective area can be captured in the circuit. Therefore, the maximum value of the collection efficiency of the rectenna array, $\eta_{\text{cap, max}}$, is shown by the following equation:

$$\eta_{\text{cap, max}} = C_{\text{r}} \frac{A_e}{A} \tag{1.11}$$

where C_{r}, A are the antenna element spacing factor and the physical area of the antenna, respectively. In array antennas, the efficiency decrease depending on the distance between antennas has been reported (Han et al. 2016), and C_{r} estimates the efficiency decrease. A flying object changes its direction of motion depending on its attitude. When the rectenna array is directly attached to the bottom surface of the flying object, η_{cap} decreases according to the radiation pattern of the receiving antenna. If the gain reduction due to the attitude change is taken into account in Eq. 1.10, η_{cap} is expressed by the following equation:

$$\eta_{\text{cap}} = \eta_{\text{cap, max}} \times f_E(\alpha, \phi) \times f_H(\beta, \theta) \times f_{\text{pol}}(\psi) \tag{1.12}$$

where f_E, f_H, f_{pol}, ϕ, θ, ψ are the E-plane radiation pattern, H-plane radiation pattern, polarization efficiency, roll angle, pitch angle, and yaw angle, respectively. Equation 1.12 assumes roll angle variation for the E-plane and pitch angle variation for the H-plane. In the case of tracking, beam angle should also be considered as a variable of E- and H-planes, and Eq. 1.12 assumes that the E-plane is the beam angle variation in the x-direction and the H-plane is the beam angle variation in the y-direction.

Antenna Transmission Efficiency η_{tra} *and RF-DC Conversion Efficiency* η_{RF-DC}

The power reflected due to mismatch among the collected power in the circuit is reradiated from the heat loss and antenna and does not reach the rectifier circuit. First, considering the reflected power of the antenna, the antenna transmission efficiency $\eta_{tra,\ antenna}$ is given by the following:

$$\eta_{tra,antenna} = 1 - 10^{\frac{S11,antenna}{10}} \tag{1.13}$$

where $S_{11,\ antenna}$ is the antenna S_{11} and the unit is dB. Next, we consider the reflected power of the rectifier circuit. The reflected power of the rectifier circuit depends on the power obtained by multiplying the collected power by the antenna transmission efficiency due to the diode nonlinearity. Therefore, the component of transmission efficiency of the rectifier circuit is added to Eq. 1.13, which is shown in the following equation:

$$\eta_{tra} = 1 - 10^{\frac{S11,antenna}{10}} - 10^{\frac{S11,rectifier\left(\eta_{tra,antenna} P_{cap}\right)}{10}} \tag{1.14}$$

where $S_{11,\ rectifier}$ is S_{11} of the rectifier circuit and the unit is dB. Similarly to the transmission efficiency, the rectifier efficiency is subject to diode nonlinearity and becomes nonlinear depending on the transmitted power. Since the characteristics of rectification efficiency also depend on the terminating load, the rectification efficiency $\eta_{RF\ -\ DC}$ is given by the following equation with the transmitted power P_{tra} and the terminating load R_L.

$$\eta_{RF-DC} = \eta_{ind} f_{RF-DC}(P_{tra}, R_L) \tag{1.15}$$

η_{ind}, $f_{RF\ -\ DC}$are fitting functions of rectenna individual difference efficiency and rectification efficiency, respectively.

Operation Frequency Dependency: 5.8 GHz vs. 28 GHz

For example, using the efficiency definition in section "Feasibility of Microwave Powered Flight of UAVs", a comparison of 28 GHz and 5.8 GHz MPTs is made using UAVs. The AR.Drone 2.0 used in Suganuma et al. (2021) has a payload of 100 g and can carry a rectenna array of up to 9 cm × 15 cm on its bottom surface. The beam collection efficiency is calculated by using the maximum number of rectenna arrays in the installed area. In the following, the UAV is hovering just above the transmission antenna and the antenna is square, i.e., $\omega_{0x} = \omega_{0y} = \omega_0$, and the spot size at distance z from the antenna is ω_z. The beam waist of the transmission system to be swept is considered to be 1 m as an upper limit from the viewpoint of portability. $\eta_{\text{beam, max}}(z)$ is a monotonically decreasing function depending on the distance z. To realize the microwave power transmission, it is necessary to design a separate beam waist for each distance z in the transmission system that maximizes $\eta_{\text{beam, max}}(z)$. For other efficiencies, $\eta_{\text{cap, max}} = 33\%$, $\eta_{\text{tra}} = 99\%$, and $\eta_{RF-DC} = 55\%$ at 28 GHz. The effect of atmospheric attenuation can be almost neglected in the microwave power transmission of about 100 m using 5.8 GHz and 28 GHz.

Figure 1.11 shows $\eta_{\text{theory, max}}$ at altitude 0 ~ 10 m for 28GHz MPT to the drone. First, $\eta_{\text{theory, max}} = 12.0\%$ was calculated at the flight altitude of 80 cm and beam waist of 5.0 cm in this study. To cover the power consumption, the power transmission is estimated to be 500 W. The AR.Drone 2.0 can be powered by a single oscillation source and TWT amplifier at an altitude of 80 cm. Similarly, $\eta_{\text{theory, max}} = 3.5\%$ at 5 m. In this case, the power transmission is 1715 W and the beam waist is 13 cm. At 10 m altitude, it is necessary to design with $\eta_{\text{theory, max}} = 1.9\%$, transmission power of 3157 W, and transmission beam waist of 19 cm. Compared

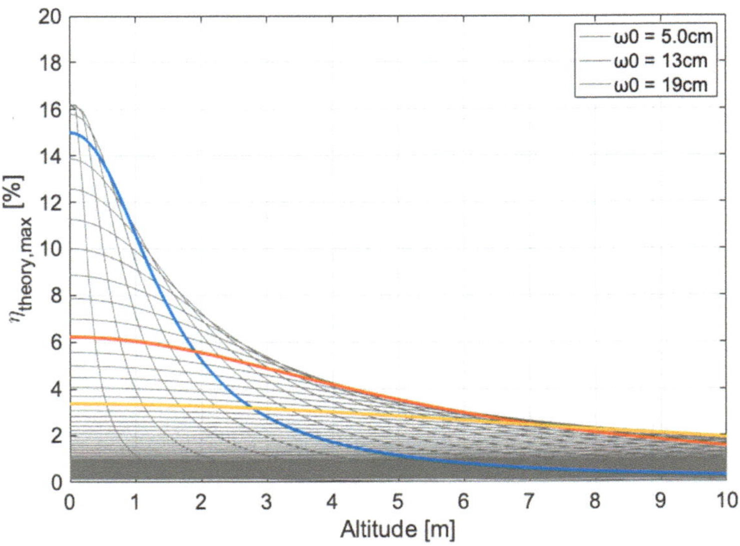

Fig. 1.11 Estimated $\eta_{\text{theory, max}}$ of AR.Drone system with 28 GHz mmW for 10 m altitude flight

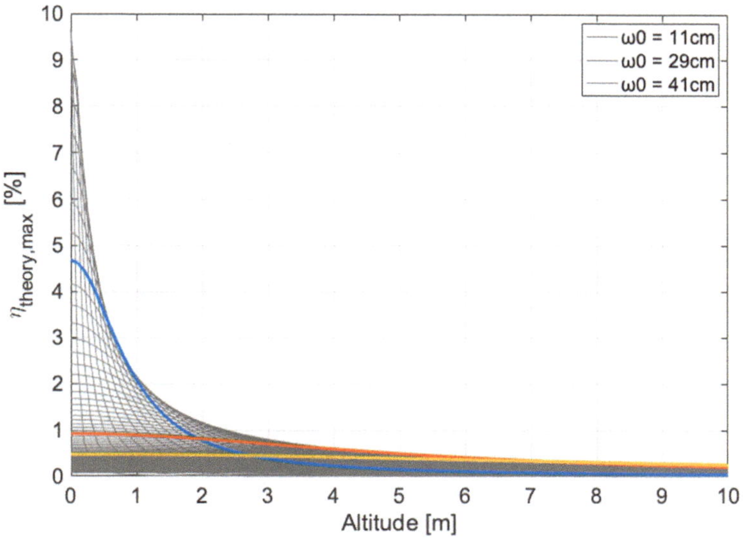

Fig. 1.12 Estimated $\eta_{\text{theory, max}}$ of AR.Drone system with 5.8 GHz MW for 10 m altitude flight

Table 1.6 28 GHz vs. 5.8 GHz in terms of $\eta_{\text{theory, max}}$ and beam waist ω_0

	28 GHz		5.8 GHz	
	$\eta_{\text{theory, max}}$ (%)	ω_0 (cm)	$\eta_{\text{theory, max}}$ (%)	ω_0 (cm)
80 cm	12.0	5	2.6	11
5 m	3.5	13	0.45	29
10 m	1.9	19	0.25	41

to 28 GHz, $\eta_{\text{theory, max}}$ at an altitude of 0–10 m is shown in Fig. 1.12 when the rectenna created in the previous study (Shimamura et al. 2017) is used for 5.8 GHz MPT. At 5.8 GHz, $\eta_{\text{theory, max}} = 2.6\%$ at 80 cm altitude; in this case, the power transmission beam waist is 11 cm. In comparison, 28 GHz is found to be about 4.6 times superior with respect to $\eta_{\text{theory, max}}$. At 5 m altitude, 28 GHz is 7.7 times superior to $\eta_{\text{theory, max}} = 0.45\%$. At 10 m altitude, 28 GHz is 7.6 times superior to $\eta_{\text{theory, max}} = 0.25\%$. A comparison of the obtained results is shown in Table 1.6.

Consider $\eta_{\text{beam, max}}$ of 28 GHz MPT to a rotary-wing UAV with power consumption of 4000 W. Based on previous studies, the bottom area of a 4000 W UAV can be estimated to be 1.79 m^2 (Fujino and Fujita 1998). Assuming that the rectenna mounting area is a quarter of the base area, a rectenna array of 67 cm on a side can be mounted. Figure 1.13 shows $\eta_{\text{beam, max}}$ calculated in the same way as the assumption for AR.Drone 2.0. The rotary-wing UAV, which is considered to be used for logistics, is assumed to fly at 3000–4000 W and an altitude of 20–50 m, considering its payload and cruising speed. At this altitude, $\eta_{\text{theory, max}}$ of more than 10% can be obtained at altitudes below 50 m by proper design of the transmission beam waist. Considering that the maximum transmission beam waist is 41 cm at 50 m, the 28 GHz MPT power transmission system is a portable system (Table 1.7).

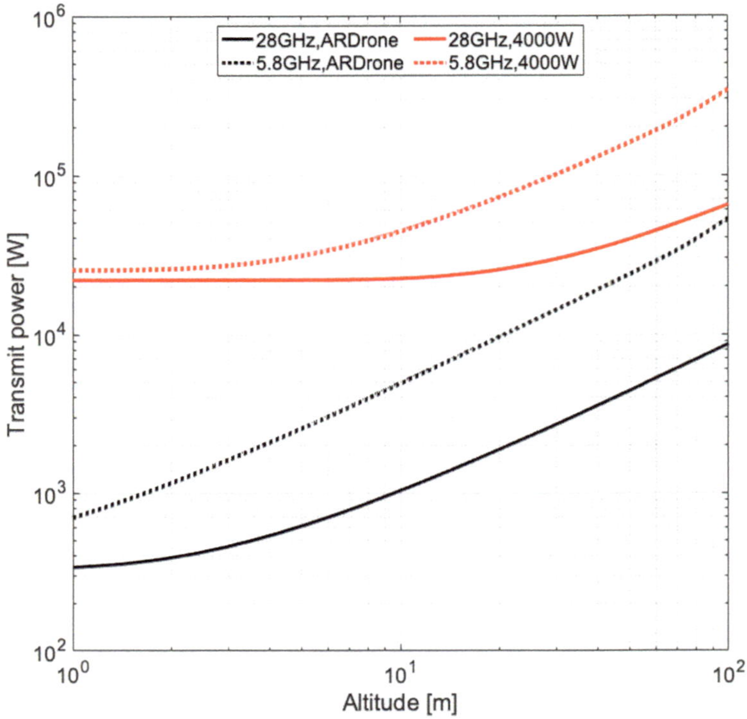

Fig. 1.13 Estimated $\eta_{\text{theory, max}}$ of 4000 W-class UAV 28 GHz mmW for 0 to 100 m altitude flight

Table 1.7 $\eta_{\text{beam, max}}$ and beam waist for each altitude for 28 GHz mmW power transmission to 4000 W class UAV

	$\eta_{\text{theory, max}}$	ω_0 (cm)
20 m	15.7	26
50 m	10.1	41
100 m	6.2	58

Figure 1.14 compares the altitude and power at 28 GHz and 5.8 GHz for the two compared UAVs when the power transmission system is designed with the optimal beam waist. In the case of the 28 GHz MPT, the AR.Drone 2.0 can fly up to an altitude of 3.5 m when the power transmission system is configured with a TWT amplifier capable of 500 W output. On the other hand, for 5.8 GHz MPT, magnetrons are considered as a possible power transmission system capable of 1 kW output. Even in this case, AR.Drone 2.0 can only fly to an altitude of 1.6 m. For a 4000 W-class UAV, the difference between 28 GHz and 5.8 GHz is small up to an altitude of about 3 m. This is because the efficiency loss due to beam dissipation is suppressed due to the low altitude and the large area of the UAV's rectenna loading. It is shown that at altitudes above 10 m, the 28 GHz MPT can fly with less than twice the transmission power of the 5.8 GHz MPT. On the other hand, the beam radius of

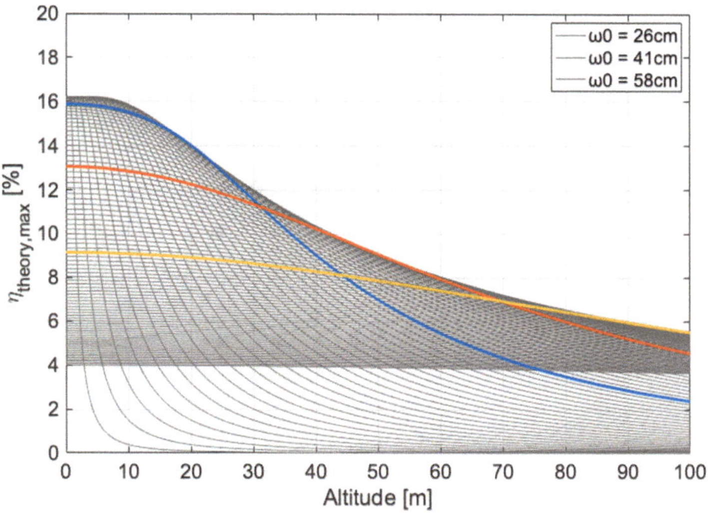

Fig. 1.14 Transmitted power vs. flight altitude for 28 GHz and 5.8 GHz power transmission using small (AR.Drone) and high-power (4 kW) drone

28 GHz MPT is smaller than that of 5.8 GHz MPT, which means that a highly accurate flight control is necessary for the 28 GHz MPT.

Figure 1.15 shows the DC power received in this study and the previous 5.8 GHz study at an altitude of 50 m and a transmission antenna diameter of 1 m, as well as the rotary-wing UAV used as a reference. The 28 GHz MPT is calculated from the η_{theory} of each rectenna area and the transmission power to obtain the DC power that can be received by the rectenna array of the UAV. To validate the 5.8 GHz MPT for comparison and under the same conditions as in the previous study, we assume that the UAV hovers on a circle with a beam angle of 9° at an altitude of 50 m without any deviation in position or attitude. In this study, the case in which power consumption can be met by less than 1 MW is considered feasible. Figure 1.15 shows that it is impossible to meet the power consumption of a UAV at 5.8 GHz even with 1 MW of power transmission. In contrast, at 28 GHz, it is possible to obtain more DC power than the power consumption of the UAV with 200 kW transmission at an altitude of 50 m. Compared to the 1 MW power transmission at 5.8 GHz, the 28 GHz power transmission can provide about ten times the power consumption, and it is considered possible to recharge the battery while maintaining the power consumption of the flight.

A comparison is made between the weight of the rectenna and the battery for 28 GHz microwave power transmission and the current battery. The weight of the rectenna and the battery is shown in Fig. 1.16. It is assumed that the rectenna fabricated in this study is installed in the area of one-fourth of the base area calculated by the equation of the previous study (Song et al. 2018). When the power input to the antenna exceeds the maximum input power of the rectifier circuit,

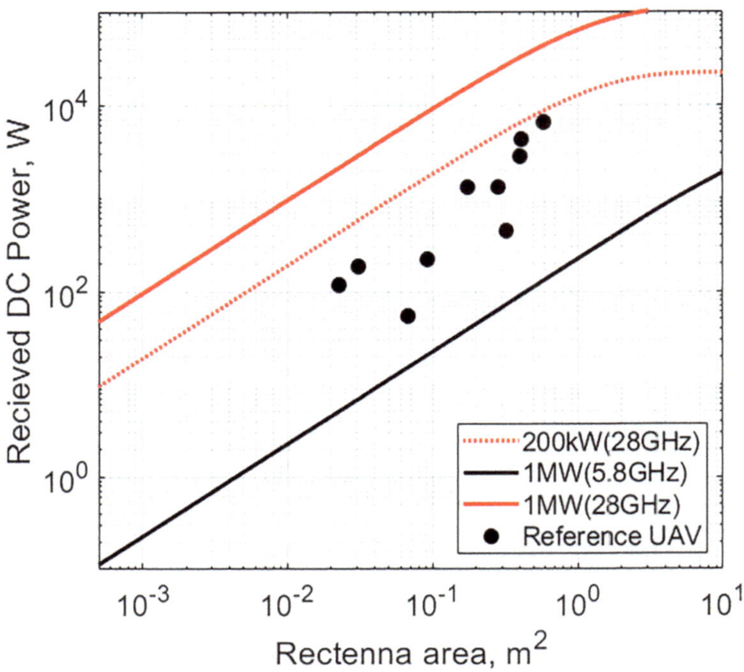

Fig. 1.15 Received DC power and UAV power consumption for each transmission power at an altitude of 50 m and a transmission antenna diameter of 1 m. Light gray indicates infeasible areas at 5.8 GHz, while dark gray indicates infeasible areas at 28 GHz

the number of rectifier circuits that can handle the input power is connected in parallel according to the previous 5.8 GHz study. As for the battery, the weight of the battery is generally 0.3 times the weight of the aircraft including the battery (Traub 2011). This is close to the case for the rotor-wing UAV of less than 1 kg, which is used as a reference in Fig. 5.5, but the percentage of battery weight tends to decrease as the aircraft weight increases. The diodes used in this study were made of silicon nitride, and in this case, the weight of the diodes was about the same as that of the battery for aircraft weighing 90 g or more. Furthermore, when a GaN diode (Mizojiri 2019) capable of 20 W input is used in the rectifier circuit, the weight of the rectifier is less than 0.3 times of the aircraft weight at 65 g or more, indicating that the weight of the rectifier can be significantly reduced compared to the reference UAV. Considering the weight of the unit that integrates and stably supplies the power obtained by the rectenna, the 28 GHz MPT should adopt the rectenna with GaN diodes.

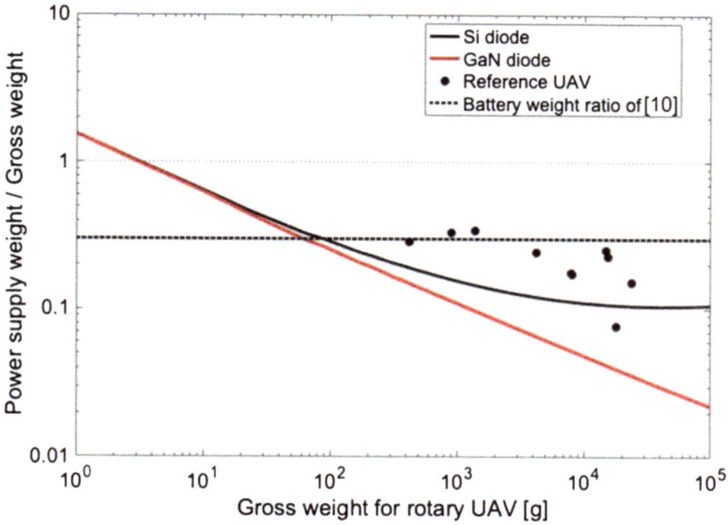

Fig. 1.16 Ratio of power supply weight to the gross weight (Traub 2011)

Flight Demonstration of Microwave Power Transmission with Drone

Mobilities Design

A demonstration will be given using a study using a drone as an example of rectenna based wireless power transfer. Drone 2.0 (Parrot, Inc.) was used for the power supply experiment because of its accessibility to sensor information, flight control from a PC, and sufficient payload to ensure stable flight even when payloads such as rectenna increase. This drone is equipped with an inertial measurement unit (IMU) and is capable of measuring flight altitude, Euler angle, and acceleration. The drone transmits IMU data to the PC and flight commands from the PC to the drone via a Wi-Fi connection with the drone as the parent unit. The drone power feeding experiment will be conducted in a shielded room from the perspective of Japanese Radio Law. Since GPS signals cannot be acquired indoors, it is not possible to control the drone's position or acquire its position. Therefore, we constructed an indoor position acquisition system to control the drone's position and estimate the beam direction angle. Figure 1.17 shows a diagram of the indoor position acquisition system. In this case, Beacon HW v4.9 (Marvelmind Robotics) is installed in the four corners of the shielded room for the indoor position supplement system, and Beacon Mini-TX (Marvelmind Robotics) is mounted on top of the drone. Modem HW on PC v4.9 (Marvelmind Robotics) is connected via USB, and the drone's location information is obtained by processing it on a PC using the dashboard software provided by Marvelmind Robotics. MATLAB and Simulink (Mathworks) are used for real-

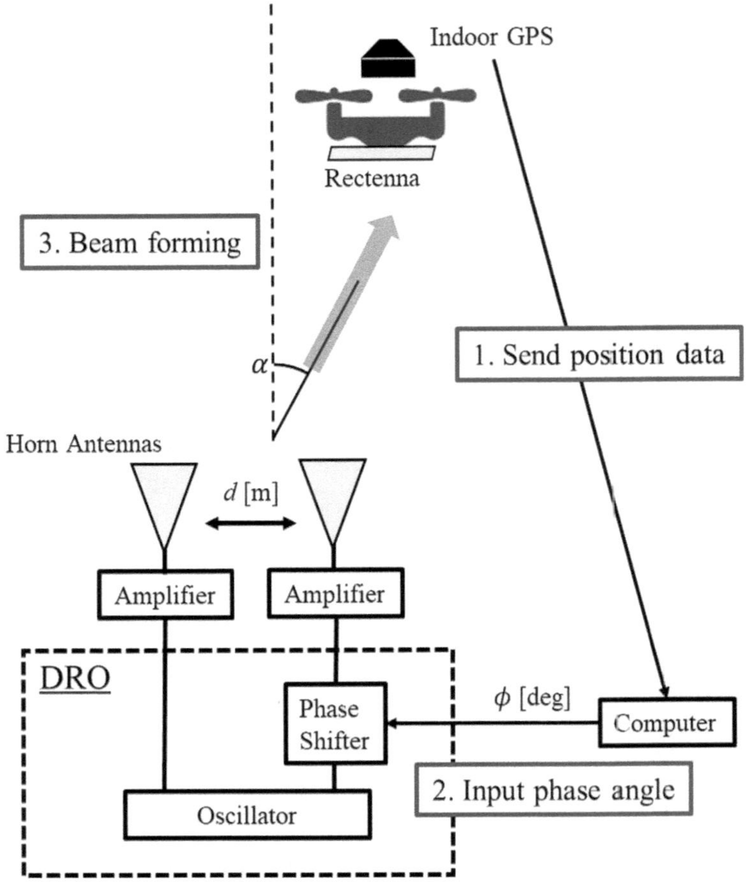

Fig. 1.17 Sample of wireless power transmission system for flying mobility

time drone control, phase angle calculation, and phase angle input to the phasor. The PID controller is configured in Simulink to ensure stable drone flight, and the drone is given pitch, roll, and yaw angles, as well as velocity in the height direction, as operating variables.

A Dielectric Resonator Oscillator (DRO, VG14-28G, Vega Technology) is used for the 28 GHz oscillator on the transmission side. This DRO has two output terminals, one of which has a built-in phasor (SKP-0631831036-SFSF-A1, SAGE Millimeter, Inc.), which is connected to a PC via USB to input the phase angle. A block diagram inside the DRO is shown in Fig. 1.18. The direction angle of the transmission beam and the phase angle are as follows:

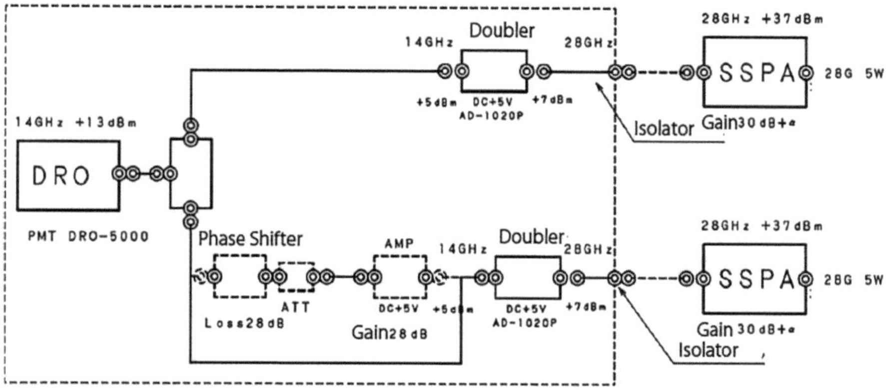

Fig. 1.18 Block diagram inside the DRO

$$\alpha = \sin^{-1}\left(\frac{\lambda\phi}{2\pi d}\right) \tag{1.16}$$

where α, ϕ, λ, d are the beam direction angle, phase angle, wavelength, and distance between transmission antennas, respectively. In this experiment, a power amplifier (WPAC172PG, Advanced Microwave Inc.) is used to amplify the transmitted power by connecting it to the two input terminals of the DRO with a coaxial K-connector. The power amplifier is connected to the transmission antenna via an isolator, coaxial K-connector, coaxial waveguide converter, and waveguide flange. A horn antenna (LB-28-20-C-KF, A-INFO) is used as the power transmission antenna. In the power feeding experiment, this antenna is arrayed in the x-direction as a phased array antenna.

The gain of this antenna is 19.2 dBi. The radiation characteristics of the power feed array antenna and the variation of the center gain are shown in Fig. 1.19. From Fig. 1.19, it can be confirmed that with tracking enabled, the reduction in gain can be reduced compared to the case without tracking.

The receiving antenna used in this experiment was designed as a 16-patch array antenna, with a single-shunt rectifier circuit on the back side. In addition, an F-class load is inserted in the output side filter to reflect harmonics. The power receiving antennas are arranged in four arrays and placed on the bottom of the drone as shown in Fig. 1.20. The diode used in this rectifier circuit is MA4E1317 (MACOM), which operates with a rectification efficiency of 55.5% at 245 mW input power when the termination resistance is 130 Ω. The conversion efficiency of this rectifier circuit with respect to input power is shown in Fig. 1.21.

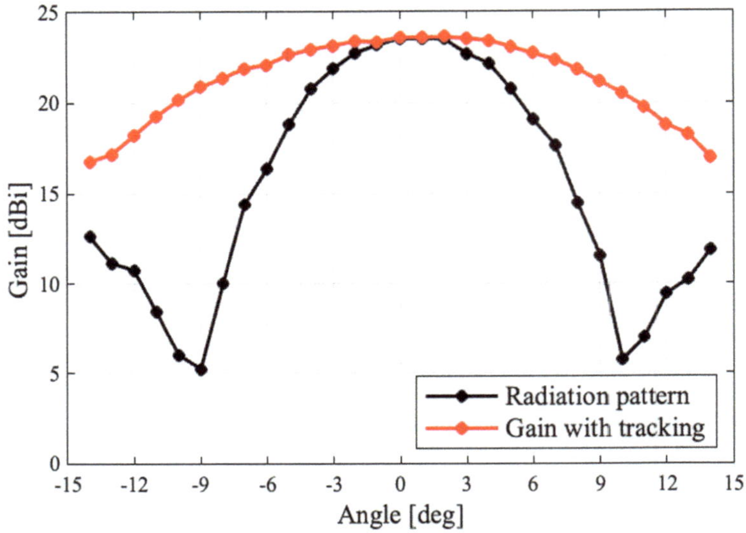

Fig. 1.19 Radiation characteristics of the power feed array antenna and the variation of the center gain

Flight Demonstration

We conduct power supply experiments on drones flying autonomously along predetermined flight paths. In other words, all flight commands other than takeoff and landing are given automatically on a PC, and beamforming is performed according to the drone's position. A total of 3.8 W is radiated from the antenna on the power transmitting side. To verify the effectiveness of beamforming, we compared the results when the phase angle is input to the DRO (with beamforming) and when the phase angle is not input (without beamforming). The drone is set to hover in place for a total of 1 s after takeoff, move directly over the transmission antenna for a total of 40 s, and then move back to the point where it took off for landing. The drone's flight altitude is set at 1.2 m.

First, the flight trajectory of the drone is shown in Fig. 1.22. These results confirm that there is no difference in the state of drone flight with and without beamforming.

Next, Fig. 1.23 shows the output power from the power receiving rectenna attached to the drone. The power acquisition is intermittent without beamforming. On the other hand, the power output with beamforming is continuous. The flight condition is almost the same, indicating that the beam control in the x-direction is working well. Continuous power feed without beamforming lasted about 5 s. Beamforming enabled continuous power was received for 48 s while hovering just above the transmission antenna. The average value of power acquisition was −4.06 dBm for the former and 1.42 dBm for the latter, a 3.6 times average improvement of 5.48 dBm.

Fig. 1.20 Photo of the power receiving rectennas attached to the bottom of the drone

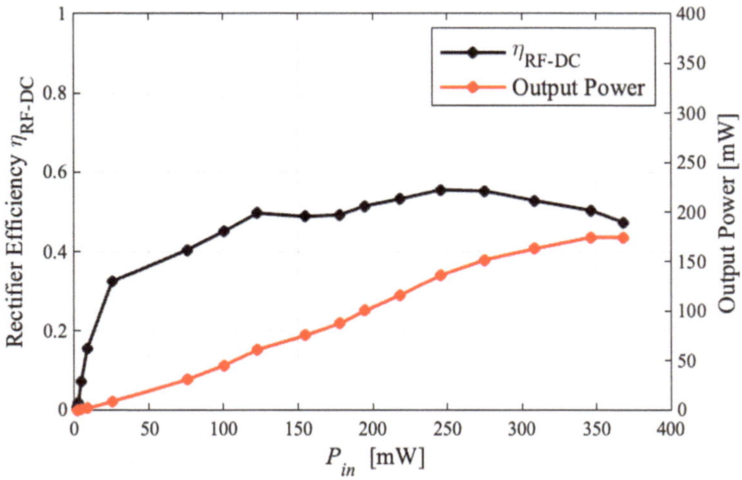

Fig. 1.21 Conversion efficiency and output power of the rectifier circuit with respect to input power

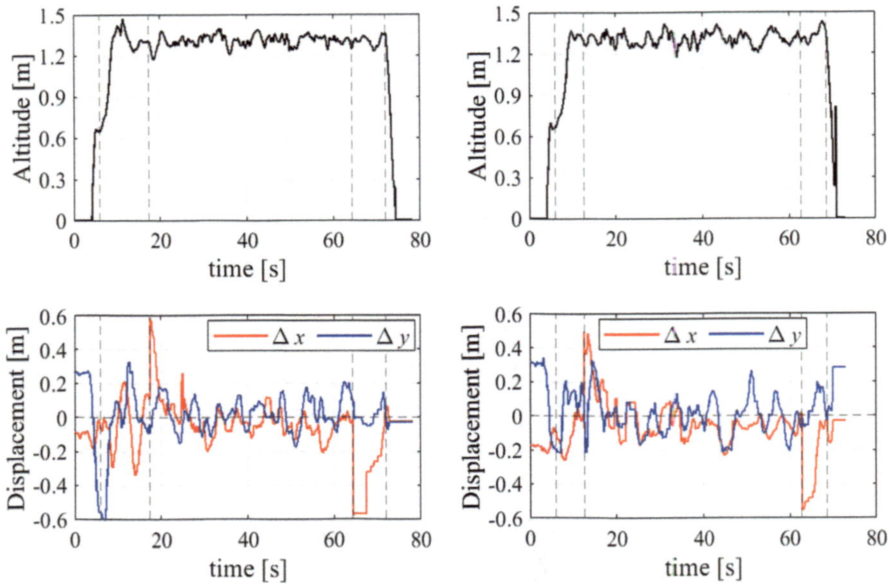

Fig. 1.22 Trajectories of altitude and xy-coordinates with and without drone beamforming

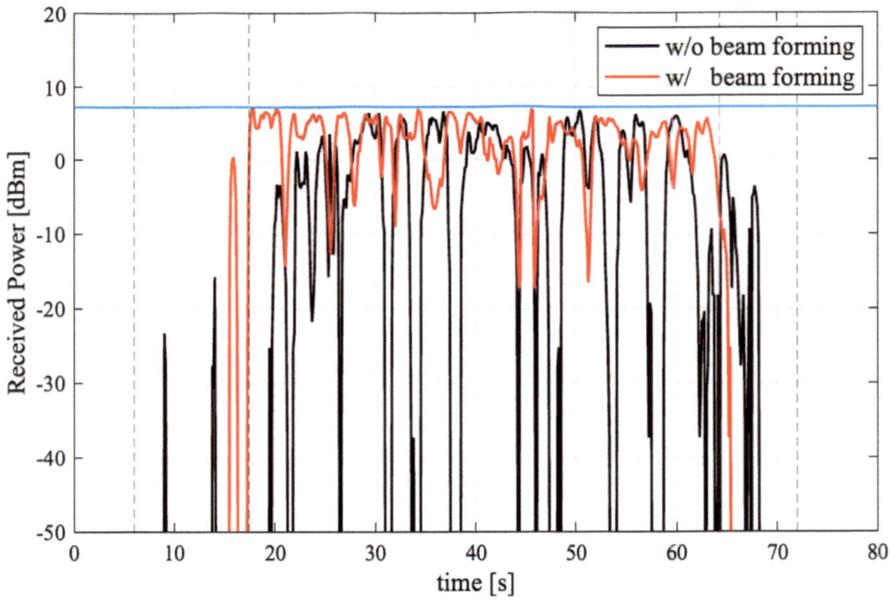

Fig. 1.23 Output power from the power receiving rectenna attached to the drone with and without beamforming

References

Avanzini G, Giulietti T (2013) Maximum range for battery-powered aircraft. J Aircr 50(1):304–307

Blackwell T (2005) Recent demonstrations of laser power beaming at DFRC and MSFC. AIP Conf Proc 766:73

Brown W (1969) Experiments Involving a Microwave Beam to Power and Position a Helicopter. IEEE Transact Aerospace Electron Sys AES-5(5):692–702

Brown WC (1977) Electronic and mechanical improvement of the receiving terminal of a free-space microwave power transmission system. Raytheon Co, Wayland, MA

Cabrera T, Alicia G-G, José M, Sánchez A, José A (2020) Wireless power transfer for electric vehicles: foundations and design approach. Springer, Cham

Chiou HK, Chen IS (2010) High-efficiency dual-band on-Chip Rectenna for 35- and 94-GHz wireless power transmission in 0.13-μm CMOS technology. IEEE Trans Microw Theory Tech 58(12)

Fujino Y, Fujita M (1998) Development of a high-efficiency Rectenna for wireless power transmission–application to a microwave-powered airship experiment. Rev Commun Res Lab 43(3):367–374

Gur O, Rosen A (2009) Optimizing electric propulsion Systems for Unmanned Aerial Vehicles. J Aircr 46(4):1340–1353

Gutmann RJ, Borrego JM (1979) Power combining in an array of microwave power rectifiers. IEEE Transac Microwave Theory Tech 27(12):958–968

Han Y, Han KK, Lin JJ, Fard GM (2016) Visual monitoring of civil infrastructure systems via camera-equipped Unmanned Aerial Vehicles (UAVs): a review of related works. Visual Eng 4(1):1–8

Hashimoto Y, Yuan Q, Aoki T (2018) Drone driven by microwave. 2018 Asia-Pacific microwave conference (APMC), Kyoto. 1444–1446

Hatano K, Shinohara N, Mitani T, Seki T, Kawashima M (2013) Development of 24GHz-Band MMIC Rectenna. Radio and Wireless Symposium (RWS) IEEE 50:199–201

He P, Xu J, Zhao D (2020) A W-band Rectenna using on-Chip CMOS switching rectifier and on-PCB tapered slot antenna achieving 25% effective-power-conversion efficiency for wireless power transfer. IEEE/MTT-S International Microwave Symposium (IMS) 2020:1055–1058

Hemour S, Lorenz CHP, Wu K (2015) Small-footprint wideband 94GHz rectifier for swarm micro-robotics. 2015 IEEE MTT-S Int Microw Symp I:5–8

Hu B, Li H, Li T, Wang H, Zhou Y, Zhao X, Hu X, Du X, Zhao Y, Li X, Qi T (2020) A long-distance high-power microwave wireless power transmission system based on asymmetrical resonant magnetron and cyclotron-wave rectifier. Energy Rep 7:1154

Kawashima N, Takeda K (2008) Laser energy transmission for a wireless energy supply to robot. In: Balaguer C, Abderrahim M (eds) Robotics and automation in construction. Intech, Rijeka

Kazemi H (2022) 61.5% efficiency and 3.6 kW/m2 power handling Rectenna circuit demonstration for radiative millimeter wave wireless power transmission. IEEE Trans Microw Theory Tech 70(1):650–658

KR Li, KY See, WJ Koh, and JW Zhang (2017) Design of 2.45 GHz microwave wireless power transfer system for battery charging applications. 2017 Progress in electromagnetics research symposium-Fall (PIERS–FALL). pp 2417–2423

Litchfield M, Schafer S, Reveyrand T, Popović Z (2014) High-efficiency X-Band MMIC GaN power amplifiers operating as rectifiers. 2014 IEEE MTT-S International Microwave Symposium (IMS2014):1–4

Matsui K et al (2017) Microstrip antenna and rectifier for wireless power transfar at 94 GHz. Proceedings of IEEE Wireless Power Transfer Conference (WPTC), Taipei, pp 1–3

Mizojiri S et al. (2019) Demonstration of sub-terahertz coplanar rectenna using 265 GHz gyrotron, 2019 IEEE MTT-S wireless power transfer conference (WPTC), London

Mizojiri S et al (2018) Subterahertz wireless power transmission using 303-GHz Rectenna and 300-kW-class Gyrotron. IEEE Microw Wireless Comp Lett 28(9):834–836
Mukherjee R, Heiges M, Matthies L (2010) Review of microautonomous vehicle sizing and power modelling. Micro- Nanotechnol Sensors Sys Applications II 7679:268–282
Nakamura M et al (2016) A 5.8 GHz-band high efficiency rectifier with a low resistance and high breakdown voltage GaAs Schottky Barrier Diode. IEICE Tech Rep 116(12):11–15
Nariman M, Shirinfar F, Pamarti S, Rofougaran A, Flaviis FD (2017) High-efficiency millimeter-wave energy-harvesting systems with Milliwatt-level output power. IEEE Transon circuits and systems-II 64(6):605
Nex F, Remondino F (2014) UAV for 3D mapping applications: a review. Appl Geomat 6(1):1–15
Nguyen DH, Suganuma S, Shimamura K, Mori K (2020) Millimeter wave power transfer to an autonomously controlled micro aerial vehicle. Transact JSASS 63(3):101–108
Nugent TJ, Kare JT (2010) Laser power for UAVs. In: Laser motive white paper-power beaming for UAVs. NWEN, Seattle, WA
Ohira T (2013) Power efficiency and optimum load formulas on RF rectifiers featuring flow-angle equations. IEICE Electronics Express (ELEX) 10(11):1–9
Pines DJ, Bohorquez F (2006) Challenges facing future micro-air-vehicle development. J Aircr 43: 290–305
Reveyrand T, Ramos I, Popovic Z (2012) Time-reversal duality of high-efficiency RF power amplifiers. Electron Lett 48(25):1607–1608
Rim CT, Mi C (2017) Wireless power transfer for electric vehicles and mobile devices. Wiley, New York
Roberg M, Reveyrand T, Ramos I, Falkenstein EA, Popovic Z (2012) High-efficiency harmonically terminated diode and transistor rectifiers. IEEE Trans Microw Theory Tech 60(12):4043–4052
Rodenbeck C et al (2021) Microwave and millimeter wave power beaming. IEEE J Microwave 1(1):229
Sakai N, Noguchi K, Itoh K (2020) 5.8 GHz band high efficiency 1 W rectenna with a short stub connected dipole antenna. IEICE Tech Rep 120(172):7–12
Schafer S, Coffey M, Popović Z (2015) X-band wireless power transfer with two-stage high-efficiency GaN PA/rectifier. IEEE Wireless Power Transfer Conference (WPTC) 2015:1–3
Schlesak J, Alden A, Ohno T (1988) A microwave powered high altitude platform. In: Proceedings of the IEEE MTT-S International Microwave Symposium Digest. IEEE, New York, pp 283–286
Shaulov E, Jameson S, Socher E (2017) W-band energy harvesting rectenna array in 65-nm CMOS. IEEE MTT-S International Microwave Symposium (IMS) 2017:307–310
Shimamura K et al (2017) Feasibility study of microwave wireless powered flight for micro air vehicles. Wireless Power Transfer 4(2):146–159
Shinohara N (2013) Beam control technologies with a high-efficiency phased Array for microwave power transmission in Japan. Proc IEEE 101(6):1448–1463
Song BD, Park K, Kim J (2018) Persistent UAV delivery logistics: MILP formulation and efficient heuristic. Comput Ind Eng 120:418–428
Suganuma S et al (2021) 28 GHz microwave-powered propulsion efficiency for free flight demonstration. J Spacecr Rocket 59:342
Sugimori K, Oosaki K, Ueda H (1999) RF to DC energy converter electron tube for SPS system application (Part2) -comparison with the simulation results. Memoirs of National Institute of Technology Kurume College 15(1):5–10. (in Japanese)
Suh YH, Chang K (2002) A high-efficiency dual-frequency rectenna for 2.45- and 5.8-GHz wireless power transmission. IEEE Trans Microw Theory Tech 50(7):1784–1789
Suh I-S et al (2015) Wireless charging technology and the future of electric transportation. SAE international, Warrendale, PA
Sun H, Guo Y, He M, Zhong Z (2012) Design of a high-efficiency 2.45-GHz rectenna for low-input-power energy harvesting. IEEE Antennas Wirel Propag Lett 11:929–932

Tornerro JLG, García MP, Quirós RG, Arnause JCS (2016) Design of Ku-band wireless power transfer system to empower light drones. 2016 IEEE wireless power transfer conference (WPTC) Aveiro. pp 1–4
Torresan C, Berton A, Caronenuto F, Gennaro SFD, Gioli B, Matese A, Miglietta F, Vagnoli C, Zaldei A, Wallace L (2017) Forestry applications of UAVs in Europe: a review. Int J Remote Sens 38(8–10):2427–2447
Traub LW (2011) Range and endurance estimates for battery-powered aircraft. J Aircr 48(2): 703–707
Vanke VA, Savvin VL, Boudzinski IA, Bykovski SV (1995) Development of cyclotron wave converter. Abstracts of the Second International Wireless Power Transmission Conference WPT'95, October, Japan
Wang C, Yang B, Shinohara N (2021a) Study and design of a 2.45-GHz rectifier achieving 91% efficiency at 5-W input power. IEEE Microw Wirel Compon Lett 31(1):76–79
Wang Y, Wei G, Dong S, Dong Y, Yu X, Li X (2021b) A high-efficiency self-synchronous RF–DC rectifier based on time-reversal duality for wireless power transfer applications. Electronics 11: 90
Watson DC, Grow RW, Johnson CC (1971) A cyclotron-wave rectifier for S-band and X-band. IEEE Trans Electron ED-18(1):3
N Weissman, S Jameson, and E Socher (2014) Wband CMOS on-chip energy harvester and rectenna. IEEE MTT-S Int Microw Symp Dig. pp 1–3
Wu P, Chen Y, Zhou W, Ren ZH, Huang SY (2020) A wide dynamic range rectifier array based on automatic input power distribution technique. IEEE Microw Wirel Compon Lett 30(4):437–440
Yang X et al (2008) X-band circularly polarized rectennas for microwave power transmission applications. J Electron 25:389
Ye J, Yang C, Zhang Y (2016) Design and experiment of a rectenna array base on GaAs transistor for microwave power transmission. 2016 IEEE international conference on microwave and millimeter-wave technology (ICMMT), Beijing, pp 323–326
Yoo TW, Chang K (1992) Theoretical and experimental development of 10 and 35 GHz rectennas. IEEE Transac MTT 40(6):1259–1266
Yoshida S, Nishikawa K, Kawasaki S (2019) 10W class high power C-band rectifier using GaN HEMT. IEEE Wireless Power Transfer Conference (WPTC) 2019:595–598
Zhang C, Kovacs JM (2012) The application of small unmanned aerial systems for precision agriculture: a review. Precis Agric 13(6):673–712
Zhao X, Tuo X, Ge Q, Peng Y (2017) Research on the high power cyclotron-wave rectifier. Phys plasma 24(7):073117
Zhao D, He P, Wang X (2019) Millimeter-wave rectenna and rectifying circuits for far-distance wireless power transfer. 2019 12th Global Symposium on Millimeter Waves (GSMM). pp 90–92

Chapter 2
Microwave Rocket

Yasuhisa Oda and Toshikazu Yamaguchi

Abstract This comprehensive study further extends into the concept of microwave rockets, elucidating their working principle and thrust generation mechanisms. Experimental applications of high-power microwave sources are explored alongside thrust measurements and performance trends observed with single- and multi-pulse operations. It delves into the role of millimeter wave plasma in the thruster and investigates the impact of beam profile and frequency. The chapter also presents demonstration experiments, including single- and multi-pulse flight experiments, as well as a thrust generation experiment involving long-distance transmission-type thruster with beam optical conversion. An analysis of the microwave rocket system and launch system beam source concludes the study. This work provides vital insights into the advancement of microwave power technologies and their potential in propulsion systems.

Keywords Microwave rockets · Thrust generation mechanisms · High-power microwave sources · Single pulse · Multi-pulse · Millimeter wave plasma · Beam profile · Flight experiment · Long-distance transmission thruster · System analysis

What Is a Microwave Rocket?

A propulsion system that receives propulsive energy from outside of a vehicle remotely using a microwave beam is called microwave propulsion, and several methods have been proposed.

One of the oldest studies on microwave propulsion is a method which utilizes stationary plasma generated by a continuous wave (CW) microwave. Plasma in a

Y. Oda (✉)
Department of Mechanical Engineering, Setsunan University, Osaka, Japan
e-mail: yasuhisa.oda@mec.setsunan.ac.jp

T. Yamaguchi
Department of Communication and Business, Edogawa University, Chiba, Japan
e-mail: tyamaguc@edogawa-u.ac.jp

K. Mori et al. (eds.), *Beamed-mobility Engineering*, Springer Tracts in Electrical and Electronics Engineering, https://doi.org/10.1007/978-981-99-4618-1_2

chamber heats the propellant and exhaust jet. Some research work verified plasma ignition and heating of the propellant by microwave power.

A microwave thermal propulsion is the system in which steady microwave power heats propellant without plasma. Microwave power is injected into ceramic pipes in which propellant flows and heats up propellant up to several thousand K and exhaust jet through nozzle to generate thrust force. This type of microwave propulsion system was studied in NASA, and a demonstration experiment of propellant heating using a gyrotron for fusion heating was also planned.

On the other hand, a method of generating thrust by the shock wave induced by the discharge plasma which was generated by pulsed microwave power was proposed by the University of Tokyo and the Japan Atomic Energy Research Institute (JAERI), currently the National Institute for Quantum Science and Technology (QST). The research was carried out experimentally using a gyrotron with 1 MW output, including the indoor flight experiments as demonstration. Since this method can generate a thrust force larger than the thruster weight because the propeller can have a very simple structure consisting only of a reflector that also serves as a nozzle, we named this system as "microwave rocket," aiming a launch vehicle to space.

This chapter summarizes studies on microwave rocket representing remarkable experimental results.

Thrust Generation Mechanism of Microwave Rocket

The principle of microwave rocket that receives a pulsed microwave and generates thrust is explained as follows:

The microwave beam is focused by a reflector, and focused power induces a discharge of the propellant gas. The discharge plasma absorbs microwave power and expands rapidly, generating shock waves in front of microwave plasma. The physical drive mechanism of this shock wave will be described in detail in another chapter of this book.

The shock wave forms high pressure in the thruster, which gives a force to the reflector that also serves as a fluid dynamic nozzle.

Microwave rocket utilizes the beam-shaped microwave which has advantage in plasma growth in one-dimensional direction. Microwave rocket has a cylindrical nozzle to convert the energy absorbed by one-dimensionally developed plasma which is generated by ms-order long pulses into thrust. Since it is intended for launch applications, its operation is premised on being performed in the atmosphere, and air acts as a propellant.

The air plasma ignited near the focused reflector induces a shock wave propagating toward the outlet of the cylindrical nozzle. At this time, the pressure behind the shock wave is maintained for a certain period of time inside the thruster, similar to the detonation engine. As the plasma growth reaches the end of the cylindrical nozzle, the pressure behind the shock wave is exhausted from the nozzle, and as the

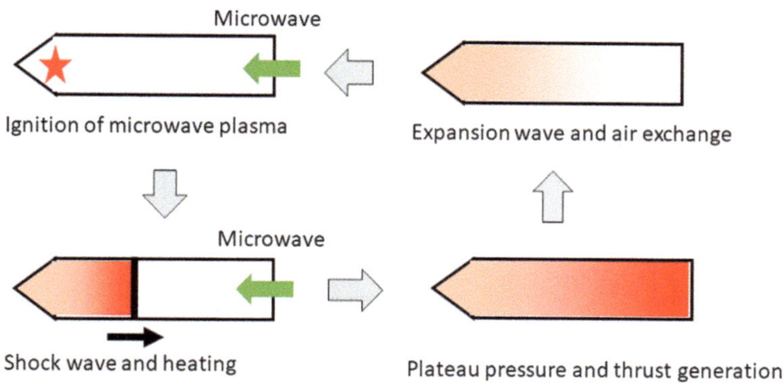

Fig. 2.1 Thrust generation cycle of microwave rocket

Fig. 2.2 Typical pressure history at thruster wall through a cycle

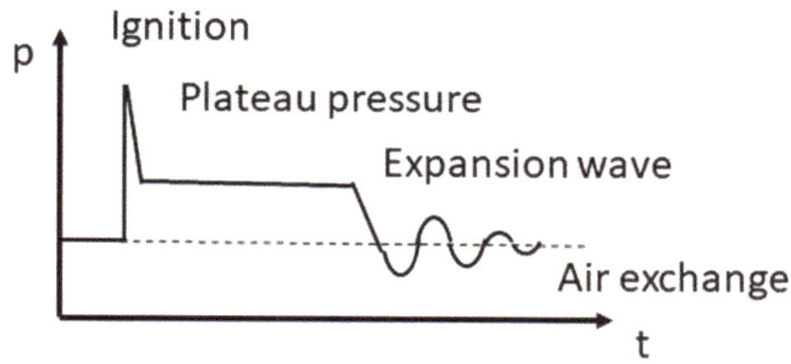

expansion wave propagates from the end of the nozzle, the pressure inside the propeller decreases.

Since the pressure inside the thruster after expansion wave propagation is lower than the ambient air pressure, new air is taken in from the periphery and returns to the initial state. When the next pulse is incident here, a new thrust generation cycle is started (Oda et al. 2006a).

Shock wave driven by microwave plasma in tube forms microwave supported shock wave, which consists of shock adiabatic compression, heating region, and rarefaction wave as shown in Figs. 2.1. The pressuare at closed end of the tube keeps constant, namely, plateau pressure, then pressure deline by expansion wave follows as shown in Fig. 2.2. The relation of gas parameters, pressure p and temperature T, in each region as shown in Fig. 2.3 is represented in the following way:

Region 1 to region 2 is shock relation:

$$\frac{p_2}{p_1} = 1 + \frac{2\gamma}{\gamma + 1}\left(M_{\text{shock}}^2 - 1\right)$$

Here γ and M_{shock} are specific heat ratio and Mach number of shock wave, respectively.

Region 2 to region 3 is heating relation:

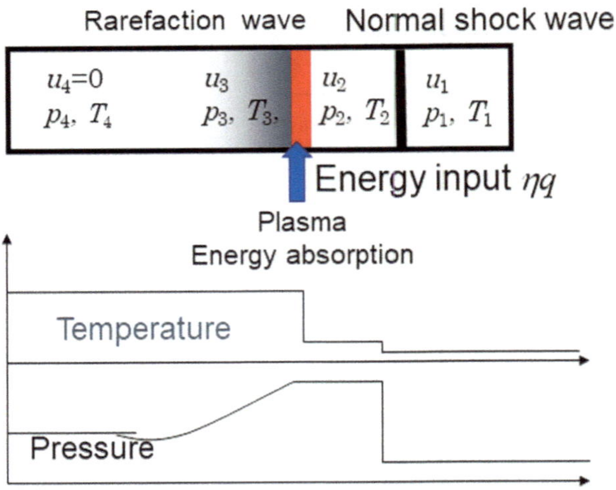

Fig. 2.3 Distribution of microwave supported shock wave

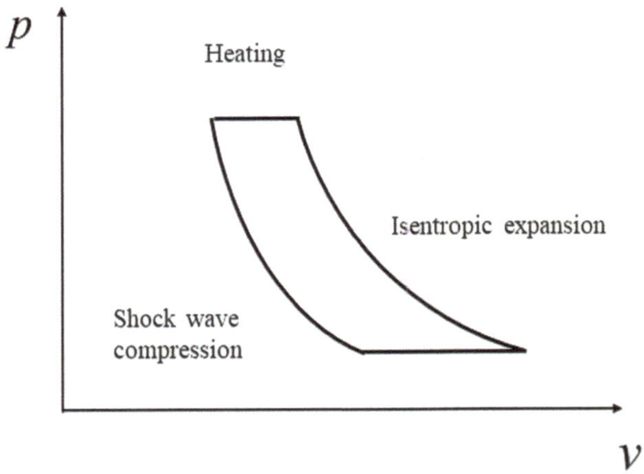

Fig. 2.4 p-V diagram of microwave rocket

$$p_2 = p_3, T_3 = T_2 + \frac{\eta S}{c_p u \rho}$$

Here S, c_p, ρ, u, and η are the average power density, isobaric specific heat, fluid density, local fluid velocity, and absorption efficiency of millimeter-wave power in the plasma heating region, respectively.

Region 3 to region 4 is rarefaction wave relation:

$$\frac{p_4}{p_1} = \frac{p_3}{p_1}\left[1 - \frac{\gamma - 1}{2} M_{3c}\right]^{2\gamma/(\gamma-1)}, \left(M_{3c} = \frac{u_3 - U_{\text{shock}}}{\sqrt{\gamma R T_3}}\right)$$

Here, M_{3c} is the Mach number of the rarefaction wave head relative to the tube end (Oda et al. 2009).

A series of cycles consists of adiabatic compression by shock waves, heating by plasma, adiabatic expansion by nozzles, and intake/exhaust processes, which can be represented as follows on the p-V diagram as shown in Fig. 2.4.

Thrust Generation Experiment and Performance Trend of Microwave Rocket

Experimental Application of High-Power Microwave Sources

Experiments of microwave rockets require the use of microwave beams in the 100 GHz band at the level of several 100 kW to 1 MW. A joint research between the University of Tokyo and the Japan Atomic Energy Research Institute utilized a 170 GHz/1 MW long-pulse gyrotron, which was developed for the RF power source for the electronic cyclotron heating and current drive system in ITER. Since this gyrotron was developed for fusion heating applications, it required some addition to apply to microwave rocket experiments (Oda et al. 2019).

Gyrotrons in the plasma heating system provide their output to fusion plasma in a fusion device or a dummy load through evacuated transmission lines. Since microwave rocket experiment requires to radiate microwave power into the atmosphere, microwave power is delivered to a shield room, a dedicated experimental facility. The shield room was originally installed for the purpose of developing a power injection device for plasma, namely, a launcher, and it is designed so that microwaves radiated inside the shield room do not leak to the outside, and the door of the shield room has interlock system for human safety. During the experiment, access to the shield room is restricted, and all measurements are assumed to be performed remotely.

In order to extend the transmission line to the shield room, a transmission line with a total length of 40 m composed of corrugated waveguides with an inner diameter of 63.5 mm was utilized. In order to keep the inside of the transmission line in 10–3 Pa level of vacuum, a diamond window or sapphire window is installed at the end of the transmission line. The same diamond window or sapphire window is used as the gyrotron output window. It was difficult to completely suppress the influence of reflection from the window to radiate microwave to the atmosphere, and it was limitation of the operating conditions of the gyrotron (Oda et al. 2018).

Further, to produce repeating pulses of the gyrotron power, high-voltage power supply of gyrotron produced modulated output for the anode voltage or the body voltage to control the gyrotron oscillation. In the microwave rocket experiment, it was necessary to find the operating parameters of the gyrotron while observing the behavior of the plasma in the microwave rocket in order to completely stop the microwave power between pulses.

Thrust Measurement Experiment and Performance with a Single Pulse

Thrust performance of microwave rocket under single pulse with 1 ms or less pulse width was observed as a first step. An aluminum parabolic mirror and a conical mirror combined with a cylindrical body, namely, "rocket style" model, are utilized as thruster models for thrust measurement as shown in Fig. 2.5.

The thrust of the aluminum parabolic mirror thruster was measured by the vertical flight test since the thruster could generate a sufficiently large thrust impulse to drive vertical motion of a thruster with weight of 100 g. The flight motion of the thruster was measured using a laser displacement gauge, and the thrust impulse was deduced from the initial velocity. Typical result of the aluminum parabolic mirror thruster is presented in Figs. 2.6 and 2.7. By increasing pulse duration of microwave, thrust impulse was increased. Indeed, the momentum coupling coefficient, the ratio of

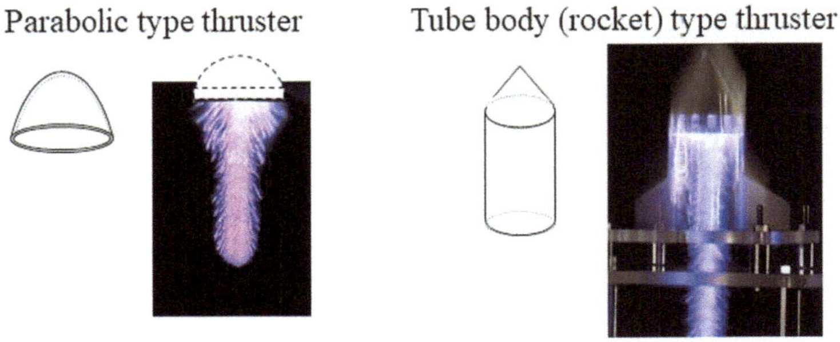

Fig. 2.5 Thruster models applied for thrust generation test

Fig. 2.6 Dependence of thrust impulse of parabolic thruster model on pulse width and power of microwave

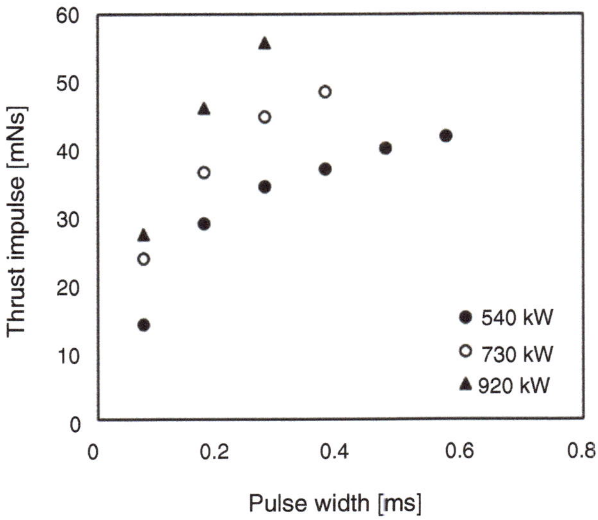

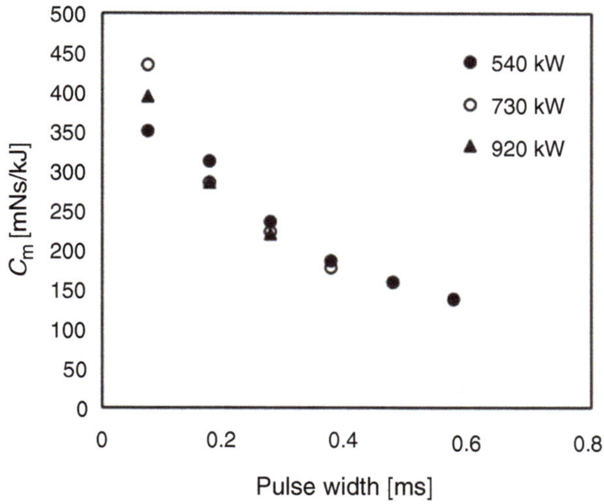

Fig. 2.7 Dependence of momentum coupling coefficient of parabolic thruster model on pulse width and power of microwave

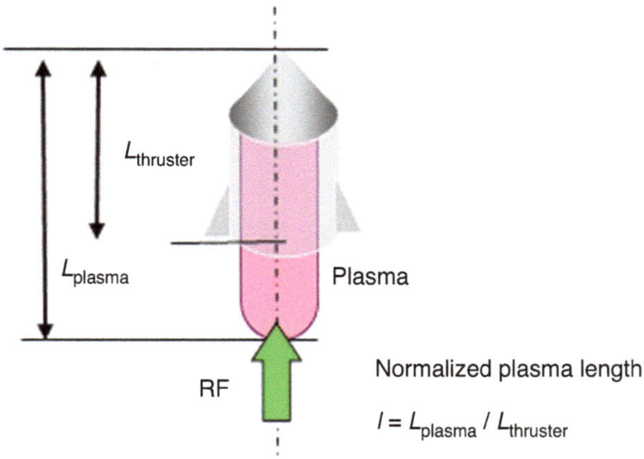

Fig. 2.8 Definition of normalized plasma length

thrust impulse and injected energy, was declined by increment of pulse duration. At the peak, coupling coefficient exceeds 400 mNs/kJ (Nakagawa et al. 2004).

The thrust of "rocket style" thruster models which combines a conical mirror with a cylindrical body was measured by vertical flight test or pressure measurement in the thruster, verifying the internal shock wave behavior to calculate the thrust force from the pressure history. Especially thruster model with long body was too heavy for vertical flight test; pressure measurement is useful for this case. Typical result of rocket style thruster model is presented in Figs. 2.8, 2.9, and 2.10. The thrust impulse rises linearly with the increase of the pulse duration, and its trend changes to reach a plateau at some point. The momentum coupling coefficient kept a constant value for shorter pulse duration than starting plateau of thrust impulse. The maximum pulse duration of constant coupling coefficient increases with the length of the cylindrical body. In this case, when the gyrotron output is 850 kW and the cylinder length is 180 and 380 mm, the length of the plasma in the thruster body was found as 40% of

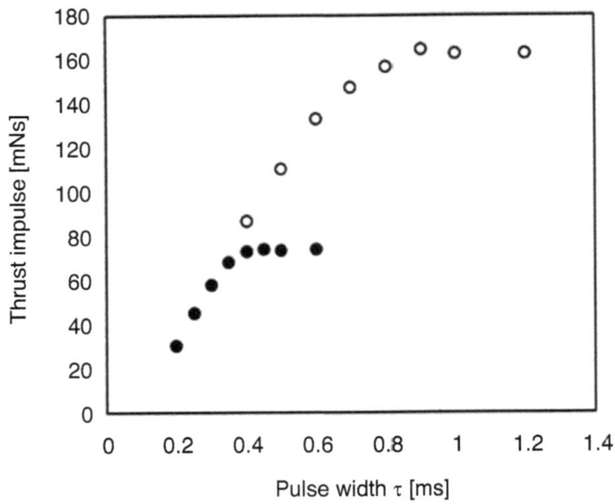

Fig. 2.9 Dependence of thrust impulse of rocket-type thruster model on pulse width and power of microwave. ● $L = 180$ mm model at 680 kW, ○ $L = 380$ mm model at 680 kW

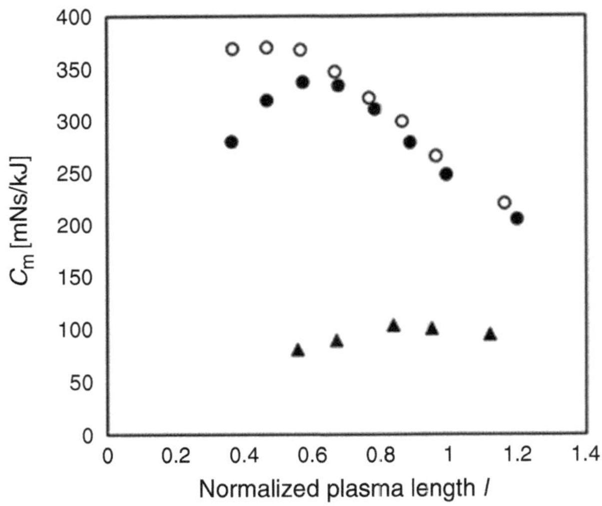

Fig. 2.10 Dependence of momentum coupling coefficient of rocket-type thruster model on pulse width and power of microwave. ● $L = 180$ mm model at 680 kW, ○ $L = 380$ mm model at 680 kW, ▲ $L = 380$ mm model at 320 kW

the length of the thruster body. For both cases, the momentum coupling coefficient is about 350 mNs/kJ, and the momentum coupling coefficient decreases at a long-pulse duration (Oda et al. 2006b).

As conclusion of single pulse operation, the thruster models with the cylindrical body can maintain the momentum coupling coefficient for extended pulse duration so that the greater thrust impulse can be generated with longer thruster models. Since then, studies using "rocket-type" thruster model have become the mainstream of microwave rocket.

Millimeter-Wave Plasma in the Thruster

The key of the thrust generation mechanism of microwave rockets is the characteristics of atmospheric discharge of 100 GHz band high-power microwaves (hereinafter referred to as “millimeter waves” only in this section).

In a “rocket-type” thruster model, the length of the plasma changes depending on the width of the pulse. Photographs of plasma development taken by a high-speed camera present that the ionization front of plasma propagates at a constant speed. Its propagation speed depends on the power density of the millimeter wave and increases with increasing power level. The speed is order of sonic speed in atmospheric air. From point of thrust generation, this speed range is suitable for shock wave driving in the air.

The shock wave accompanying by propagation of plasma can be visualized by the shadow graph method. The visualized wave front in the closed space, such as the cylindrical body part of the thruster model, confirmed that the vertical shock wave was formed in front of the ionization front of plasma (Oda et al. 2011).

A time-integrated photograph of the millimeter-wave atmospheric plasma represents the structure on the fine filament around to the columnar growth history of the plasma. It has been clarified that this structure is caused by the electromagnetic field distribution near the ionization front. The electromagnetic field distribution has the millimeter-wave wavelength scale structure, resulting in the behavior of a fine spot plasma motion, and their trajectories form the macroscopic filament structure (Oda et al. 2020).

The behavior of plasma in the focusing reflector represents that ionization front of plasma propagates toward direction of millimeter-wave source. The millimeter wave is almost absorbed by the ionization front of plasma, and there is no energy behind it.

The propagation velocity of plasma and its shape change greatly with ambient pressure. From atmospheric pressure (1 atm) to about 0.1 atm, plasma has filament structure, and the propagation speed gradually increases as the pressure decreases. For lower pressure, less than 0.1 atm condition, the filamentous structure disappears. In such plasma condition, the thruster cannot generate thrust (小田 小紫 2006).

The pressure threshold of the filament structure formation depends on the gas, and the filament shape was not formed with argon and helium at 1 atm condition (Figs. 2.11, 2.12, 2.13, and 2.14).

Multi-Pulse Thrust Generation Experiment and Performance

Thrust under multi-pulse injection was measured and its thrust performance was analyzed. For repetitive pulse oscillation of gyrotron, voltage modulation of the body power supply is utilized. Modulation at a maximum of 5 kHz is available on JAEA gyrotron operation system. Through the thrust measurement of repetitive pulsed operation, the thrust force was evaluated mainly by the method of estimating

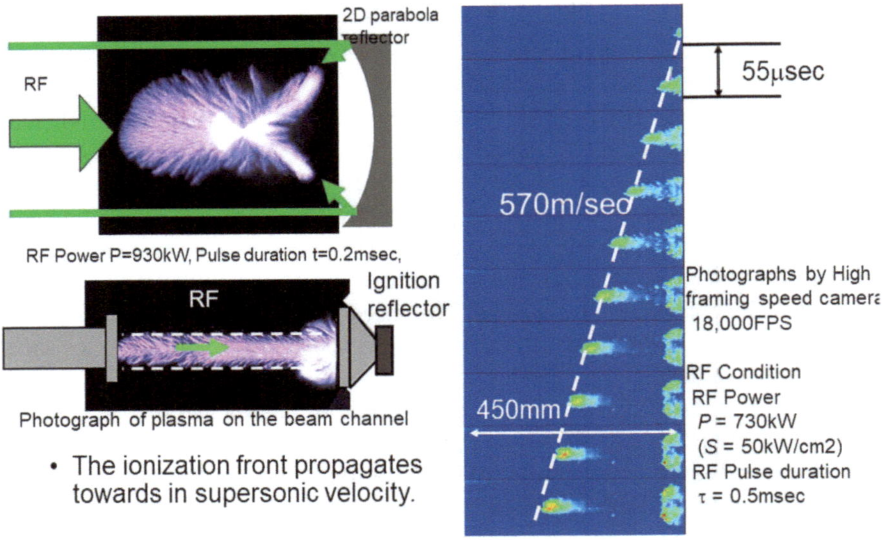

Fig. 2.11 Integrated and fast-framing photograph of millimeter-wave plasma. Left-up, plasma in 2D parabola mirror; left-down, plasma on the RF beam from corrugated waveguide; right series of fast-framing photographs of plasma development on the RF beam

Fig. 2.12 Shadow graph photo of shock wave generated by millimeter-wave plasma in a tube

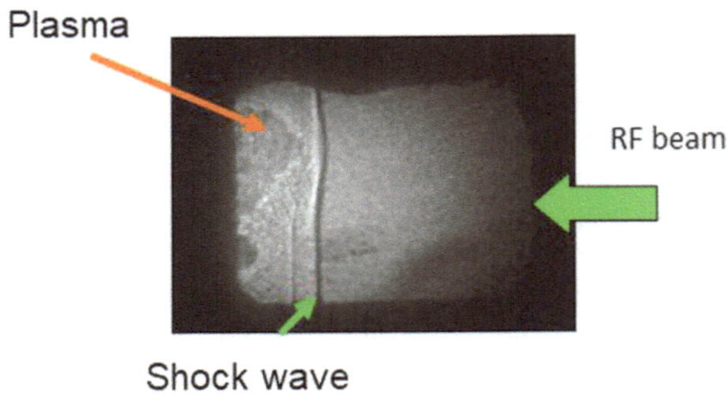

from the pressure history inside the thruster. For the condition that continuous thrust generation can be expected, the thrust measurement using the pendulum type thrust stand was performed.

In multi-pulse operation, the thrust impulses after the second pulse became smaller than the impulse of the single pulse operation, and the momentum coupling coefficient tended to decrease pulse by pulse. When the air exchange between the pulses is insufficient, the temperature of the air in the thruster remains high. In such condition, the propagation speed of the shock wave increases and the peek pressure decreases, and then the thrust impulse decreases. In order to clarify the effect of air exchange between pulses, we conducted an experiment using a thruster with a forced air exchange system that supplies compressed air from the focusing mirror, and it forcibly exhausts air through thruster tail. The experiment verified the change of

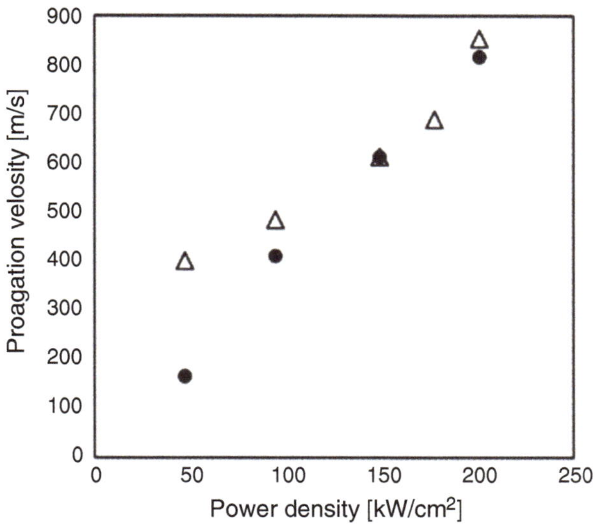

Fig. 2.13 Dependence of propagation velocity of ionization front of millimeter-wave plasma and shock wave on millimeter-wave power density. Millimeter-wave frequency is 170 GHz

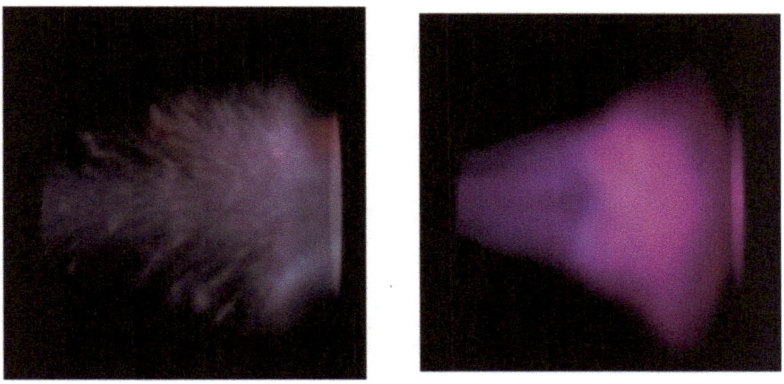

Fig. 2.14 Photograph of millimeter-wave breakdown plasma in reduced pressure condition: left, 1.0 atm; right, 0.1 atm

thrust performance by air exchange. As a result, the momentum coupling coefficient had a tendency to recover by increasing amount of supplied air during the interval of pulses. Figure 2.15 presents this result using a rate of air replacement in the thruster, namely, partial filling rate (Oda et al. 2008) (Figs. 2.16 and 2.17).

From this result, we obtained the following design restriction of microwave rocket: in the single pulse operation, a large thrust impulse can be obtained with a thruster with a longer cylindrical part; in the multi-pulse operation, since it takes time to replace the air in the thruster, there is a limitation of the minimum interval between the pulses; otherwise, the decrease in the momentum coupling coefficient is caused. Then the thruster with longer cylindrical part works as disadvantage in multi-pulse operation. For the generation of continuous thrust force, appropriate design

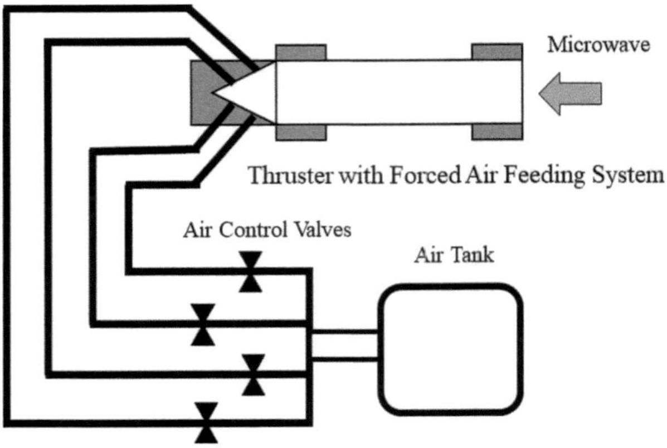

Fig. 2.15 Diagram of force air feeding system for measurement of partial filling rate effect under multi-pulse operation of microwave rocket

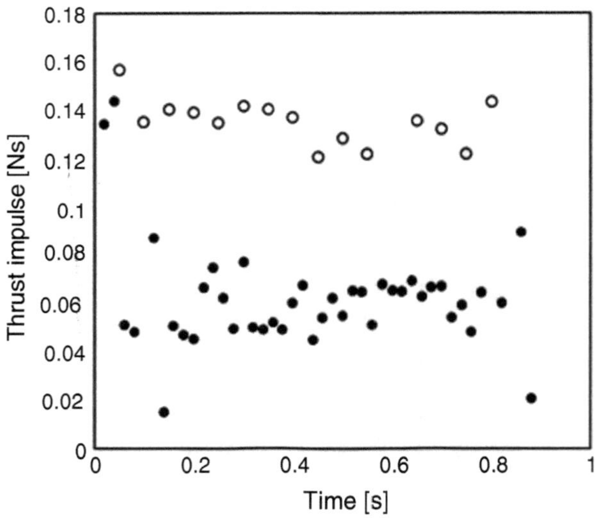

Fig. 2.16 Thrust impulse of each pulse under multi-pulse operation. ● repetition rate 50 Hz, ○ repetition rate 20 Hz

conditions shall exist. Moreover, the effective mechanism that promotes air exchange between pulses, like an active vane system, is the key to improving performance.

In the pulse repetition experiment without active air exchange system, a thrust of 30 N was generated under the repetition condition of 200 Hz with a pulse width of 0.8 ms at a gyrotron output of 570 kW. This result is the maximum thrust level obtained by beaming propulsion. It is expected to realize a system with practical performance in combination with active air exchange mechanism (小松怜史, 山口敏和 2012).

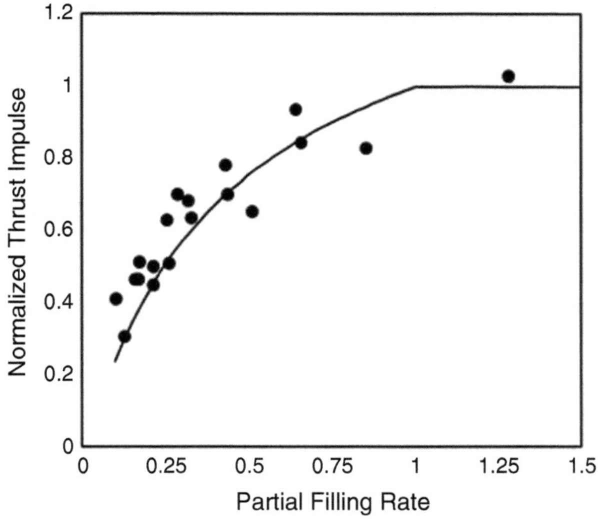

Fig. 2.17 Dependence of normalized thrust impulse on partial filling rate

Impact on Beam Profile and Frequency

The MW-class high-power microwave system utilizes corrugated waveguides for power transmission. Radiation from the corrugated waveguide has a beam profile with a single peak since the fundamental transmission mode (HE11 mode) in the corrugated waveguide is coupled to a Gaussian beam in free space. The effect of beam profile to thrust performance was evaluated. Using phase correcting mirror, the flat top beam profile which is superposition of the Gaussian-like beam and hollow beam is applied for thrust measurement, and thrust force is generated as similar to Gaussian-like beam profile.

In addition, we conducted a thrust measurement experiment at 134 GHz in addition to 170 GHz using the multifrequency gyrotron. As a result, the same level of thrust generation can be obtained. This is preliminary result for frequency dependence of microwave rocket performance study, and further observation with wide frequency range condition is necessary (Fig. 2.18) Oda et al. (2010b).

Demonstration of Microwave Rocket

Flight Experiment with a Single Pulse

The thrust measurement represented that the discharge caused by the gyrotron output power can generate thrust force; we conducted a flight experiment of a “rocket-type” thruster model that combined a conical focusing mirror and a cylindrical body in 2003. To reduce the weight of the model, the thruster model is made of a plastic sheet, and the focusing part is made of aluminum foil. The weight of the thruster

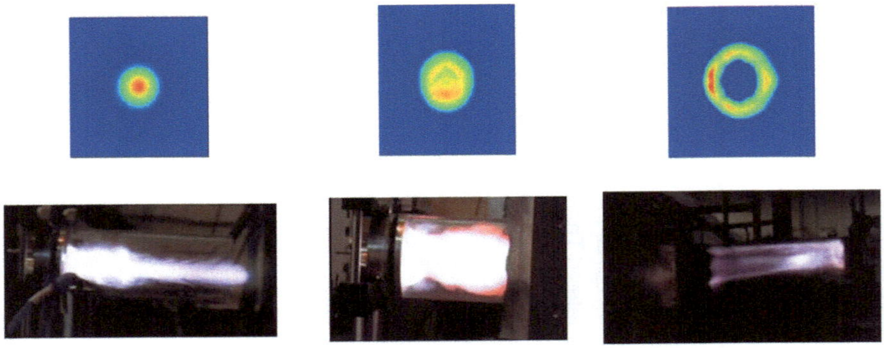

Fig. 2.18 Microwave plasma on different beam profile

Fig. 2.19 Schematic of flight test setup of microwave rocket

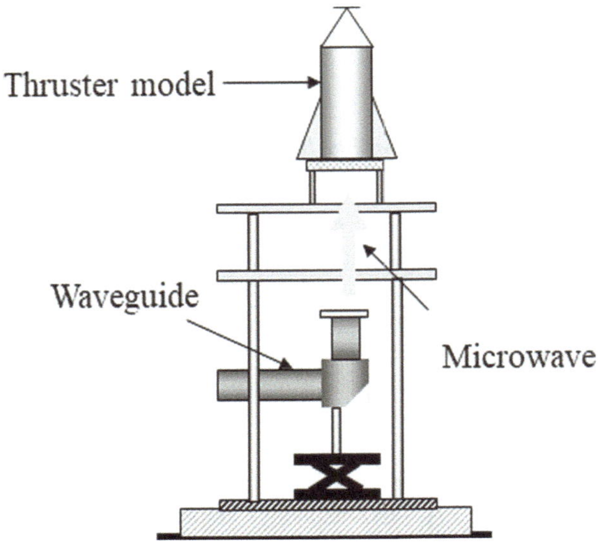

model was 10 g. In the flight experiment, single pulse is injected into the thruster. The pulse width was set to obtain the maximum thrust of the model according to thrust measurement result. Plasma was generated in the thruster with the microwave pulse injection accompanying exploration-like sound of shock wave. The thruster model takes parabolic motion under gravity, and flight reaches to the ceiling 2 m above the initial position by the thrust impulse. It was a demonstration that a sufficiently large thrust force could be generated with respect to the mass of the thruster model (Figs. 2.19 and 2.20). Obviously this was an important achievement as a principle demonstration of a microwave rocket (Nakagawa et al. 2004).

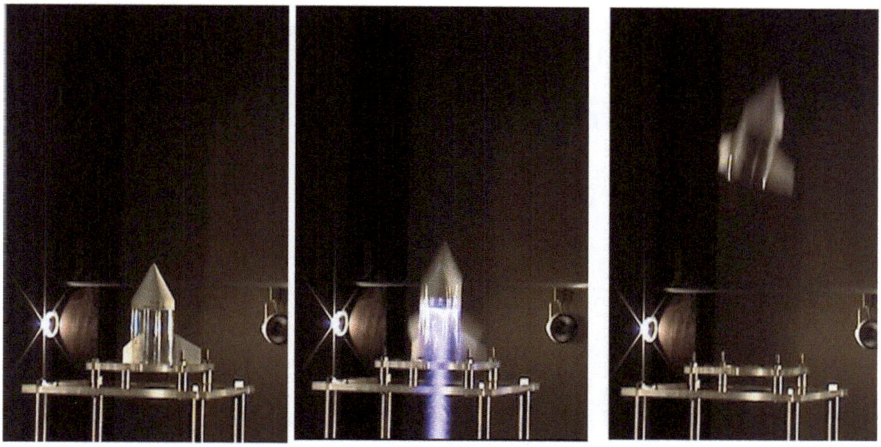

Fig. 2.20 Flight test of microwave rocket with single pulse injection

Flight Experiment with Multi-Pulse Operation

With progress of gyrotron system study, the repetitive pulse operation of gyrotron becomes available. This technology opened the door to a flight experiment of a 100 g "rocket-type" thruster model in 2009. The thruster model is made of a thin aluminum sheet, and the total length is 300 mm and the inner diameter of the cylindrical part is 100 mm, which is larger than that of the single pulse test. To prevent interference between the launch stage and the microwave beam, the thruster is hung with thin wire and placed in the initial position.

In the experiment, a pulse of about 600 kW and 1.25 ms was injected into the thruster model at 100 Hz for about 0.7 s. The thruster model flew to a height of about 1.2 m while continuously generating thrust and then climbed to a point of 2 m by parabolic motion. Movie of a high-speed camera captured the light emitted by plasma ignition inside the thruster model and the deformation vibration of the entire thruster by air exchange process. The beam output system is fixed and no active attitude control is performed by the thruster model, but the autonomous attitude stability of the "rocket-type" propulsion model keeps the attitude almost straight and upward during powered flight. The generation of continuous thrust was successfully demonstrated (Figs. 2.21 and 2.22). There are only a limited number of cases of launch demonstration in free space with repetitive pulses in beaming propulsion system; this demonstration was a great achievement in this field (Oda et al. 2010a).

For higher flight test, there remains an issue of quasi-optics of microwave beam. The microwave beam from the waveguide outlet expands greatly by the beam propagation of 1 m or more. Therefore, when the thruster model is installed at the 2 m point of vertical flight test system, expanded beam could not generate sufficient thrust although the plasma ignited in the thruster. This result indicates that proper beam optics is mandatory for beaming propulsion.

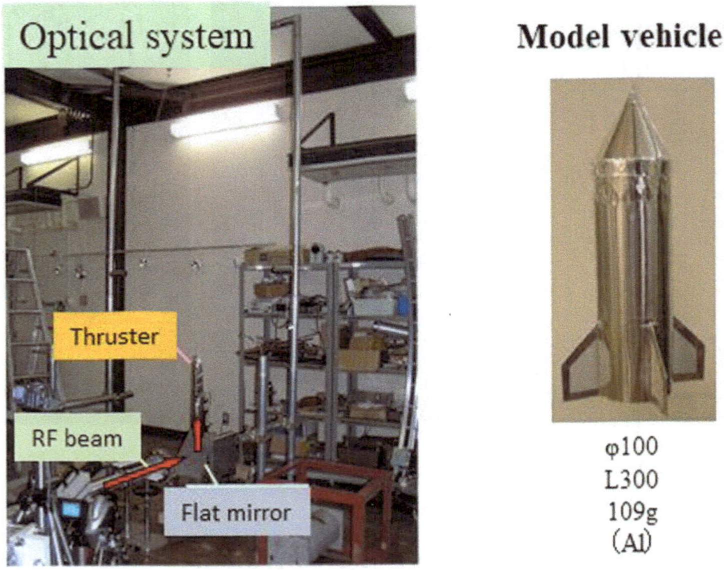

Fig. 2.21 Flight test setup and photograph of model vehicle

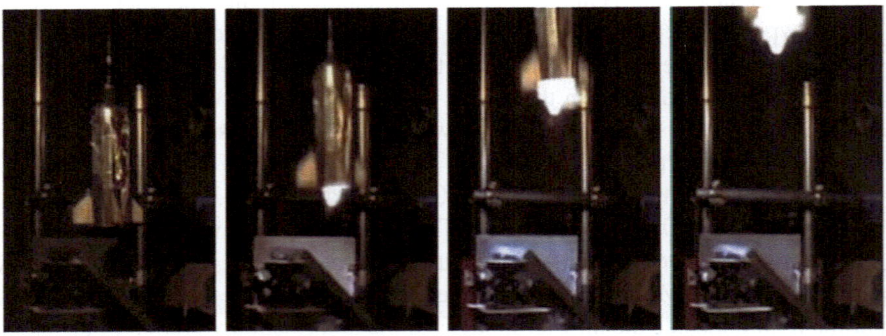

Fig. 2.22 Flight of model vehicle propelled by repetitive microwave pulses

Thrust Generation Experiment of Long-Distance Transmission Type Thruster with Beam Optical Conversion

Proper control of the microwave beam profile is required to provide a constant power to the moving thruster. For example, the beam profile from a waveguide output expands with short distance propagation due to the diffraction, and the power density of the profile is reduced to less than half of that of the initial profile. As a method of mitigating the diffraction, by increasing the beam waist diameter of the propagating beam, it is possible to maintain a power density of almost a certain level over a long distance. However, if the beam waist diameter is increased, the power density is

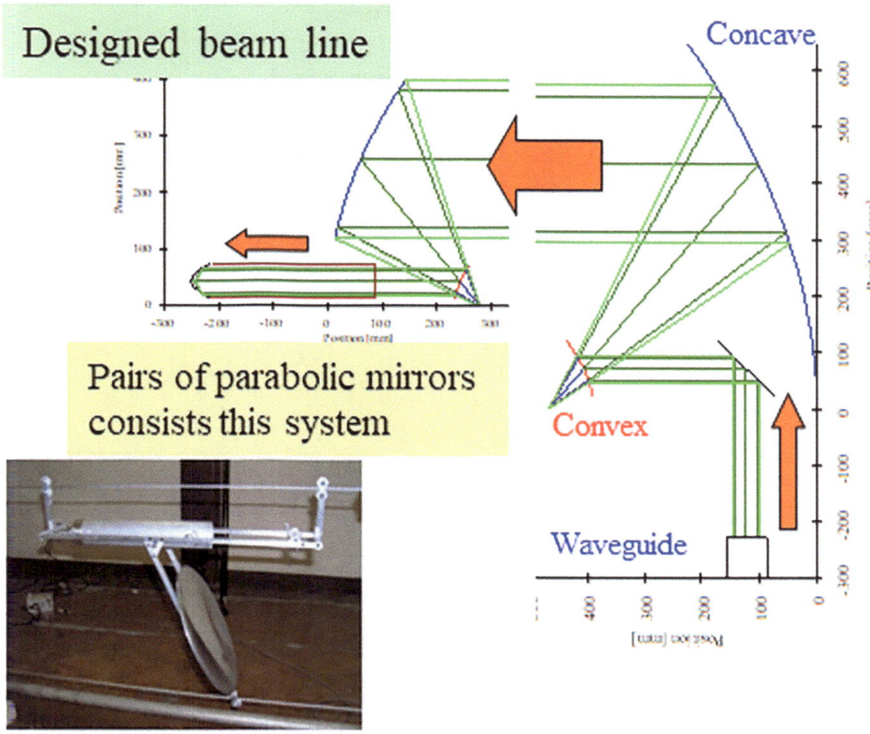

Fig. 2.23 Microwave optics for 5-m-distance beam transmission line

lowered as a whole, which makes it difficult to ignite the plasma, and the efficiency of shock wave driving by the plasma is greatly reduced, so that thrust cannot be generated well. Therefore, it is necessary to use an optical system that converts the beam diameter on the thruster side as well.

For the demonstration of continuous thrust generation of microwave rocket with long-distance traveling, quasi-optics for 5 m distant power transmission is developed in the experimental site at Japan Atomic Energy Agency.

The microwave output from the gyrotron is delivered to the experimental site which is RF shielded room through corrugated waveguide system, and microwave beam is converted into a Gaussian beam with 240 mm of beam waist, of which profile is less affected by spreading with a beam transmission of about 5 m. The beam from corrugated waveguide with 63.5 mm inner diameter is converted to three mirror system which consists of a planar mirror, a machined convex mirror, and a pressed φ750 mm offset parabolic mirror (Fig. 2.23).

On the thruster, a lightweight mirror machine with a diameter of 280 mm is used as the primary mirror to focus the transmission beam, and another mirror with a convex surface provides a reconversion to a beam with the high enough power density to ignite plasma for thrust generation. With usage of an offset parabolic mirror, this beam conversion system is off-axis system, and it could maintain beam

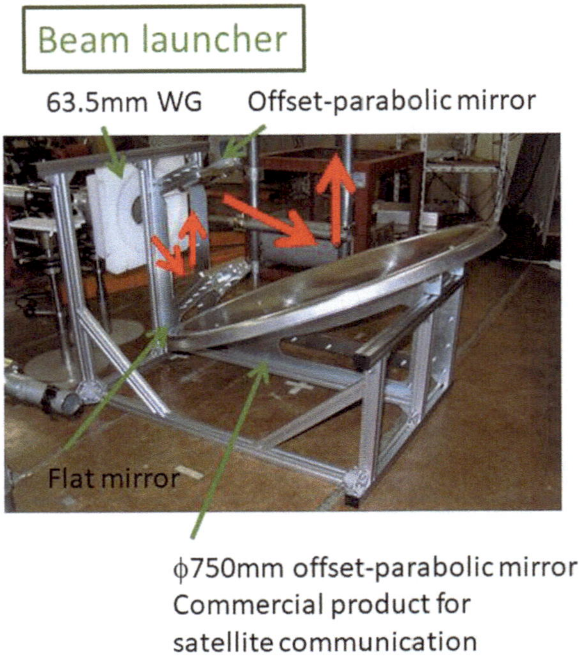

Fig. 2.24 Launcher mirror setup

conversion with respect to movement in the beam axis direction. To catch up the thruster unit motion, the mirror system is fixed on the unit. In addition to mirror system, an air tank is mounted on the thruster unit, and a forced injection system is onboard to increase the thrust. In order to maintain the position of the thruster on the horizontally settled beamline in the shield room, the thruster unit put on the rope and the unit can slide freely by the thrust force (Figs. 2.24, 2.25, and 2.26).

The demonstration represented the thruster model with a beam converter slides stably on the beamline over 5 m by injecting a pulse of about 400 kW/1 ms at 100 Hz repetition for 3 s (Fig. 2.27).

The ignition status of the plasma was compared with the location of the thruster, and the plasma was ignited in the same way at the start point, at middle way, and at a position 5 m away.

The generated thrust at a 2 m position from the start point is estimated from the acceleration motion of the thruster by repetitive pulse injection for 1 s. The dependence of thrust force on repetition rate was measured. The measured momentum coupling coefficient is lower than that of single pulse operation, and the generation of continuous thrust of 3 N over 1 s with forced air injection is demonstrated (Figs. 2.28, 2.29, and 2.30). The dependence of thrust performance was similar to those results obtained without beam conversion system, and we concluded the beam conversion system can provide sufficient support to the thrust generation of microwave rocket for long-distance operation (Oda et al. 2010a).

This demonstration was a significant remark of beaming propulsion which contains all the necessary components, a thruster, air feed system, and beam optics

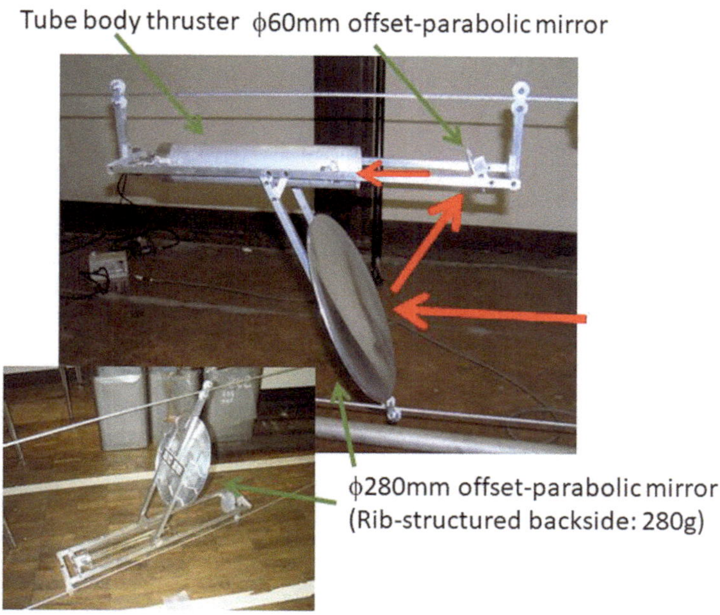

Fig. 2.25 Model vehicle with beam focusing system

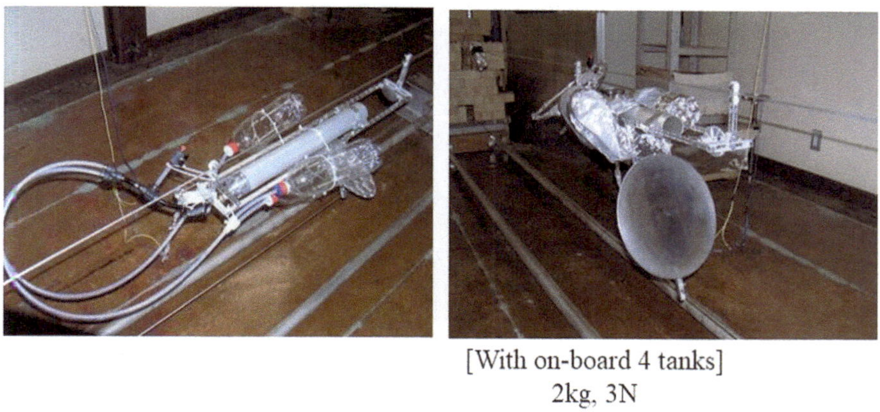

Fig. 2.26 Model vehicle with onboard force air feeding system

for long-distance transmission, onboard, namely system level demonstration of beaming propulsion.

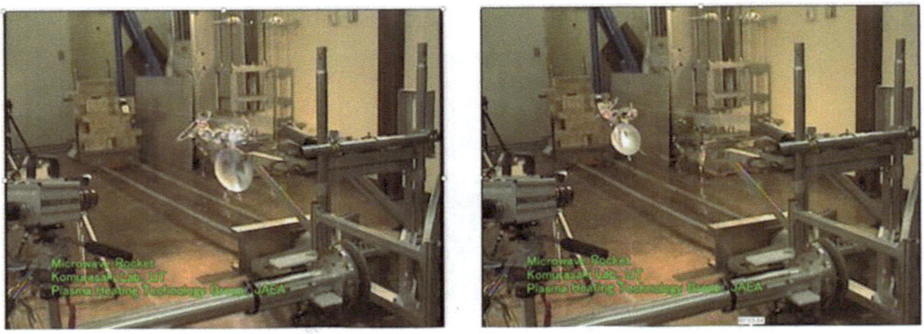

Fig. 2.27 Horizontal acceleration test with multi-pulse operation

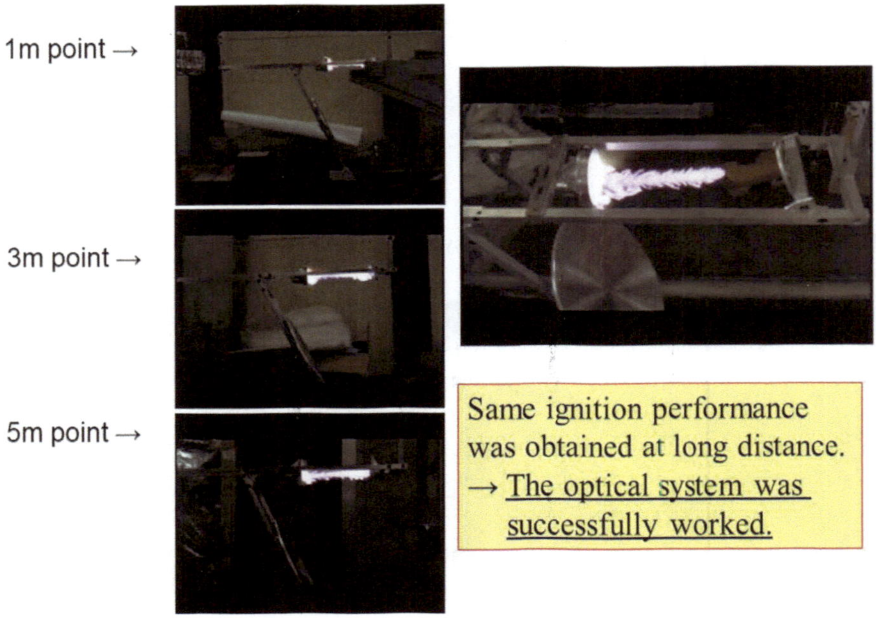

Fig. 2.28 Plasma ignition at distant position from beam launcher

Microwave Rocket System Analysis

Based on the thrust performance parameters obtained in the experiment using the gyrotron, the evaluation of the feasibility of the microwave rocket as an actual launch system was also observed. Beaming propulsion such as microwave rockets aims to build a system that is significantly cheaper than the conventional space transportation system by covering the initial construction cost of the beam source with a large number of launches. Therefore, it is important to estimate the scale of the required beam source.

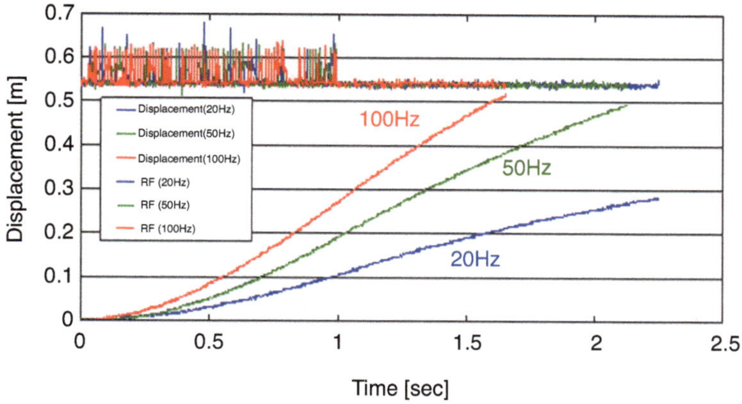

Fig. 2.29 Measured accelerating vehicle motion under different repetition rate

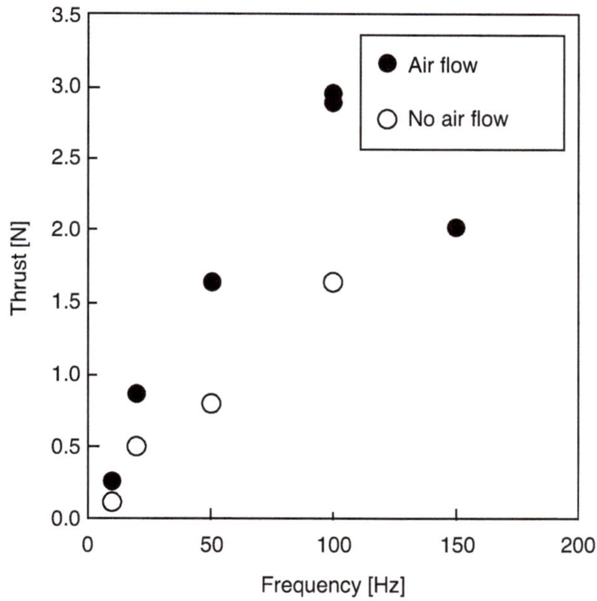

Fig. 2.30 Measured thrust force under multi-pulse operation with long-distance beam transmission

Focusing on the fact that the microwave rocket can generate thrust in the atmosphere, the system study assumed to replace the first stage of the space launch rocket which uses chemical rocket propulsion, and the beam source power was assumed exceeding 100 GW. Although power level is far higher than current system, this can be achieved by a cluster of a large number of gyrotrons. In addition, the antenna scale required for beam transmission within the atmospheric altitude of 10 km, which is the first stage operating area, was analyzed. As a result, a beam of several meters is required, and scale of transmission antenna was assumed as several tens of meters.

For more advanced idea, an antenna system using active phased array system can realize dynamic focusing system on the beam line, and small-scale thruster can be applied. Further study of system analysis is strongly demanded.

Launch System Beam Source

The launch beam source of the microwave rocket is expected to be at least several tens of MW to several GW class and will be configured as a cluster system combining a large number of gyrotrons. A system that combines multiple gyrotrons in this way is common in fusion heating systems. For example, the international fusion project, ITER, delivers 20 MW of power for its electron cyclotron heating system utilizing 24 gyrotrons for power source.

The same method can be used for the beam source of the microwave rocket. To consider the scale of beam source plant, several components play dominant role in design. For instance, the gyrotron needs to have a distance to mitigate the mutual interference of the superconducting magnets required to drive gyrotron, and the high-voltage power supply which produces 50 kV/50 A class for single tube is another dominant device for plant design. Power source which utilizes large number of gyrotron requires a large site. Furthermore, in order to introduce the gyrotron output to the antenna, a transmission system exceeding 100 m is required, and an energy control system for the transmission system is also required. In addition, a large number of devices such as auxiliary equipment required for cooling such as a gyrotron are also required (Fig. 2.31). The figure is an assumed layout of a 1 GW class beam source, but it is considered that a beam source site much larger than the antenna scale is required. Since it is a very large-scale system, the cost will be high, but various aspects are considered in system study for the realization of a microwave rocket (小田, 今井 2017).

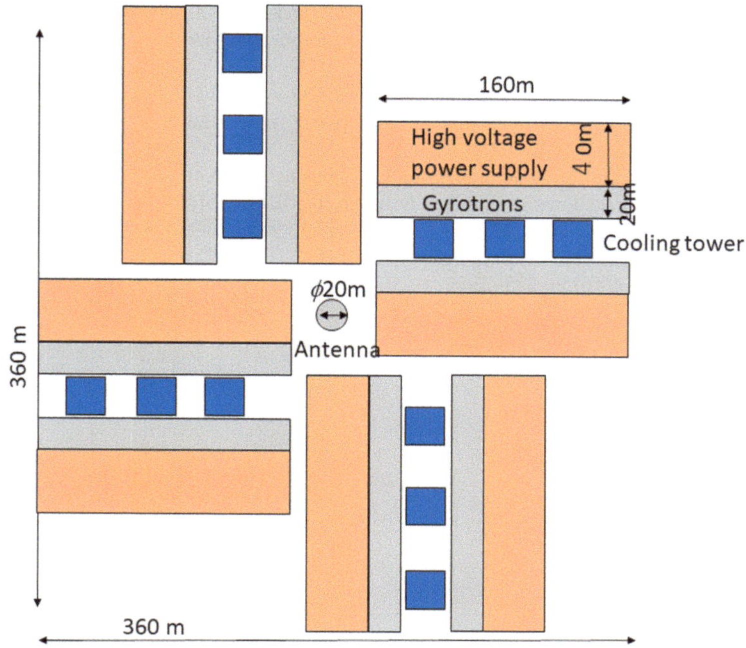

Fig. 2.31 Land plan of ground beam source for microwave rocket

References

Nakagawa T, Mihara Y, Komurasaki K, Takahashi K, Sakamoto K, Imai T (2004) Propulsive impulse measurement of a microwave-boosted vehicle in the atmosphere. J Spacecr Rocket 41(1):151–153

Oda Y, Komurasaki K, Takahashi K, Kasugai A, Sakamoto K (2006a) A thrust generation model of microwave rocket. J Space Technol Sci 22(2):30–37

Oda Y, Komurasaki K, Takahashi K, Kasugai A, Sakamoto K (2006b) Plasma generation using high-power millimeter wave beam and its application for thrust generation. J Appl Phys 100: 113307

Oda Y, Shibata T, Komurasaki K, Takahashi K, Kasugai A, Sakamoto K (2008) Thrust performance of microwave rocket under repetitive pulse operation. J Propuls Power 25(1):118–122

Oda Y, Kajiwara K, Takahashi K, Kasugai A, Sakamoto K, Komurasaki K (2009) In-tube shock wave driven by atmospheric millimeter-wave plasma. Japan J Appl Phys 48(11):116001

Oda Y, Yamaguchi T, Shimada Y, Komurasaki K, Kajiwara K, Takahashi K, Sakamoto K Study of high-power millimeter-wave beam transmission for microwave beaming propulsion. Proceedings of 35th international conference on infrared, September 2010a

Oda Y, Yamaguchi T, Shiraishi Y, Komurasaki K, Kajiwara K, Takahashi K, Kasugai A, Sakamoto K (2011) A one-dimensional propagation of shock wave supported by atmospheric millimeter-wave plasma. J Infrared Millimeter Terahertz Waves 32(6):877–882

Oda Y, Anderson J, Doane J, Kajiwara K, Ikeda R, Takahashi K, Sakamoto K (2018) A study on multi frequency operation of RF transmission system for EC H&CD system. Nuclear Fusion 58(10):106041

Oda Y, Ikeda R, Kajiwara K, Kobayashi T, Hayashi K, Takahashi K, Moriyama S, Sakamoto K, Eguchi T, Kawakami Y, Mitsunaka Y, Darbos C, Henderson MA (2019) Development of 1st ITER Gyrotron in QST. Nuclear Fusion 59(8):086014
Oda Y, Takahashi M, Ohnishi N, Komurasaki K, Sakamoto K, Imai T (2020) A study on the macroscopic self-organized structure of high-power millimeter-wave breakdown plasma. Plasma Sources Sci Technol 29:075010
Oda Y, Kajiwara K, Takahashi K, Sakamoto K, Yamaguchi T, Komurasaki K. Dependence of millimeter wave breakdown structure on beam profile. IEEE. 2010b.
小松怜史, 山口敏和, 小田靖久, 齋藤翔平, 福成雅史, 小紫公也, 梶原健, 高橋幸司, 坂本慶司 (2012) 大電力・高デューティ比作動によるマイクロ波ロケットの推力向上. 日本航空宇宙学会論文集 60(6) pp 235–237
小田, 今井 (2017) ギガワット出力ミリ波ビーム源開発のための技術課題. プラズマ核融合学会誌 93(10):473–477
小田 小紫, 高橋 春日井, 坂本 (2006) 大電力ジャイロトロンを用いたマイクロ波ロケットの推進性能の大気圧力依存性. プラズマ核融合学会誌 83(3) pp 296–299

Chapter 3
Laser Propulsion and Demonstrations

Hiroshi Katsurayama, Yusuke Nakamura, and Akihiro Sasoh

Abstract This comprehensive research explores the potential of laser propulsion as a cost-effective alternative for space launches and as a solution for space debris removal. The first part delves into pulsed laser orbital launchers, introducing their principles, examining different vehicle types, and assessing past studies for feasibility in mass space transportation. Detailed trajectory analyses and cost evaluations show significant potential for cost reduction, especially with high-power lasers like laser diodes or fiber lasers in arrays with coherent coupling. The second part demonstrates laser propulsion's practical applications, detailing the operation of laser-driven in-tube accelerators (LITAs) and wall-ablative, laser-driven in-tube accelerators and presenting experimental procedures for satellite spin breaking using TEA CO2 and Nd:YLF lasers. The chapter concludes by proposing the use of laser ablation impulses for deorbiting nonfunctional satellites, presenting a promising strategy for space debris removal.

Keywords Laser propulsion · Pulsed laser orbital launchers · Vehicle types · Mass space transportation · High-power lasers · Laser-driven in-tube accelerator · Transversely excited atmospheric CO2 laser · Satellite spin breaking · Nd:YLF laser · Space debris removal

Supplementary Information The online version contains supplementary material available at https://doi.org/10.1007/978-981-99-4618-1_3.

H. Katsurayama (✉)
Department of Mechanical and Physical Engineering, Tottori University, Tottori, Japan
e-mail: katsurayama@tottori-u.ac.jp

Y. Nakamura · A. Sasoh
Department of Aerospace Engineering, Nagoya University, Aichi, Japan
e-mail: yusuke.nakamura@mae.nagoya-u.ac.jp; akihiro.sasoh@mae.nagoya-u.ac.jp

K. Mori et al. (eds.), *Beamed-mobility Engineering*, Springer Tracts in Electrical and Electronics Engineering, https://doi.org/10.1007/978-981-99-4618-1_3

Pulsed Laser Orbital Launcher

Introduction

Some innovative space programs, such as solar-powered satellites, are not economically feasible because of the high cost of conventional chemical rocket launch systems. For example, to build a solar-powered satellite with an output of 1 GW into geostationary Earth orbit (GEO), transporting the required 10^4 tons of material using a Japanese H2A rocket would cost 400 billion dollars. Therefore, a new low-cost launch system is needed, and one promising candidate is a pulsed laser launcher (Birkan 1992) (Fig. 3.1). This system can achieve a high payload ratio and reduce costs because energy can be supplied from a laser base on the ground and atmospheric air is available as propellant.

This chapter first introduces the concept of pulsed laser launchers and past related research. Next, it presents a trajectory analysis of a pulsed laser launcher into GEO and discusses its feasibility as a mass space transportation system. Finally, it discusses the types of lasers suitable for launch and the necessary technological innovations.

What Is Pulsed Laser Propulsion?

The concept of laser propulsion (Fig. 3.1), which uses laser energy transmitted from a laser base on the ground or in space to generate thrust, was proposed by Kantrowitz (Kantrowitz 1972) in 1972, more than 10 years after the invention of the laser.

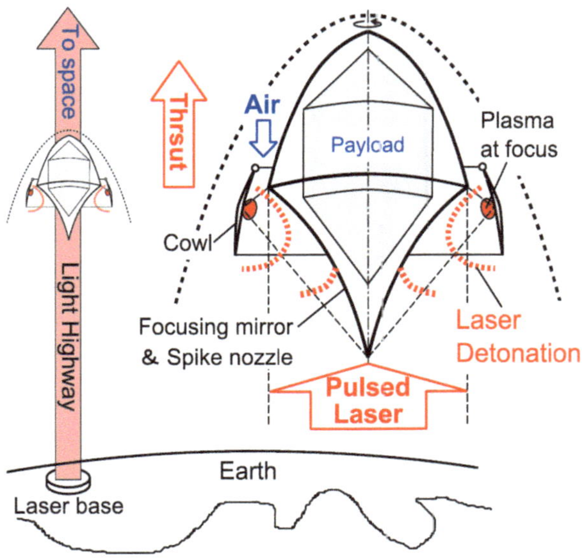

Fig. 3.1 Laser launching system

Around the same time, Raizer (Raizer 1977) showed that energy conversion from a pulsed laser to a fluid occurs via laser-supported detonation waves, and experimental and analytical studies began. This section introduces the research conducted on pulsed laser propulsion systems.

Propulsion Principle of the Pulsed Laser Propulsion System

Figure 3.2 shows the concept of a pulsed laser propulsion system (Lightcraft type (Myrabo 2001, 2003). In pulsed laser propulsion, the transmitted laser beam is focused by the vehicle wall, causing a gas breakdown in the propellant. At the leading edge of the generated plasma, electrons absorb the laser beam by inverse bremsstrahlung radiation. The laser absorption layer and the shock wave propagate together in the form of a high-pressure wave called a laser-supported detonation (LSD) (Raizer 1977). The LSD process is very similar to that of chemical detonation.

The LSD expands rapidly during a single laser pulse, driving a blast wave. The blast wave propagates through the nozzle and exerts a reaction force on the vehicle wall. The vehicle gains thrust by repeating this process.

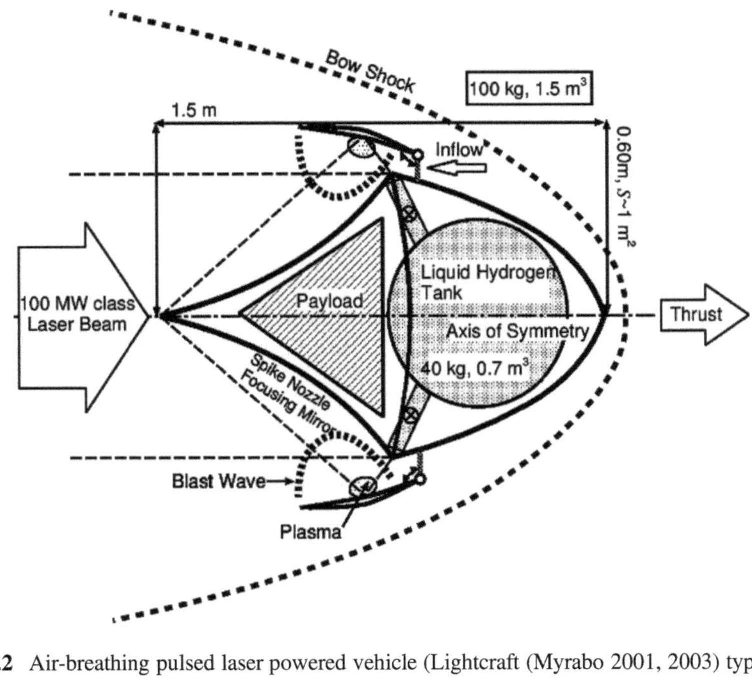

Fig. 3.2 Air-breathing pulsed laser powered vehicle (Lightcraft (Myrabo 2001, 2003) type)

Various Vehicle Types and Their Research Trends

The first propulsion system proposed by Kantrowitz (Kantrowitz 1972) was to obtain thrust by heating ablators mounted on the vehicle with a laser beam. However, when laser propulsion is used as a launch system, air-breathing laser propulsion is advantageous because it can achieve a high specific impulse and large payload ratio. This section presents several types of air-breathing pulsed laser propulsion that have been proposed. A detailed review of ablative laser propulsion is also available in Phipps et al. (2010). The first air-breathing pulsed laser propulsion system was the pulse-jet type with a bell-shaped nozzle (Fig. 3.3a) proposed by Pirri and Monsler (1974), which was studied during the 1970s and early 1980s. The blast wave induced by the LSD propagates through the nozzle. After the blast wave is exhausted from the nozzle, the nozzle breathes the air from its exit, and the next second pulse is irradiated. Thrust measurements have been performed on this bell-shaped type using a CO_2 pulsed laser of about 5–50 J (Pirri and Monsler 1974; Ageev et al. 1980), and Ageev et al. (1980) have shown a scaling law between the laser energy and the nozzle length. Myrabo et al. (1998) proposed a thrust model of the bell-shaped type using an analytical solution and investigated the specific impulse in the case of the rocket type with onboard propellant. However, at that time, no lasers were powerful enough to enable flight experiments, and the research went downhill.

Later, in the 1990s, kW-class lasers became available, and in 1998, Myrabo et al. (1998) conducted the first flight demonstration using a 10 kW-class CO_2 pulsed laser. This flight demonstrator, called Lightcraft (Fig. 3.2), is designed to fly in the pulse-jet mode, which breathes air from the nozzle exit, during subsonic flight and fly in the ramjet mode, which takes compressed air from the inlet, during supersonic flight. The spike nozzle of the vehicle's rear part serves as a receiver that focuses the laser beam over the cowl and receives the driven blast wave. In 2001, they launched the Lightcraft, whose diameter is 12 cm and weight is 50 g, to 71 m in height. Additionally, Salvador et al. (Salvador et al. 2013) have attempted to test the thrust generation of the Lightcraft in supersonic airflow. The demonstration by Myrabo et al. brought pulsed laser propulsion back into the spotlight, and pulsed laser propulsion research has been active in the USA, Germany, and Japan.

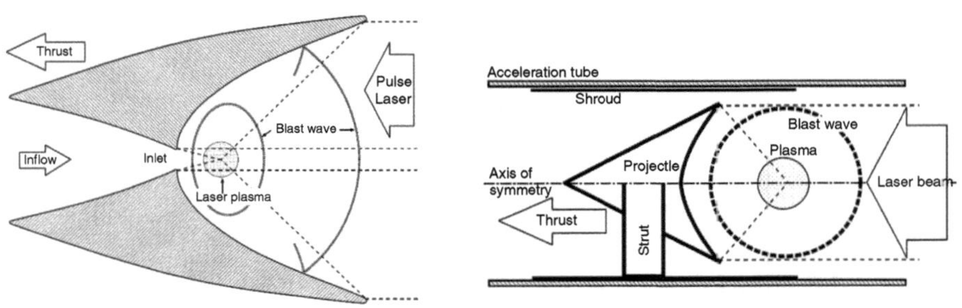

(b) Bell nozzle type (a) Laser in–tube accelerator (LITA)[14)]

Fig. 3.3 Several types of air-breathing laser powered vehicle (Not-Lightcraft type)

In Germany, Schall et al. (2000) conducted a launch demonstration of a bell-shaped type with a 50 g weight using an 8 kW pulsed laser and succeeded in launching it to a height of 60 cm. In Japan, Sasoh (2000) proposed and demonstrated the laser-driven in-tube accelerator (LITA) (Fig. 3.3b), which accelerates a projectile in a tube like a ram accelerator by blast waves driven by pulsed lasers. Mori et al. (2004; Mori and Katsurayama 2010) measured the thrust of a laser pulse-jet with a conical nozzle and proposed an optimal nozzle scale for the input pulsed laser energy. Mori (2017; Chongfa et al. 2020) also proposed a system in which a pulsed laser with an annular cross-sectional area is used to launch a spherical-shaped object while maintaining stable beam riding, and they measured its thrust impulse. Komurasaki et al. have also conducted various experiments (e.g., Shimano et al. 2017; Shimamura et al. 2017, 2014; Wang et al. 2013) to elucidate the propagation mechanism of laser-supported detonation.

Past Studies of Pulsed Laser Launch Systems

As mentioned above, various pulsed laser propulsion systems have been proposed, and many experiments and analyses have been conducted to elucidate their propulsion performance and physical mechanisms. However, there has been a little concrete study of the feasibility of this system as a launch system. In the 1970s, Moeckel discussed transportation to GEO by laser propulsion in a series of papers (Moeckel 1972a,b, 1975), which was the beginning of the system consideration. He performed a simple analysis using the rocket equations and showed that a payload ratio of about 0.2–0.8 is achievable. However, his analysis was impractical because it did not consider the effects of aerodynamic drag at launch or the efficiency of converting laser power to thrust.

In the 1990s, based on Moeckel's analysis, launch analyses into LEO (low Earth orbit) were conducted, considering the switch from the ramjet to rocket mode in propulsion mode (Toki 1991), laser transmission efficiency, and thrust conversion efficiency (Kare 1995; Humble and Pierson 1995). These analyses indicate that the payload mass m_{pl} capable of being launched per laser power P_L is about $m_{pl}/P_L = 1$–5 kg/MW. (Note that this value is about two orders of magnitude larger than the value obtained from Katsurayama et al. (2009)'s modern analysis, which uses a realistic thrust model and considers laser power optimization, as described below.)

These analyses of the 1990s assumed a conversion efficiency from laser energy to thrust. However, there has been little discussion of the realism of this assumption. There was also no discussion of parameters necessary for cost evaluation, such as the minimum laser power that would make launch feasible and the conditions that would maximize the launchable weight of the payload per MW.

Single-stage launches using pulsed laser propulsion require several to several tens of megawatts of laser power per kilogram of vehicle weight. Moreover, at least a 100 MW-class laser base must be constructed to make this a realistic launch system.

Since the laser price is estimated at least $10/W (Kare 2005), constructing a 100 MW-class laser base would cost as much as $1 billion. Therefore, it was necessary to carefully consider how to achieve a single-stage launch while suppressing laser power.

Feasibility of Mass Space Transportation by Laser Propulsion

As mentioned above, past system studies of pulsed laser launchers have been inadequate and have not reached the point of feasibility studies. Therefore, taking a launch to GEO as an example, Katsurayama et al. (2009) conducted a launch analysis using a realistic thrust model based on numerical analysis and experiments, estimated the achievable payload ratio and cost, and examined the feasibility. This section introduces its results. A pulsed laser launch system requires a laser base with average power in the several 100 MW to several GW class. Therefore, the cost reduction of the laser oscillator is essential to establishing the system, and it is realistic to launch from one laser base.

Transportation to GEO Via Light Highway

A pulsed laser launch system requires a laser base with average power in the several 100 MW to several GW class. To obtain aerodynamically dynamic stability of the vehicle, several researches (Libeau and Myrabo 2002; Takahashi and Ohnishi 2012) propose a vehicle shape that can generate thrust vector and torque in the direction where the vehicle axis always coincides with the beam axis. Using this technique, the beam and the fuselage can maintain their flight direction perpendicular to the ground. As a result, the vehicle tracking and beam pointing are unnecessary, and we carry the vehicle into space along a single light highway established vertically from the ground by laser (Fig. 3.4).

On this highway, the vehicle initially flies in "pulse-jet" mode, which closes the foreside intake, and breathes and exhausts air from the rear side. Next, when the vehicle has accelerated sufficiently to take advantage of ram compression, it breathes air from the foreside intake and shifts the mode to "ramjet." Finally, when the vehicle reaches a high altitude and cannot take in enough air, it closes the air intake and switches the mode to "rocket," using the onboard hydrogen as a propellant. The vehicle accelerates to orbital velocity in a single stage using these three modes. Considering the requirements for beam transmission, a launch trajectory, as shown in Fig. 3.5, in which the vehicle is accelerated vertically over a short distance along the optical highway and put into a long elliptical orbit, is desirable. In this trajectory, the vehicle is launched near the equator by laser propulsion and is rapidly accelerated. To take full advantage of this acceleration, the vehicle once crosses the GEO through inertial flight and reaches the apogee point. Finally, an onboard kick motor puts the vehicle into GEO by accelerating and decelerating at the apogee and the

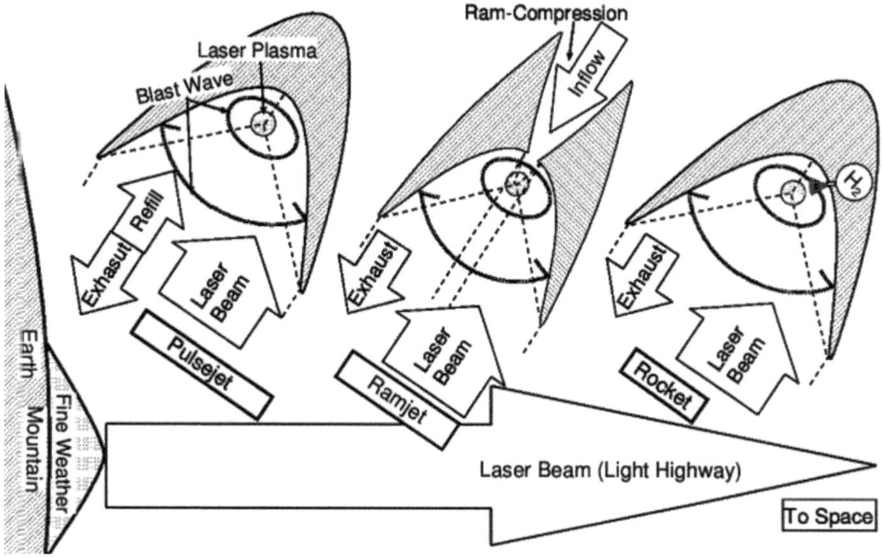

Fig. 3.4 Light highway (Katsurayama et al. 2009)

Fig. 3.5 Proposed super-synchronous transfer orbit (Katsurayama et al. 2009)

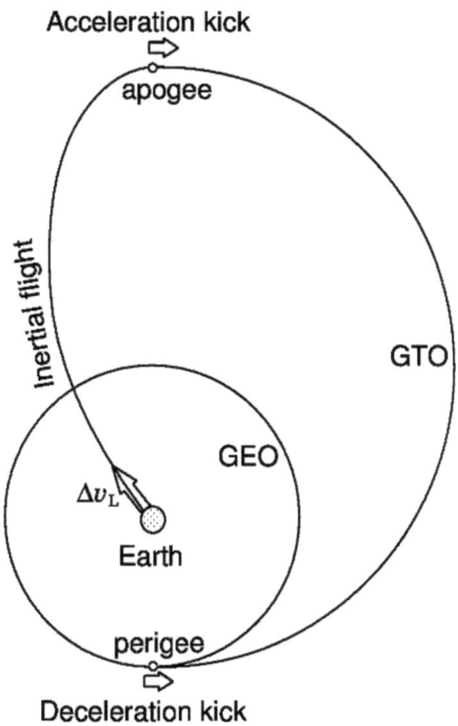

perigee points on the geostationary transfer orbit (GTO). Such an orbital insertion is generally called a super-synchronous transfer orbit (Walters et al. 1995). From the viewpoint of gravity loss, one might think that it would be advantageous to launch the rocket not vertically, as shown in Fig. 3.5, but horizontally to the Earth's surface from the middle of the trajectory, as in the case of a typical chemical rocket. However, a vertical launch is more realistic for a pulsed laser launcher from the points described below. First, to prevent beam degradation (attenuation, fluctuation, spreading, and so on.) due to the atmosphere, a vertical launch is desirable because of its shortest beam transmission distance. In addition, inclining the beam in the horizontal direction requires the generation of lift and precise control for the vehicle to follow the beam, but it is not practical to mount a mechanism to achieve this on the vehicle. In a vertical launch, on the other hand, the vehicle is placed on the light highway and launched autonomously along it. As noted above, the vertical launch does not require capturing the fast-moving vehicle or directing the beam toward it. If we launch the vehicle in the horizontal direction, a system that switches beams emitted from multiple bases may be necessary to reduce the beam transmission distance. However, such a system is impractical for laser propulsion because only one single laser base is realistic to suppress its construction cost.

Flight Trajectory and Payload Ratio

Katsurayama et al. (2009) set the requirement that the acceleration and deceleration by the kick motor be limited to 2 km/s on the trajectory of Fig. 3.5. And they performed a trajectory analysis assuming vertical acceleration up to $\Delta v_L = 10.85$ km/s by laser propulsion to satisfy this requirement. Unlike previous studies, the study modeled the thrust for each flight mode realistically using experimental values, numerical fluid dynamics analysis, and analytical solutions. The details can be found in Katsurayama et al. (2009, 2006, 2003, 2008, 2004). Figure 3.6 shows the flight trajectories and achievable payload ratios λ for characteristic laser power $P_L/m_{v,0}$, where $m_{v,0}$ is the initial vehicle mass. The λ and the flight time of the ramjet mode decrease with decreasing $P_L/m_{v,0}$, and the ramjet mode becomes unavailable at $P_L/m_{v,0} < 1.47$ MW/kg. At this condition, the beam transmission is maximized for distance and time, requiring the transmission of 400 km and 160 s.

Figure 3.7 shows the payload weight per MW of beam power m_{pl}/P_L and λ versus $P_L/m_{v,0}$. The λ increases with $P_L/m_{v,0}$, but m_{pl}/P_L takes its maximum value $(m_{pl}/P_L)_{max} = 0.085$ kg/MW at $P_L/m_{v,0} = 2.5$ MW/kg. Therefore, $(P_L/m_{v,0})_{opt} = 2.5$ MW/kg is the optimal characteristic laser power to minimize the cost of the laser oscillator without wasting laser power.

Cost Evaluation

Katsurayama et al. (2009) evaluated a mission to launch a 10^4-ton solar-powered satellite into GEO using a pulsed laser launcher as a cost assessment, assuming a

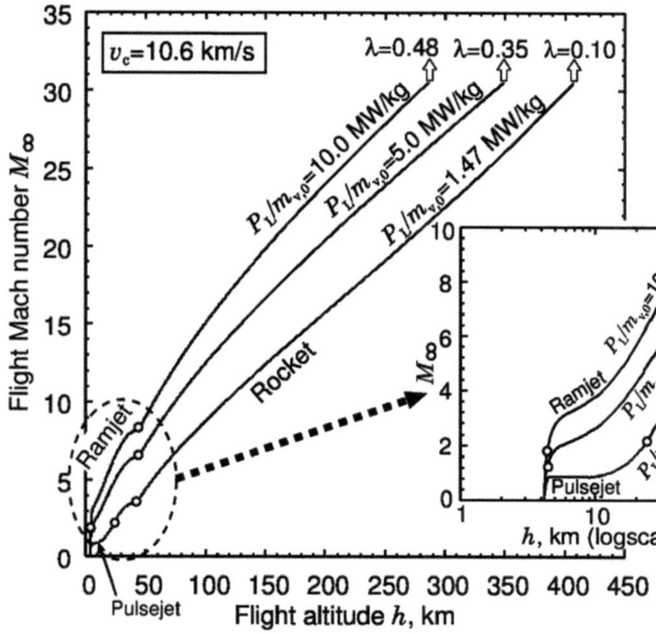

Fig. 3.6 Flight Mach number vs. flight altitude (Katsurayama et al. 2009)

Fig. 3.7 Variation of payload mass per unit beam power and payload ratio with specific beam power (Katsurayama et al. 2009)

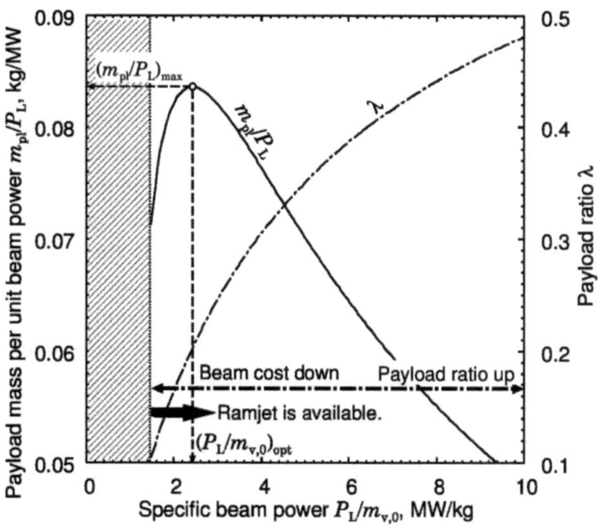

satellite with a power output equivalent to 1 GW per standard nuclear power plant (Collins 1993). For the laser oscillator, the cost of the oscillation base is assumed to be \$10/W, assuming the application of arraying technology with coherent coupling (Shirakawa and Ueda 2005) to laser diodes, which currently have the lowest manufacturing cost per unit output (Kare 2005). The costs other than the laser

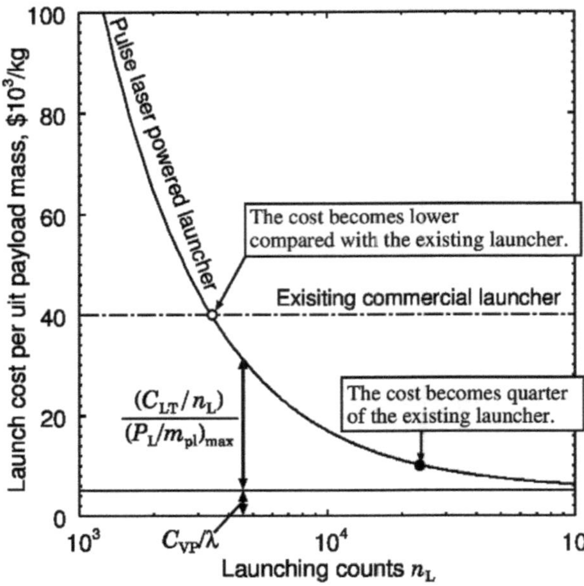

Fig. 3.8 Variation of launch cost per unit payload mass with n_L (Katsurayama et al. 2009)

oscillator include the standard Japanese electricity cost C_E(=\$0.06/kWh) and the vehicle production cost C_{vp} (=\$$10^3$/kg) (Richard et al. 1988). The costs of the laser cooling system, power supply system, and maintenance are considered negligible compared to the cost of the laser oscillator (Kare 2005) and therefore are not taken into account.

Figure 3.8 shows the variation of the launch cost per unit payload mass estimated using the optimal characteristic laser power versus the launching counts n_L. When n_L is smaller than 10^3 times, the cost of the laser oscillator is dominant. The electricity cost is negligibly small. Since the cost of the laser oscillator depreciates as n_L increases, the cost would be equivalent to the current launch system if the total payloads are divided into 3500 launches. In the case of $n_L = 24{,}000$ times, the cost becomes about one-fourth of the current launch system, and if n_L can be larger than 10^5, the cost is almost only the cost of producing the vehicle (about 2000 dollars per payload). The electricity costs are negligibly small even at this point. As a result, if a laser oscillator with a power output of 5 GW class can be used, it has a transport capacity of about 0.5 tons per launch, and it is predicted that a 10^4-ton solar-powered satellite can be launched at a quarter of the conventional transport cost by dividing it into about 20,000 launches. In addition, if the number of launches can be further increased and the C_{VP} can be reduced to one-tenth through mass production, the launch cost can be reduced by one magnitude or more from the current level.

Consideration of Laser Types and Necessary Innovations

As mentioned above, a pulsed laser launcher requires a laser with a power output of 100 MW to several GW. Let us consider what kind of lasers is available.

First, let us consider whether it is possible to build a single ultra-high-power laser with existing technology and use it for the launch. Past various pulsed laser propulsion experiments used CO_2 lasers. The reason is that (1) the CO_2 laser is capable of the relatively high output power of several 10 kW (Kalinin et al. 1998), (2) plasma quickly absorbs its 10.6 μm wavelength, and (3), however, the atmosphere does not absorb it. For example, Myrabo (2001, 2003) conducted a Lightcraft launch experiment using a 10 kW CO_2 pulsed laser (1 kJ/pulse, pulse width 30 μs, operating frequency 10 Hz). However, an output of a few 10 kW is far small than the power required for a pulsed laser launcher. In addition, the CO_2 laser tends to have poor beam quality at high power, making it difficult to use it as a single laser with a large aperture. Therefore, at the current level of technology, the CO_2 laser would not be appropriate as a laser for launches.

In the field of laser fusion, on the other hand, solid-state lasers capable of outputting high power in a single unit have been used. Typical lasers of this type are the petawatt pulsed laser with 500 J/pulse and 0.5–1 ps pulse width and the gigawatt pulsed laser with 500 J/pulse and 0.1–4 ns pulse width at the Laser Energetics Research Center of Osaka University in Japan. However, since these lasers emit only a few pulses daily, it is not easy to use them for pulsed laser launchers. High-power pulsed lasers with high repetition rates have also been developed for routine use in laser fusion. For example, a Japanese public-private research group has developed the diode-pumped solid-state lasers (DPSSL), called HALNA-10 (Kawashima et al. 2005) (wavelength ~1 μm 10 J/pulse, 10 ns pulse width, 10 Hz repetition rate) and TERU (Watari et al. 2021) (50 J/pulse, 0.5 Hz), and continues its development to achieve higher power and repetition rate (250 J/pulse at 100 Hz) (Watari et al. 2021; Mori et al. 2021; Ogino et al. 2022). However, the energy conversion efficiency of DPSSL is as low as about 10%, and the cooling system is extensive, so there are still many technical issues to achieving even higher output.

As described above, current technology makes it difficult to provide the high power required for a launch with a single laser. Therefore, an array of existing laser diodes (LDs) with low cost is promising (Kare 2005; Komurasaki et al. 2005). Laser diodes can also operate in pulsed mode, and power levels of several kW are currently available. Its conversion efficiency is as high as 50% or more, and the cooling system is simple. It is also highly reliable, with a lifetime of over 3000 h. Even when arrayed, the transmission efficiency can be as high as 70% (Komurasaki et al. 2005) if the phases are aligned, and there is little beam degradation due to arraying. However, when multiple LDs are arrayed and oscillated, the problem arises that the phases emitted from the individual LDs are not aligned (Komurasaki et al. 2005).

However, technologies for aligning the phases of laser arrays have also been developed, and coherent coupling technology is already available by arraying diode-pumped fiber lasers (Shirakawa and Ueda 2005). Fiber lasers with an output power of over 100 kW are already commercially available, and their beam quality is excellent. Fiber lasers may be the most promising at present because the energy loss from the pump source diode lasers can be kept within about 20% (Kare 2005). Kare (2005) estimated the construction cost of a laser base with an output of

100 MW, consisting of 2200 laser modules of 6 10 kW fiber lasers, and concluded that the cost per watt is the order of $10/W.

The above suggests that LDs or fiber lasers in arrays will enable the construction of the laser base from 100 MW to several GW as an extension of the existing technology.

The next problem is securing the power needed for the laser. For a pulsed laser launcher with a power output of about 5 GW, laser irradiation lasts about 100 s (Katsurayama et al. 2009), so it is necessary to secure about 500 GJ of power per launch. Superconducting magnetic energy storage (SMES) (Nitta 2004; Minami 1998) can be used for this purpose. The SMES is currently being actively researched for use in power plants and linear motor cars. This technology is expected to store the maximum power of 5 G Wh, with a speedy discharge/recharge response and an efficiency of 90% (Minami 1998). Assuming laser diodes, which have a conversion efficiency of about 50%, about 1100 GJ of energy is required for a single launch. Furthermore, assuming about 60 launches per day (20,000 launches/year), about 65 TJ per day must be stored in the SMES. This energy is smaller than the total daily energy of a standard 1 GW-class nuclear power plant (about 85 TJ) and well within the feasible range.

As for non-laser issues, it is necessary to prevent the vehicle from deviating from the light highway when it is agitated by the wind. As mentioned in section "Transportation to GEO Via Light Highway", a solution to this problem is to obtain natural aerodynamic stability by devising the vehicle shape (Libeau and Myrabo 2002; Takahashi and Ohnishi 2012), but Takahashi et al. (Libeau and Myrabo 2002) also pointed out the necessity of beam control, and further research is needed.

Summary

This chapter introduced previous research on pulsed laser launchers and the system analysis of mass transport to GEO. The system analysis showed that a pulsed laser launcher could transport a payload of 0.085 kg per MW by launching through the light highway. In order to depreciate the cost of the laser base and make it smaller than the cost of the current launch system, it was found that the payload would need to be divided into 3500 launches, and if divided into 24,000 launches, a pulsed laser launcher would be feasible at about one-fourth the cost of a conventional launch system. Furthermore, if the payload is divided into 10^5 launches, a cost of about $2000 per kilogram could be realized. Furthermore, by further increasing the number of launches and reducing the cost of manufacturing the vehicle through mass production, the launch cost could be reduced by an order of magnitude or more, making it an excellent low-cost launch system.

The types of high-power lasers (100 MW to several GW class) required for the launch were also examined. Laser diodes or fiber lasers in arrays with coherent coupling are the most promising, and the laser base could be constructed as an extension of current technology.

Demonstration of Laser Propulsion and Space Debris Removal (Demonstration Movie Available)

Launch Demonstration of Laser-Driven In-Tube Accelerator (LITA) with Gas Driver

Launch operation was conducted using a highly repetitive pulse transversely excited atmospheric (TEA) CO_2 laser (wavelength, 10.6 mm; 2.6 J/pulse; repetition frequency of 60 Hz) sent through a NaCl window at the top of the launch tube (Sasoh 2001). The schematics of this experiment and the picture of the launched projectile are shown in Fig. 3.9. The launch tube was made of acrylic and is 0.5 m in length. The projectile, made of aluminum alloy A7075-T6, was initially located near the bottom of the launch tube. The laser beam incident on to the forefront of the projectile was designed to be reflected on the projectile twice and be focused behind it. At the focal point, the laser energy was input to the driver gas, Xe at 100 kPa, filled in the launch tube. A laser-induced blast wave yielded a thrust to lift up the projectile.

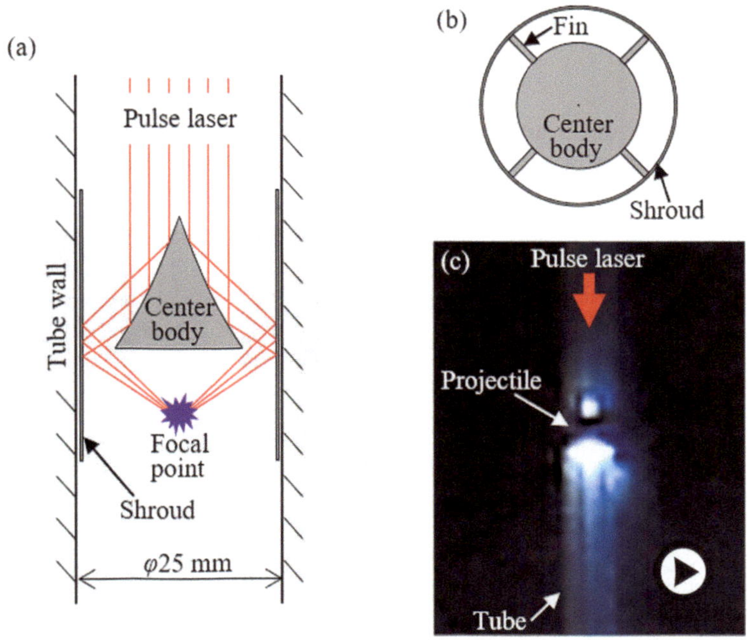

Fig. 3.9 (**a**) Schematic illustration of the experiment of LITA. (**b**) Top view of the projectile. (**c**) Picture of the projectile taken during lifting up (▶ https://doi.org/10.1007/000-bn2)

Launch Demonstration of Wall-Ablative, Laser-Driven In-Tube Accelerator

The wall-ablative, laser-driven in-tube accelerator (WA-LITA) had an outer brass frame and a 25-mm-side square inner cross section and a 500 mm effective length (Sasoh et al. 2009). The schematics of this experiment and the picture of the launched projectile are shown in Fig. 3.10. A pair of opposing walls was made of polyacetal (POM) copolymer as ablators. The other sides were acrylic windows. The projectile had a cross section fitted to the square tube. The projectile was made of aluminum alloy (A7075-T6). A collimated laser beam was sent into the launch tube through a ZnSe window at the bottom of the launch tube, reflected on the projectile parabolas, and then focused onto the ablator walls. The upper end of the launch tube was plugged with an acrylic window. On a side wall, two circular ducts near the top and bottom were connected to a vacuum chamber; the vacuum chamber pressure was kept lower than 20 Pa. Laser pulses were irradiated from a transversely excited atmospheric (TEA) CO_2 laser (ML205E, Selective Laser Coating Removal Co.; wavelength of 10.6 μm, 5.3 J/pulse, repetition frequency of 50 Hz).

Demonstration of Satellite Spin Breaking Using a TEA CO_2 Laser (2016)

The spin motion of an artificial satellite model was stopped by repetitive laser ablation impulses. A picture of the satellite model is shown in Fig. 3.11a. Ablator

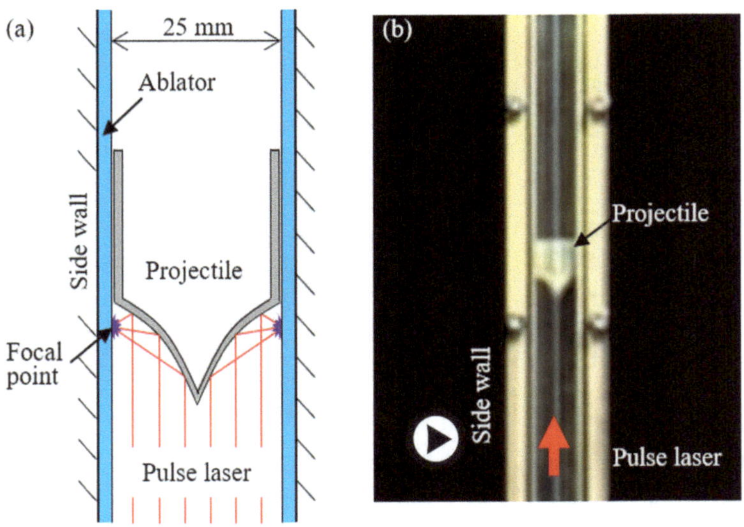

Fig. 3.10 (**a**) Schematic illustration of the experiment of WA-LITA. (**b**) Picture of the projectile taken during lifting up (▶ https://doi.org/10.1007/000-bn1)

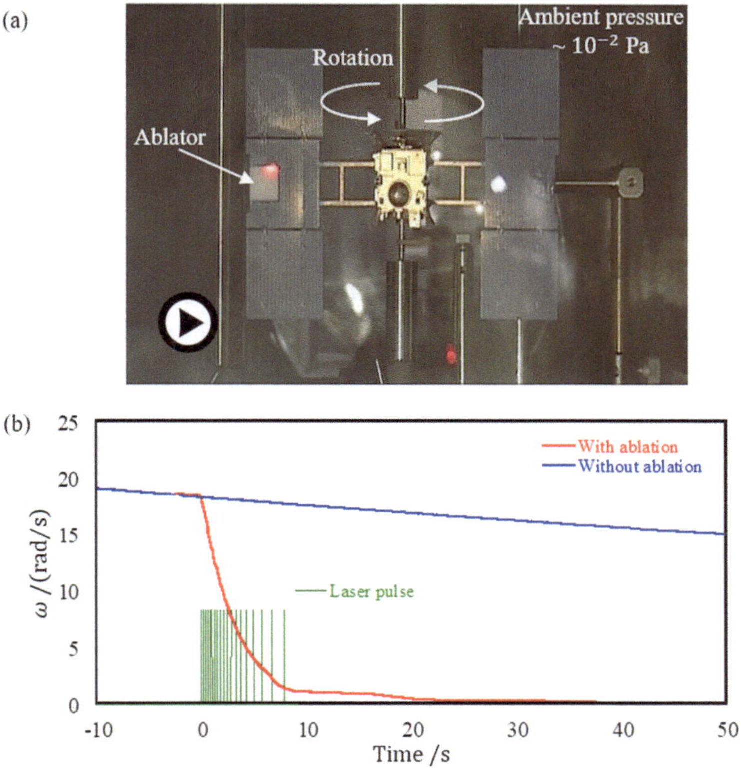

Fig. 3.11 (**a**) A picture of the artificial satellite model used in this experiment. (**b**) Histories of angular velocity ω with and without ablation (▶ https://doi.org/10.1007/000-bn0)

sheets made of polyacetal (POM) were placed at solar panel models. The irradiation of laser pulses from a transversely excited atmospheric (TEA) CO_2 laser (ML205E, Selective Laser Coating Removal Co.; wavelength 10.6 μm, 5.3 J/pulse) was synchronized to the model spin motion. Histories of the angular velocity of the model with and without ablation are shown in Fig. 3.11b. The angular velocity with ablation decreased faster than that without ablation, which was braked only by drag.

Demonstration of Satellite Spin Breaking Using an Nd: YLF Laser (2021)

The spin motion of a model with eight panels was stopped by laser ablation impulses with high-repetition-rate laser (Nd: YLF laser, INNOSLAB IS81-E, Edgewave; wavelength 1047 nm). A schematic illustration of this experiment and a picture taken during the experiment are shown in Fig. 3.12a, b, respectively. Ablator sheets made of polycarbonate (PC) were placed at every panel. Laser pulses with the pulse energy of 10 mJ were shot on the ablator sheet with the repetition frequency of 1.2 kHz during one of the panels in the angular range ϕ. Histories of the angular velocity with and without ablation are shown in Fig. 3.12c.

The use of laser ablation impulses for deorbiting space debris was proposed (Phipps et al. 1996). In recent years, deorbiting nonfunctional satellites with repetitive pulsed laser irradiations from a rendezvoused satellite with an onboard pulsed laser has been planned (Fukushima et al. 2021).

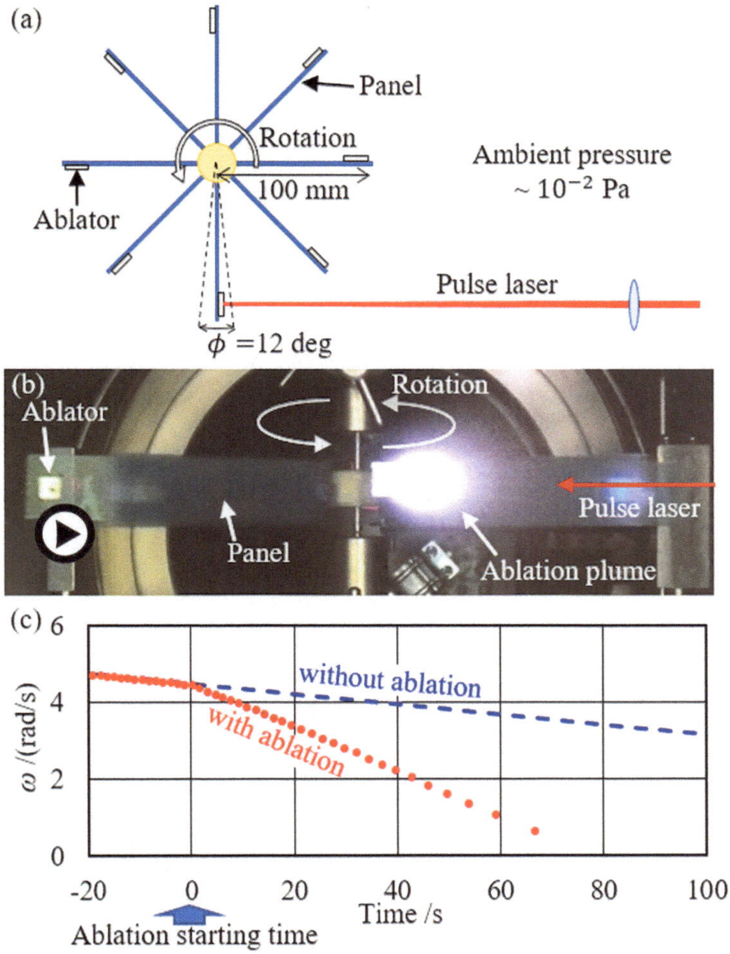

Fig. 3.12 (**a**) A schematic illustration of the experiment. (**b**) A picture of the model during this experiment. (**c**) Histories of angular velocity ω with and without ablation (▶ https://doi.org/10.1007/000-bn3)

References

Ageev VP, Barchukov AI, Bunkin FV, Konov VI, Korobeinikov VP, Putjatin BV, Hudjakov VM (1980) Acta Astronaut 7:79

Birkan MA (1992) J Propuls Power 8:354

Chongfa X, Tran DT, Mori K (2020) J Propuls Power 36:415

Collins P (1993) Proc 5th Int Energy Conf 3:50

Fukushima T, Hirata D, Adachi K, Itaya Y, Yamada J, Tsuno K, Ogawa T, Saito N, Sakashita M, Wada S, Ebisuzaki T (2021) End-of-life deorbit service with a pulsed laser onboard a small satellite. In: Proceedings of 8th European conference on space debris. The ESA Space Debris Office, Darmstadt, Germany, ESA

Humble WE, Pierson BL (1995) J Guid Control Dyn 18:1259

Kalinin VP, Goryachkin DA, Gromovenko VM, Romanov NA, Rodionov AY, Dimakov SA, Irtuganov VM, Kuprenyuk VI, Sherstobitov VE, Rezunkov YA (1998) Proc SPIE 3574:187

Kantrowitz A (1972) Astronaut Aeronaut 10:74

Kare JT (1995) J Propuls Power 11:535

Kare TJ (2005) AIP Conf Proc 766:128

Katsurayama H, Komurasaki K, Momozawa A, Arakawa Y (2003) Trans Japan Soc Aero Space Sci Space Tech Japan 1:9

Katsurayama H, Ushio M, Komurasaki K, Arakawa Y (2004) AIP Conf Proc 766:117

Katsurayama H, Komurasaki K, Arakawa Y (2006) J Japan Soc Aeronaut Space Sci 54:63. (in Japanese)

Katsurayama H, Komurasaki K, Hirooka Y, Mori K, Arakawa Y (2008) J Propuls Power 24:999

Katsurayama H, Komurasaki K, Arakawa A (2009) Acta Astronaut 65:1032

Kawashima T et al (2005) IEEJ Tras EIS 125:233. (in Japanese)

Komurasaki T, Nakagawa S, Ohmura YA (2005) Japan Soc Aeronaut Space Sci, Space Tech Japan 3:7

M.A. Libeau, and L.N. Myrabo, AIAA Paper 02–3781, 2002

Minami M, Mitsubishi heavy industries, Ltd (1998) Technical Rev 35:39

Moeckel WE (1972a) J Spacecr Rocket 9:863

Moeckel WE (1972b) J Spacecr Rocket 9:942

Moeckel WE (1975) J Spacecr Rocket 11:700

Mori K (2017) J Spacecr Rocket 54:1180

Mori K, Katsurayama H (2010) J Propuls Power 26:1305

Mori K, Komurasaki K, Arakawa Y (2004) J Spacecr Rocket 41:887

Mori Y et al (2021) J Plasma Fusion Res 97:352. (in Japanese)

LN Myrabo (2001) AIAA Paper 2001–3798

Myrabo LN (2003) AIP Conf Proc 664:49

LN Myrabo, DG Messitt, and FB Jr Mead (1998) AIAA paper 98–1001

Nitta T (2004) J Plasma Fusion Res 80:578. (in Japanese)

Ogino J et al (2022) Laser Rev 50:377. (in Japanese)

Phipps CR, Albrecht G, Friedman H, Gavel D, George EV, Murray J, Ho C, Priedhorsky W, Michaelis MM, Reilly JP (1996) ORION: clearing near-earth space debris using a 20-kW, 530-nm, earth-based, repetitively pulsed laser. Laser Particle Beams 14(1):1–44

Phipps C, Birkan M, Bohn W, Eckel H-A, Horisawa H, Lippert T, Michaelis M, Rezunkov Y, Sasoh A, Schall W, Scharring S, Sinko J (2010) J Propuls Power 26:609

Pirri AN, Monsler MJ (1974) AIAA J 12:1254

Raizer YP (1977) Laser-induced discharge phenomena. Consultants Bureau, New York and London

J.C. Richard, C. Morales, and L.N. Myrabo, AIAA Paper 88–2970, 1988

Salvador II, Myrabo LN, Minucci MAS, de Oliveira AC, Toro PGP, Channes JB Jr, Rego IS (2013) J Propuls Power 29:718

Sasho A (2000) J Phys IV France 10:41

Sasoh A (2001) Laser-driven in-tube accelerator. Rev Scientific Instruments 72:1893

Sasoh A, Suzuki S, Matsuda A (2009) Wall-propelled, in-tube propulsion with repetitive-pulsed laser ablation. J Propulsion Power 25(2):540–542

Schall WO, Bohn WL, Eckel H-A, Mayerhofer W, Riede W, Zeyfang E (2000) Proc. Of the SPIE: high-power laser ablation III 3065. International Society for Optical Engineering, Bellingham, WA, p 472

Shimamura K, Komurasaki K et al (2014) IEEE Trans Plasma Sci 41:3121

Shimamura K, Matsui K, Ofosu J, Yokota I, Komurasaki K (2017) Appl Phys Lett 110:134104

Shimano T, Ofosu JA, Matsui K, Komurasaki K, Koizumi H (2017) Trans Japan Soc Aero Space Sci 60:378

Shirakawa A, Ueda K (2005) Laser Rev 33:254. (in Japanese)

Takahashi M, Ohnishi N (2012) AIAA J 11:2600

Toki K (1991) J Space Technol Sci 8:23
Walters M, Baker JF, Shurmer IM (1995) Proc. of Flight Mechanics/Estimation Theory Symposium. NASA Goddard Space Flight Center, p 435. Document ID: N95–27804
Wang B, Komurasaki K, Arakawa Y (2013) J Phys D 46:375201
Watari T et al (2021) Laser Rev 49:173. (in Japanese)

Chapter 4
Spectroscopy of Air Discharge Plasmas Induced by a Gyrotron Beam

Kuniyoshi Tabata

Abstract This comprehensive study explores the practicalities of beamed energy propulsion (BEP) systems, focusing on the spectroscopic measurements of air discharge plasmas within a microwave rocket's nozzle. It unpacks the mechanism by which millimeter-wave energy is transformed into thrust via plasma generation and millimeter-wave supported detonation. Special attention is given to the non-equilibrium dynamics of neutral particles and ions in atmospheric millimeter-wave discharges. A practical method to investigate these dynamics is introduced through optical emission spectroscopy (OES), a nonintrusive technique capable of studying plasma without disturbing it. The study features real-world examples, adding practical value and feasibility to the theoretical aspects. This exploration substantiates BEP's potential in revolutionizing space transportation, bringing substantial cost reduction and high efficiency.

Keywords Beamed energy propulsion · Microwave Rocket · Millimeter-waves · Gyrotron · Plasma · Millimeter-wave supported detonation, Optical emission spectroscopy, Non-equilibrium dynamics, Air discharge plasmas, Atmospheric millimeter-wave discharges

Introduction

This chapter describes the spectroscopic measurements of air discharge plasmas inside *Microwave Rocket*'s nozzle, which is one of beamed energy propulsion (BEP) systems. Part of this chapter is based on the author's studies during the Ph.D. course (Tabata 2022).

BEP is a space transportation system that converts a wirelessly transferred energy (lasers or microwaves) into propulsion energy. Because the air around a flight vehicle is used as its propellant, no onboard propellant is necessary. The specific

K. Tabata (✉)
Institute of Space and Astronautical Science (ISAS), Japan Aerospace Exploration Agency (JAXA), Sagamihara, Japan
e-mail: tabata.kuniyoshi@jaxa.jp

K. Mori et al. (eds.), *Beamed-mobility Engineering*, Springer Tracts in Electrical and Electronics Engineering, https://doi.org/10.1007/978-981-99-4618-1_4

impulse I_{sp} is infinite in contrast to the value for chemical rockets, which is less than 450 s. Moreover, because BEP requires no complicated mechanisms such as turbopump systems used for chemical rockets, the manufacturing costs of the vehicle itself become much lower than those of conventional rockets. Therefore, BEP is one means of drastically reducing space transportation costs.

Microwave Rocket (Komurasaki and Tabata 2018; Komurasaki et al. 2017; Shimamura et al. 2020; Nakagawa et al. 2004; Oda et al. 2006, 2009a; Fukunari et al. 2017, 2014; Kakinuma et al. 2016) uses microwaves of millimeter-order wavelength, which we designate as "millimeter-waves," for atmospheric energy beaming. One benefit of using millimeter-waves instead of lasers is the cost of oscillators. Reportedly, a gyrotron, which is a high-power millimeter-wave beam oscillator, only costs one-tenth as much as a laser oscillator.

Microwave Rocket converts the millimeter-wave energy into thrust impulse via plasma. First, at the closed end of a thruster, plasma is ignited by focusing the millimeter-wave and thereby creating a strong electric field called millimeter-wave discharge plasma. The ionization wave front of the generated plasma propagates toward a millimeter-wave source, with an accompanying shock wave propagating at supersonic speed: millimeter-wave supported detonation (MSD). The air inside the thruster is pressurized and exhausted from the thruster's nozzle, which produces thrust force.

First, the relation between the propagation velocity of the discharge front and the millimeter-wave beam power density is explained as a key factor in terms of thrust performance of *Microwave Rocket*. The following equations are Euler equations that determine the parameters of the fluid associated with detonation in a coordinate fixed to discharge front: the detonation wave front.

$$\rho_2 V_2 = \rho_1 V_{\mathrm{MSD}} \tag{4.1}$$

$$\rho_2 V_2^2 + p_2 = \rho_1 V_{\mathrm{MSD}}^2 + p_1 \tag{4.2}$$

$$\frac{1}{2} V_2^2 + C_{\mathrm{p}} T_2 = \frac{1}{2} V_{\mathrm{MSD}}^2 + C_{\mathrm{p}} T_1 + q \tag{4.3}$$

Herein, ρ, V, p, T, and C_{p}, respectively, stand for mass density, fluid velocity, pressure, temperature, and specific heat. The number 1 and 2 indicate the detonation wave's upstream and downstream states, respectively. Besides, the equations are, respectively, equations of continuity, momentum conservation, and energy conservation. The third equation includes the following source term q, which expresses the heat input to neutral particles via plasma.

$$q = \frac{S_{\mathrm{beam}}}{\rho_1 V_{\mathrm{MSD}}} \tag{4.4}$$

Here, S_{beam} and V_{MSD}, respectively, denote the beam power density of incident millimeter-wave and the propagation velocity of millimeter-wave discharge's front, indicating that the relation between S_{beam} and V_{MSD} affects the source term q. For

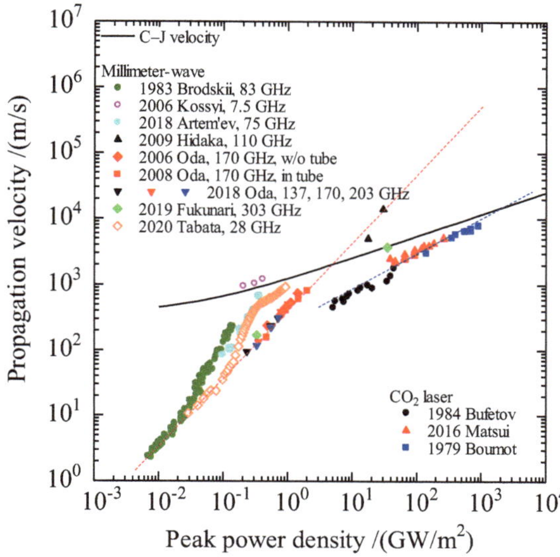

Fig. 4.1 Dependence of propagation velocity of ionization front on the peak power density for various beam frequencies including microwaves, millimeter-waves, and CO_2 laser (Tabata et al. 2020a; Oda et al. 2006, 2018, 2009b; Kossyi n.d.; Artem'ev et al. 2019; Brodskii et al. 1984; Hidaka et al. 2009; Fukunari et al. 2019; Bufetov et al. 1987; Bournot et al. 1979; Matsui et al. 2017)

this reason, many earlier studies have investigated the relation at widely various beam power densities, as Fig. 4.1 presents (Tabata et al. 2020a; Oda et al. 2006, 2018, 2009b; Kossyi n.d.; Artem'ev et al. 2019; Brodskii et al. 1984; Hidaka et al. 2009; Fukunari et al. 2019; Bufetov et al. 1987; Bournot et al. 1979; Matsui et al. 2017). In the figure, it is particularly interesting that the propagation velocity of millimeter-wave discharges is higher than that of laser discharges at a certain beam power density. Furthermore, it is noteworthy that the breakdown threshold power density in millimeter-wave frequencies is about 20 GW/m^2, which is higher than the beam power densities used for experimentation, as shown in Fig. 4.1. This fact indicates that millimeter-wave discharges can be sustained and indicates that the discharge front can propagate even at a beam power density that is lower than the threshold level, which we designate as an "under-critical" condition. On the other hand, "over-critical" condition indicates the one where the local beam power density is larger than the critical threshold level.

It is still unclear why an ionization front can propagate toward a millimeter-wave source in under-critical conditions. Furthermore, many researchers are working diligently on the reproduction of the ionization wave propagation velocity of air millimeter-wave discharges. According to Bogatov et al. and Borodacehva et al., the ultraviolet radiation generated a pre-ionization region in front of the bulk plasma called "halo," and that discharge can be induced in the precursor region (Bogatov et al. 1983; Borodacheva and Semenov 1985). However, this model yields no estimate of the propagation velocity in the range where the propagation velocity is greater than the speed of sound. Takahashi (Takahashi and Komurasaki 2018) and Hamasaki (Hamasaki and Ohnishi 2018) solved a diffusion equation of electrons and Euler equations on neutral particles. Reportedly, the Townsend number *E/N*, which is related to the ionization frequency, becomes larger because of gas heating and thereby the low number density of heavy particles. However, because of the insufficient ionization frequency, the propagation velocity remained lower than experimentally obtained results. In the simulation, the shock wave propagated faster than the ionization wave did. Nakamura et al. assumed a hypothetical ionization frequency higher than the values by BOLSIG+ and reproduced the "comb-shaped" structure propagation using a two-dimensional simulation (Nakamura et al. 2018). The filamentary plasma structure like comb-shaped one is typically observed in millimeter-wave discharges. Furthermore, to reproduce the velocity without hypothetical ionization frequency, the non-equilibrium of millimeter-wave discharge plasma was specifically examined by a one-dimensional simulation (Nakamura and Komurasaki 2020). Electron diffusion equation and conservation equations of electrons' kinetic energy and neutral particles' vibrational energy were solved. In the model, Nakamura incorporated the effects of excitation of the neutral particles in the precursor region via photon diffusion from the bulk plasma. Results showed that the existence of highly excited neutrals increased the ionization frequency. Accordingly, the simulation yielded a greater velocity than those of Takahashi and Hamasaki's studies. The results indicated the importance of the non-equilibrium characteristics of air millimeter-wave discharge plasmas.

Here, the non-equilibrium temperature characteristics in atmospheric discharge plasmas are explained. The temperatures are characterized by electron temperature T_e, electronic excitation temperature T_{exc}, vibrational temperature T_{vib}, rotational temperature T_{rot}, and translational temperature T_{trans}. Relations between these temperatures generally satisfy the following in both AC and DC discharges at atmospheric pressure (Staack et al. 2006).

$$T_{\mathrm{e}} > T_{\mathrm{exc}} > T_{\mathrm{vib}} > T_{\mathrm{rot}} = T_{\mathrm{trans}} \tag{4.5}$$

Figure 4.2 depicts the expected spatial temperature-structure of millimeter-wave discharge plasma. There, E, P, T_e, T_{vib}, and $T_{tr,rot}$, respectively, stand for the amplitude of the electric field, pressure, electron temperature, vibrational temperature, and translational-rotational temperature. In the figure, a millimeter-wave beam propagates from the right to the left. In the region (a), T_e and T_{vib} increase because of the millimeter-wave absorption via inverse bremsstrahlung. As one report of the literature (Park and Griffith 1991) presents, the characteristic time of energy transfer from electrons to vibrational excitation is expected to be much short at electron densities of around 10^{21} m^{-3}. Furthermore, the region (b), energy relaxation from vibrational to translational-rotational excitations, follows the region (a), in which gas pressure has a peak because of isometric heating. Here, the rotational and translational excitation modes are assumed to be in an equilibrium state because the reaction rate of the energy transfer is high enough at atmospheric pressure (Prevosto et al. 2016). Finally, a state of thermodynamic equilibrium is achieved. The non-equilibrium structure of millimeter-wave discharges has been presumed by two-dimensional Computational Fluid Dynamics (CFD) simulation (Tabata et al. 2023).

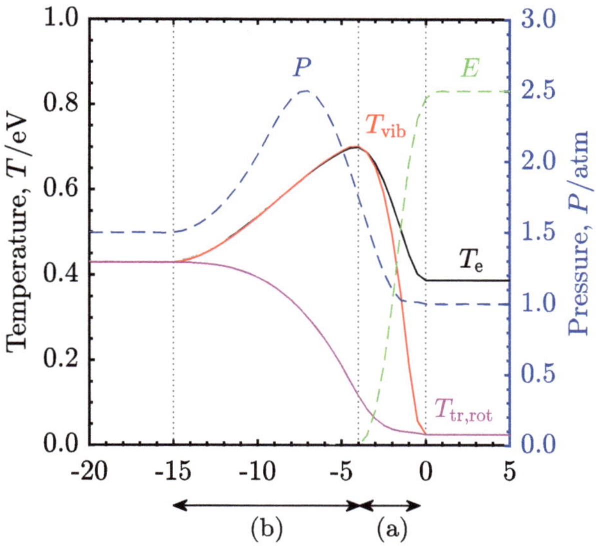

Fig. 4.2 Non-equilibrium temperature-structure of millimeter-wave discharge plasmas predicted by the author: (**a**) inverse bremsstrahlung (millimeter-wave absorption) and thereby decrease of electric field and (**b**) vibrational-translational energy relaxation and thereby pressure peak appear. In the figure, a millimeter-wave beam propagates from the right to the left. The absolute value of respective axis is meaningless because the value depends on parameters such as ambient pressure and millimeter-wave power density

In this way, one can expect that the non-equilibrium dynamics of neutral particles and ions are of importance in atmospheric millimeter-wave discharges. This chapter introduces a practical method to investigate the non-equilibrium characteristics of millimeter-wave discharge plasma, including examples of actual measurements. We specifically examine optical emission spectroscopy (OES), which enables us to take non-contact measurements without disturbing the plasma.

Optical Emission Spectroscopy (OES)

Benefits of OES for the Measurement of Millimeter-Wave Discharges

Various plasma diagnostic techniques are known in electric propulsion (EP) and plasma processes, such as probe methods and spectroscopy methods. The probe method is one in which a metal rod is inserted directly into the plasma to obtain the internal information. However, for the spectroscopy method, light emission from the plasma is measured with no contact to the plasma to obtain internal information. To investigate millimeter-wave discharges, an optical technique is desirable because placement of a needle in the discharges might induce local disturbance of the electric field or might cause errors in measurement. Optical emission spectroscopy (OES) is used widely for millimeter-wave discharge plasma measurements because it can measure the plasma temperature and density without contact. Other methods include laser absorption spectroscopy, but optical emission spectroscopy is commonly used because it accommodates a simple and low-cost experiment setup.

General Remarks on Optical Radiation of Atoms and Molecules

Emission spectra from plasma are classifiable into three types: free-free radiation, free-bound radiation, and bound-bound radiation. Figure 4.3 is an image that can help elucidate those radiations. In the free-free radiation, the free electrons are slowed as they pass near the nucleus of an atom or a molecule. The kinetic energy of the slowed electrons is emitted as luminescence. The relative velocity at the time of passage can vary continuously, which gives a continuous spectrum or Bremsstrahlung radiation. The free-bound radiation is the process of emitting the excess energy when a free electron is trapped in an atom or a molecule. This transition is also a continuous spectrum called recombination radiation. Bound-bound radiations have excess energy in an atom or a molecule that is emitted as luminescence when an excited atom or molecule is de-excited from an excited energy level to lower level. In optically thick plasmas, black body radiation appears along with those radiations. The next section presents an examination of the line spectra of bound-bound radiations and introduces methods to measure the number densities and temperatures of electrons and neutral particles.

Measurement of Electron Number Density

Fundamentals of Spectrum Broadening

The spectrum measured using a spectrometer is broadened because of several effects. One effect, the Stark effect, enables us to measure the electron number density. This effect is attributable to the fact that the change of the electron number density in a discharge plasma causes the electron orbitals of atoms or molecules to fluctuate because of Coulomb forces, resulting in a change in the wavelength of the emitted light during de-excitation. The relation between the electron number density

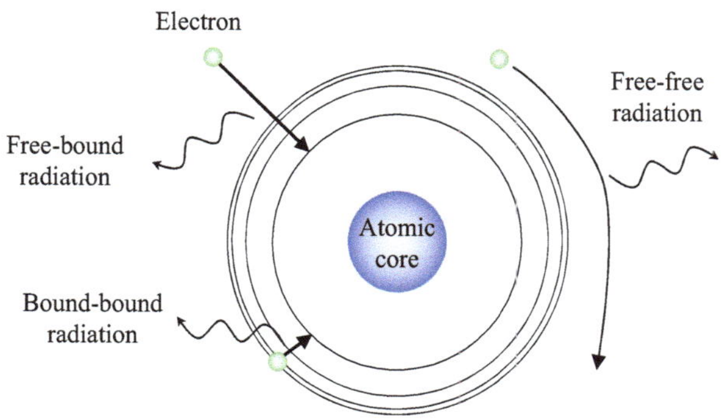

Fig. 4.3 Classification of emission spectra: free-free radiation, free-bound radiation, and bound-bound radiation

and Stark broadening is already known semi-experimentally, which enables us to obtain the electron number density.

Next, general remarks on spectrum broadening are given. The functions of broadening can be decomposed into Gaussian f_G and Lorentzian f_L functions, as the following equation presents.

$$f_G = \frac{1}{\sqrt{\pi}\Delta\lambda_G} \exp\left[-\left(\frac{\lambda-\lambda_0}{\Delta\lambda_G}\right)^2\right] \tag{4.6}$$

$$f_L = \frac{1}{\pi} \frac{\Delta\lambda_L}{(\lambda-\lambda_0)^2 + \Delta\lambda_L^2} \tag{4.7}$$

In these equations, λ, λ_0, and $\Delta\lambda_G$ and $\Delta\lambda_L$ are the wavelength, the central wavelength of a broadening spectrum, and the variables related to the spectrum's broadening of Gaussian and Lorentzian, respectively. The broadening is evaluated by the full width at half maximum (FWHM), which is relevant to the dispersion of distribution functions. The FWHMs of Gaussian and Lorentzian functions are presented below.

$$\mathrm{FHWM}_G = 2\sqrt{\ln 2}\Delta\lambda_G \tag{4.8}$$

$$\mathrm{FWHM}_L = 2\Delta\lambda_L \tag{4.9}$$

The actual spectrum is the convolution of all the broadened spectra. Voigt function f_V is generally used as the convolution integral of Gaussian and Lorentzian functions as Eq. 10 shows.

$$f_V = \frac{\Delta\lambda_L}{\pi^{\frac{3}{2}}\Delta\lambda_G} \int_{-\infty}^{\infty} \frac{\exp\left[-\left(\frac{\lambda-\acute{\lambda}_0}{\Delta\lambda_G}\right)^2\right]}{\left(\lambda-\acute{\lambda}_0\right)^2 + \Delta\lambda_L^2} d\acute{\lambda}_0 \tag{4.10}$$

The FWHM of Stark broadening, which depends on electron number density, is identifiable by decomposing actual spectra into Gaussian and Lorentzian.

Stark Broadening

Broadening of hydrogen Balmer atoms is often used to measure the electron number density. Specifically, the emissions from the electronically excited level $n = 3$ to $n = 2$ of a hydrogen atom are called H_α, from the excited level $n = 4$ to $n = 2$ are H_β, and from the excited level $n = 5$ to $n = 2$ are called H_γ. The central wavelengths of the respective spectra are 656.28 nm, 486.13 nm, and 434.05 nm. Many earlier studies have derived the relation between the electron number density and FWHMs (Konjević et al. 2012). The simulated Balmer line's broadening by Gigosos et al.

(Gigosos et al. 2003) agrees well with experimentally obtained results. Therefore, the following equation is expected to be reasonable. Attention must always be devoted to confirming the range of electron number density and electron temperature where the formula is applicable.

$$H_\alpha : n_e\left[m^{-3}\right] = 10^{23} \cdot \left(\frac{\text{FWHM [nm]}}{1.098}\right)^{1.47135} \quad (4.11)$$

$$H_\beta : n_e\left[m^{-3}\right] = 10^{22} \cdot \left(\frac{\text{FWHM [nm]}}{0.94666}\right)^{1.49} \quad (4.12)$$

$$H_\gamma : n_e\left[m^{-3}\right] = 10^{23} \cdot \left(\frac{\text{FWHM [nm]}}{4.668}\right)^{1.45826} \quad (4.13)$$

Plavcan et al. investigated the dependence of the electron number density on the FWHM of oxygen and nitrogen atoms for a case in which the emission of hydrogen atoms cannot be observed in air plasmas (Plavcan et al. 2010). The electron number density was obtained in advance from broadening of the emission spectrum of H_α. Also, ω in Eq. (4.14) was deduced, respectively, for nitrogen and oxygen atoms.

$$n_e = \frac{\text{FWHM[nm]}}{10^{-16}\omega} \quad (4.14)$$

Three spectra of excited emissions from nitrogen atoms are observed around 740 nm. The study (Plavcan et al. 2010) evaluated broadening of the spectrum at 746.83 nm. Moreover, for the oxygen atoms, spectral broadening at 844.6 nm was evaluated. Finally, the dependence of ω on the electron temperature T_e was investigated as follows.

$$\text{Oxygen line (844.6 nm)} : \omega = 0.014(\pm 0.0024) - 0.0024(\pm 0.0025) \times T_e \quad (4.15)$$

$$\text{Nitrogen line (746.83 nm)} : \omega = 0.0066(\pm 0.0014) - 0.0004(\pm 0.0014) \times T_e \quad (4.16)$$

Those equations are effective at electron temperatures from 0.7 to 1.2 eV. From the formulae above, the FWHM dependence on the electron number density was calculated assuming electron temperature T_e of 1 eV, as presented in Fig. 4.4. Equations (4.11)–(4.13) were used for the hydrogen Balmer lines. Figure 4.4 indicates that the FWHMs of nitrogen and oxygen atoms might be less than 0.01 nm when the electron number density is approximately 1.0×10^{21} m^{-3}, which is expected to make the measurement rather difficult. Depending on the physical parameters of plasma and experimental conditions, the best choice of which Stark broadening line to be used must be made.

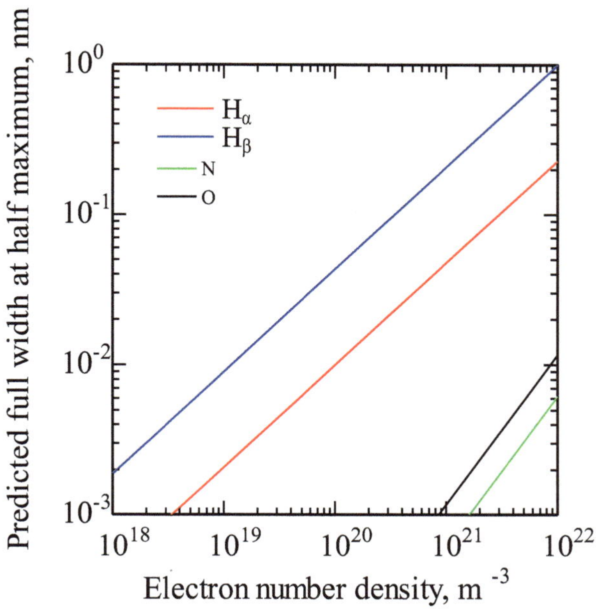

Fig. 4.4 Dependence on full width at half maximum (FWHM) of the Stark broadening of hydrogen Balmer line, oxygen, and nitrogen atoms. The electron temperature was assumed as 1 eV

Instrumental Broadening

The broadening caused by the resolution of a spectrometer is called instrumental broadening. It is desirable to measure the instrumental broadening beforehand. If a lamp with a small spectral broadening is useful, such as a low-pressure mercury lamp, then the line spectrum width can be regarded as instrumental broadening. He-Ne lamps might also be used. Other lamps for which broadening is already known can be used to measure the line spectrum by decomposition described at the beginning of this section. The instrumental broadening follows a Gaussian distribution function.

Natural Broadening

Natural broadening derives from Heisenberg's uncertainty principle, which is described by the following equation.

$$\Delta t \cdot \Delta E \geq \frac{h}{4\pi} \tag{4.17}$$

This equation shows that a shorter fluctuation time Δt is associated with greater energy fluctuation ΔE. Shorter durations for which electrons can exist at the same excited level are associated with greater fluctuation of the energy of an excited level, thereby producing greater broadening. The natural broadening follows a Lorentzian distribution function. Practically, natural broadening is usually negligible because it is slight compared to other broadening.

Doppler Broadening

Doppler broadening results from the thermal motion of particles in plasma, which induces the Doppler effect and which changes the frequency of the radiation (Djurović and Konjević 2009). If the velocity distribution of the emitting particles follows a Maxwellian, then the FWHM is expressed by the following equation.

$$\mathrm{FWHM}_{\mathrm{doppler}} = 7.162 \times 10^{-7} \lambda_0 \sqrt{\frac{T_{\mathrm{gas}}}{M}} \tag{4.18}$$

In that equation, λ_0, T_{gas}, and M, respectively, represent the central wavelength [nm], gas temperature [K], and mass per atom. FWHM has the same units as the wavelength. The Doppler broadening follows a Gaussian distribution function.

Pressure Broadening (Van der Waals Broadening)

Van der Waals broadening is pressure broadening. When the pressure is high, such as in atmospheric discharges, collisions among particles predominate. The photon emission process is disturbed. Furthermore, the spectrum is broadened by perturbations from particles passing by the particle that radiates light emission. According to a report of the literature (Balcon et al. 2007), van der Waals broadening is shown below.

$$\mathrm{FWHM}_{\mathrm{Van\ der\ Waals}} = 3.6 \times P/T^{0.7} \tag{4.19}$$

Therein, P and T, respectively, denote the pressure [atm] and gas temperature [K]. The FWHM has units of nm. The broadening follows a Lorentzian distribution function.

Practical Use of Stark Broadening in Millimeter-Wave or Microwave Discharge Plasmas

The electron number density was measured experimentally in both over-critical and under-critical conditions. In over-critical conditions, an earlier report (Schaub et al. 2016) by the group of the Plasma Science and Fusion Center at Massachusetts Institute of Technology (PSFC at MIT) described a study of the electron number density of 110 GHz millimeter-wave discharge plasmas in air in a wide range of atmospheric pressures, using Stark broadening of the H_α Balmer line. It was described that H_β line at 486.1 nm was observed but it was obscured by strong emissions from molecular nitrogen, so that Stark broadening of H_α at line 656.3 nm was chosen. Results showed that the electron number density agreed well with the following cutoff density n_{cutoff}, which is the limit of electromagnetic-wave energy absorption.

$$n_{\text{cutoff}} = \frac{m_e \varepsilon_0}{e^2}\left(\omega^2 + \nu_m^2\right) \tag{4.20}$$

Here, $\omega = 2\pi f$, ε_0, and ν_m, respectively, denote the millimeter-wave angular frequency, permittivity, and the electron-neutral particle collisional frequency.

In studies (Sidorov et al. 2016; Vodopyanov et al. 2017) reported by the group of the Institute of Applied Physics of the Russian Academy of Sciences (IAP RAS), a "point-like" discharge was generated in a nonuniform argon flow induced by a 0.67 THz and 40 kW gyrotron beam. The H_α at line 656.3 nm was observed when the minor admixture of hydrogen and electron number density was measured. The density apparently exceeded the cutoff density for a frequency of 0.67 THz. Recently, the electron number density at a frequency of 0.67 THz was measured using laser interferometer by the same group (Barmashova et al. 2023). The findings showed that the density exceeded the cutoff density.

Along with experimental investigations, the electron number density was also investigated numerically. In over-critical conditions, the experiments conducted by the PSFC at MIT indicated that a plasma filament is generated at a quarter wavelength interval in the direction of minus wavevector (toward the beam source): the "quarter-wavelength plasma structure." The numerical simulation (Boeuf et al. 2010; Zhou and Dong 2011; Kourtzanidis et al. 2014; Arcese et al. 2017) revealed that at the antinode of the standing wave generated by an incident and a reflected beam, the electric field exceeded the discharge threshold. Then, a new plasma filament was generated. In the simulation, the electron number density is approximately $0.1\text{–}1 \times 10^{22}\ \text{m}^{-3}$, which agreed with the experiment. In under-critical conditions, Nakamura (Nakamura et al. 2018) described that a plasma filament was generated under conditions where the electron number density exceeded the cutoff density, whereas the plasma structure became diffusive when the density became lower than the cutoff density.

In these ways, electron number density has an important causal role in the formation of filament structures in millimeter-wave discharge plasmas. Therefore, the measurement of electron density described in this section is useful for understanding millimeter-wave discharge plasmas in air.

Measurement of Electrons and Gas Temperature

Fundamentals of Emission Spectra of Air Discharge Plasmas

This section presents a description of the method of measuring vibrational and rotational temperatures assuming a Boltzmann equilibrium. The emission spectrum of air discharge plasmas is exceedingly complicated because the light emissions of various atoms and molecules overlap. Table 4.1 presents the electronic excitation transitions of typical molecules in air plasmas. The "Specair," commercial code, which is often used for fitting emission spectra of air discharges, can calculate spectra of NO, N_2, $N^+{}_2$, O_2, CN, OH, NH, C_2, and CO for molecules and N, O, and C for atoms, which means that various spectra appear to be mutually overlapping in air discharges. Specair and LIFBASE can calculate spectra assuming Boltzmann equilibrium for vibrational and rotational temperatures and can reveal temperatures that match the experimentally obtained spectra, by selecting the required molecular and atomic species. It is also possible to create a spectrum analysis code by yourself. In this case, one can perform spectral fitting by overlapping a continuous spectrum and band spectra. The next section presents a description of an analytical spectral calculation method.

Theoretical Scheme for Calculating Molecular Band Spectrum

This section presents a description of an example of the theoretical spectrum of molecular nitrogen, which accounts for 80% of air. Especially the second positive system (SPS) is emphasized (Akatsuka 2012; Sintsov et al. 2020). SPS is a band spectrum that appears when electronic transitions occur from C $^3\Pi_u$ to B $^3\Pi_g$. The spectrum appears at wavelengths of 300 nm to 400 nm. The SPS band spectrum is used in many reports of experimentation to elucidate air discharges and to measure the vibrational and rotational temperatures.

First, the method for calculating a molecular band spectrum is explained. The wavelength of a luminescence is determined by the energy difference between the upper excitation state and lower excitation state. There are many de-excitation processes for molecules because a molecule not only has electronic excitation but also vibrational and rotational excitations. Therefore, to obtain a completed emission spectrum, it is necessary to calculate pairs of wavelength and spectrum intensity for all processes first and then to sum all the intensities in consideration of spectral broadenings such as instrumental broadening and pressure broadening.

The vibrational and rotational temperatures of molecular nitrogen, for example, can be inferred using the spectra of the vibrational transitions $\Delta v = -2, -3$. The quantum numbers of the electronically excited, vibrational, and rotational levels are represented, respectively, as n, v, and J. Symbols ' and ", respectively, represent the upper level before the transition and the lower level after the transition. One can

Table 4.1 Names of molecular band spectra and transitions of excitation energy level (Perrin et al. 2009)

Molecule	System name	Upper state	Lower state
N_2	First positive	$B^3\Pi_g$	$A^3\Sigma_u^+$
	Second positive	$C^3\Pi_u$	$B^3\Pi_g$
	Birge – Hopfield 1	$b^1\Pi_u$	$X^1\Sigma_g^+$
	Birge – Hopfield 2	$b'^1\Sigma_u^+$	$X^1\Sigma_g^+$
	Carroll – Yoshino	$c_4'^1\Sigma_u^+$	$X^1\Sigma_g^+$
	Worley – Jenkins	$c_3{}^1\Pi_u$	$X^1\Sigma_g^+$
	Worley	$o_3{}^1\Pi_u$	$X^1\Sigma_g^+$
N_2^+	Meinel	$A^2\Pi_u$	$X^2\Sigma_g^+$
	First negative	$B^2\Sigma_u^+$	$X^2\Sigma_g^+$
	Second negative	$C^2\Sigma_u^+$	$X^2\Sigma_g^+$
NO	γ	$A^2\Sigma^+$	$X^2\Pi_r$
	β	$B^2\Pi_r$	$X^2\Pi_r$
	δ	$C^2\Pi_r$	$X^2\Pi_r$
	ϵ	$D^2\Sigma^+$	$X^2\Pi_r$
	γ'	$E^2\Sigma^+$	$X^2\Pi_r$
	β'	$B'^2\Delta$	$X^2\Pi_r$
	11000Å	$D^2\Sigma^+$	$A^2\Sigma^+$
	Infrared	$X^2\Pi_r$	$X^2\Pi_r$
O_2	Schumann – Runge	$B^3\Sigma_u^-$	$X^3\Sigma_g^-$
CO	Infrared	$X^1\Sigma^+$	$X^1\Sigma^+$
	Fourth positive	$A^1\Pi$	$X^1\Sigma^+$
	Hopfield – Birge	$B^1\Sigma^+$	$X^1\Sigma^+$
	Third positive	$b^3\Sigma^+$	$a^3\Pi$
CO^+	Comet – tail	$A^2\Pi_i$	$X^2\Sigma^+$
	Baldet – Johnson	$B^2\Sigma^+$	$A^2\Pi_i$
	First negative	$B^2\Sigma^+$	$X^2\Sigma^+$
CN	Red	$A^2\Pi_i$	$X^2\Sigma^+$
	Violet	$B^2\Sigma^+$	$X^2\Sigma^+$
	LeBlanc	$B^2\Sigma^+$	$A^2\Pi_i$
C_2	Philips	$A^1\Pi_u$	$X^1\Sigma_g^+$
	Mulliken	$D^1\Sigma_u^+$	$X^1\Sigma_g^+$
	Deslandres – D'Azambuja	$C^1\Pi_g$	$A^1\Pi_u$
	Ballik et Ramsay	$b^3\Sigma_g^-$	$a^3\Pi_u$
	Swan	$d^3\Pi_g$	$a^3\Pi_u$
	Fox – Herzberg	$e^3\Pi_g$	$a^3\Pi_u$

Table 4.2 Coefficients of second positive system of a molecular nitrogen (Akatsuka 2012)

Parameter	$N_2(C^3\Pi_u)$	$N_2(B^3\Pi_g)$
E_e, eV	11.05	7.390
ω_e, eV	2.538×10^{-1}	2.149×10^{-1}
$\omega_e\chi_e$, eV	3.526×10^{-3}	1.751×10^{-3}
B_e, eV	2.262×10^{-4}	2.030×10^{-4}
α_e, eV	2.316×10^{-6}	2.220×10^{-6}

consider the transition from n', v', J' to n'', v'', J''. First, because the energy difference between the upper and lower levels determines the light emission wavelength, it is important to calculate the electron excitation energy of each energy level correctly. The energy of each level is the sum of the electron excitation energy E_e, vibrational excitation energy $E_{vib}(v)$, and rotational excitation energy $E_{rot}(v, J)$, which can be expressed as follows.

$$E(v, J) = E_e + E_{vib}(v) + E_{rot}(v, J) \tag{4.21}$$

In that equation, the vibrational excitation energy $E_{vib}(v)$ and the rotational excitation energy $E_{rot}(v, J)$ are determined using the following equations.

$$E_{vib}(v) = \omega_e\left(v + \frac{1}{2}\right) - \omega_e\chi_e\left(v + \frac{1}{2}\right)^2 + \ldots \tag{4.22}$$

$$E_{rot}(v, J) = B_v J(J+1) - D_v J^2(J+1)^2 + \ldots \tag{4.23}$$

$$B_v = B_e - \alpha_e\left(v + \frac{1}{2}\right) + \ldots \tag{4.24}$$

$$D_v = D_e - \beta_e\left(v + \frac{1}{2}\right) + \ldots \tag{4.25}$$

Table 4.2 shows the electronic excitation energy E_e for the C $^3\Pi_u$ and B $^3\Pi_g$ levels, the constants ω_e and $\omega_e\chi_e$ for vibrational excitation, and the constants B_e and α_e for rotational excitation. Here, D_v is often negligible compared to B_v. Furthermore, in some reports of the literature, these values are expressed in units of "Kaiser" (cm^{-1}), which corresponds to 1.986×10^{-23} J.

The wavelength λ at which light emissions appear is calculable using the following equation.

$$E_{B_{v'',J''}}^{C_{v',J'}} = E_C(v', J') - E_B(v'', J'') \tag{4.26}$$

$$\lambda_{B_{v'',J''}}^{C_{v',J'}} = \frac{hc}{E_{B_{v'',J''}}^{C_{v',J'}}} \tag{4.27}$$

In those equations, h and c, respectively, denote Planck's constant and the speed of light.

The emission intensity of a line spectrum appearing at that wavelength can be calculated. The spectrum intensity is determined by the product of transition probability and the number density of an emitting particle. The luminescence energy of a photon is $h\nu = hc/\lambda$, so that the intensity [W/m^3] is described as presented below.

$$I_{\mathrm{B}_{v'',J''}}^{\mathrm{C}_{v',J'}} = \frac{hc}{\lambda} \cdot A_{\mathrm{B}_{v'',J''}}^{\mathrm{C}_{v',J'}} \cdot N_{\mathrm{C}_{v',J'}} \tag{4.28}$$

Here, $A_{\mathrm{B}_{v'',J''}}^{\mathrm{C}_{v',J'}}$ and $N_{\mathrm{C}_{v',J'}}$ denote the transition probability [s^{-1}] and the number density of the electronic excitation state C. The inverse of the transition probability is the lifetime [s] that can exist in the electronic excitation level. The transition probability is described by the following equation.

$$A_{\mathrm{B}_{v'',J''}}^{\mathrm{C}_{v',J'}} = \frac{64\pi^4 \left(v_{\mathrm{B}_{v'',J''}}^{\mathrm{C}_{v',J'}}\right)^3}{3hc^3 g_{\mathrm{C}}} \cdot \frac{1}{2J'+1} \cdot \sum_{J'J''} |R_{J'J''}|^2 \cdot q_{v'v''} \cdot S_{J'J''} \tag{4.29}$$

In that equation, g_{C}, $\sum_{J'J''} |R_{J'J''}|^2$, $q_{v'v''}$, and $S_{J'J''}$ respectively, denote the statistical weight of the electronic excitation state C, electronic transition moment, the Frank-Condon factor, and the Hönl-London factor. The number density of an electronic state C is expressed by the following equation by assuming Boltzmann equilibrium vibrational and rotational temperatures T_{vib} and T_{rot}.

$$\begin{aligned} N_{\mathrm{C}}(v',J') = (2J'+1)N_{\mathrm{C}}(0,0) \exp\left[-\frac{E_{\mathrm{C}}(v',0)-E_{\mathrm{C}}(0,0)}{k_{\mathrm{B}}T_{\mathrm{vib}}}\right] \\ \cdot \exp\left[-\frac{E_{\mathrm{C}}(v',J')-E_{\mathrm{C}}(v',0)}{k_{\mathrm{B}}T_{\mathrm{rot}}}\right] \end{aligned} \tag{4.30}$$

In this equation, k_{B} is the Boltzmann's constant. Therefore, Eqs. (4.28)–(4.30) engender the following equation using a constant K.

$$I_{\mathrm{B}_{v'',J'}}^{\mathrm{C}_{v',J'}} = \frac{K}{\lambda^4} \cdot q_{v'v''} \exp\left(-\frac{E_{\mathrm{vib}}}{k_{\mathrm{B}}T_{\mathrm{vib}}}\right) \cdot S_{J'J''} \exp\left(-\frac{E_{\mathrm{rot}}}{k_{\mathrm{B}}T_{\mathrm{rot}}}\right) \tag{4.31}$$

Equation (4.31) shows that the spectrum intensity at a wavelength is inversely proportional to λ^4.

Although countless combinations of vibrational and rotational quantum numbers exist, the following equations must be satisfied for rotational transitions in the case of second positive system.

$$\Delta J = J' - J'' = -1, 0, 1 \tag{4.32}$$

Table 4.3 Franck-Condon factors for vibrational transitions in second positive system of a molecular nitrogen (Gilmore et al. 1992)

v'	v''	Wavelength, nm	Franck–Condon factor
0	2	380.4	1.45×10^{-1}
0	3	405.8	5.12×10^{-2}
1	3	375.4	1.98×10^{-1}
1	4	399.7	1.10×10^{-1}
2	4	370.9	1.61×10^{-1}
2	5	394.2	1.39×10^{-1}
3	6	389.4	1.31×10^{-1}

Each transition is called a P-branch, a Q-branch, or an R-branch. The Hönl-London factor of a second positive system is given by the following equations for the P-, Q-, and R-branches (Phillips 1975).

$$S_{\mathrm{P}} = 6(J+1) - 10/(J+1) \tag{4.33}$$

$$S_{\mathrm{Q}} = 10/J + 10/(J+1) \tag{4.34}$$

$$S_{\mathrm{R}} = 6J - 10/J \tag{4.35}$$

Regarding the Frank-Condon factors in the second positive band, the values presented in Table 4.3 are useful (Gilmore et al. 1992).

From the expressions presented above, the wavelength and spectrum intensity corresponding to a certain transition can be calculated. The molecular band spectrum is obtained by adding all the line spectra, considering the Gaussian or Lorentzian broadenings. For example, when the broadening follows a Gaussian distribution function, the spectrum emission intensity is described by the following equation using FWHM:

$$I(\lambda) = \sum_{v', v'', J', J'} I_{\mathrm{B}_{v',J'}}^{\mathrm{C}_{v',J'}} \exp\left[-\left(\frac{\lambda - \lambda_{\mathrm{B}_{v'',J''}}^{\mathrm{C}_{v',J'}}}{\Delta\lambda}\right)^2\right] \tag{4.36}$$

where the following variable is used.

$$\Delta\lambda = \frac{\mathrm{FWHM}}{2\sqrt{\ln 2}} \tag{4.37}$$

For convolution of the Gaussian and Lorentzian distribution function, a pseudo-Voigt function or Voigt function is useful.

Practical Use of Molecular Band Spectrum Analysis for the Temperature Measurement in Air Discharge Plasmas

The group of PSFC at MIT investigated vibrational and rotational temperatures by analyzing second positive system using a commercial Specair code. The experimental conditions are over-critical conditions at a beam frequency of 110 GHz, beam power of 1–1.5 MW, and pressure range of 1–100 Torr (Hummelt et al. 2012). The results showed the rotational and vibrational temperatures, respectively, 300–500 K and 4200–6200 K, which indicate non-equilibrium.

According to results of one earlier study (Artemev et al. 2018), microwave discharge was generated using a gyrotron with 200–600 kW pulse power, 0.5–20 ms pulse duration, and 40 mm wavelength. For the experiments, a spectrum in the wavelength range of 300–400 nm was measured using a spectrometer. The minimum cross-sectional area of 9.5 cm^2 gives a peak power density of 1.3 GW/m^2 at 600 kW beam power. Analyzing the second positive system of molecular nitrogen indicated the 4000–4500 K gas temperature.

Tabata et al., in a group of the University of Tokyo, investigated the non-equilibrium characteristics using second positive system (Tabata et al. 2020a, b). The microwave frequency and atmospheric pressure are, respectively, 28 GHz and 100 kPa. The microwave beam oscillated from a 570 kW gyrotron (Kariya et al. 2007) is transmitted in a waveguide and is radiated to the atmosphere. The beam power density of the radiated microwave decreases as their beam radius increases with microwave propagation. Taking advantage of this property, the spectrum was measured at widely various beam power densities of 0.05–0.50 GW/m^2 by changing the measurement point in the microwave propagation direction. The experimental setup is shown in Fig. 4.5 (Tabata et al. 2020a). The emission spectra from 370 nm to 410 nm were measured as presented in Fig. 4.6. The molecule from which the luminescence was derived was identified as related to Park's explanation (Park and Griffith 1991). Surprisingly, N^+_2 and CN showed stronger emissions than N_2. This is true because the luminescence intensity is proportional to the product of the particle number density and transition probabilities and N^+_2 and CN have large transition probabilities. The vibrational and rotational temperatures of molecular nitrogen were deduced by decomposing the emission spectra with an analysis code based on the theoretical scheme as explained in the preceding section.

As a result, the vibrational temperature did not exceed approximately 6000 K. The rotational temperature decreased from 5500 K with the increase of beam power density. For the study (Tabata et al. 2020a), it was assumed that the electron temperature coincides with the vibrational temperature and that the rotational temperature is equal to the translational temperature. First, regarding the electron-vibrational temperature, the electron number density calculated under the assumption that Saha's equation is satisfied at around 6000 K is approximately equal to the cutoff density. An earlier report also described that the electron density becomes the cutoff density by the direct measurements using H_α line (Schaub et al. 2016). Second, as for the decrease of translational-rotational temperature with the increase

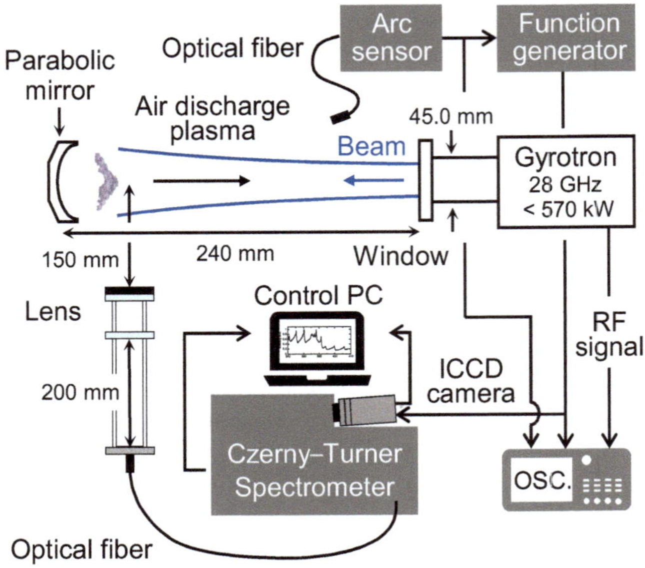

Fig. 4.5 Simplified schematic of the experimental setup for optical emission spectroscopy in air (Tabata et al. 2020a). Plasma was ignited at a parabolic mirror. The light emission from 28 GHz discharge plasma was obtained by the combination of Czerny-Turner spectrometer and ICCD camera at wavelengths of 370–410 nm

Fig. 4.6 Measured and fitted spectra of 28 GHz discharge plasmas in air at wavelengths of 370–410 nm and peak power densities of 0.12 and 0.34 GW/m^2 (Tabata et al. 2020a)

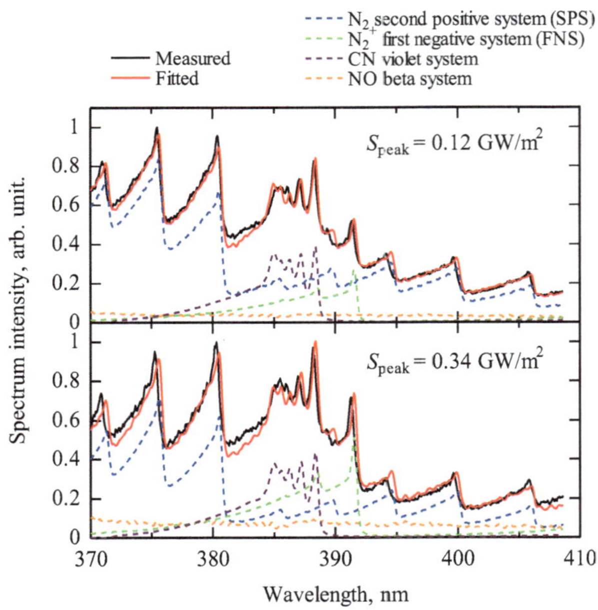

of beam power density, the increase in the propagation velocity of discharge front is regarded as attributed to the shorter energy relaxation time from vibrational to translational-rotational excitation. These measured gas temperatures are conserved thermodynamically by accounting for "plasma space occupancy." The plasma space occupancy ϕ ($0 < \phi < 1$) is a dimensionless quantity representing the volume fraction of plasma in space to account for the filamentary plasma structures observed in millimeter-wave discharges. The introduction of the plasma space occupancy modifies the energy input per unit mass in a local area where plasma is present, as described below.

$$q = \frac{1}{\phi} \cdot \frac{S_{\mathrm{beam}}}{\rho_1 V_{\mathrm{MSD}}} \tag{4.38}$$

Such energy concentration is not seen in laser discharges because there seems to be no plasma structure. The filamentary plasma structure might keep the local temperature higher in the case of microwave or millimeter-wave discharges.

Such non-equilibrium characteristics between electrons and neutral particles were discussed in a report of work by Litbak (Litbak 1991). The findings suggested that an equilibrium plasma occurs in the region of beam power density below 0.03 GW/m^2 and a non-equilibrium plasma at higher power densities. It is particularly interesting that the gas temperature shows a decreasing trend as the power density is increased, which is consistent with results reported earlier (Tabata et al. 2020a). For beam power densities that can generate non-equilibrium plasma (>0.03 GW/m^2), the propagation velocity of the discharge front is proportional to the square of the beam power density, with decreased energy input per unit mass as the power density increases. This fact might lead to a decrease of gas temperature with the increase of beam power density.

A recent study conducted using a 75 GHz gyrotron demonstrated determination of the gas temperature from the continuum spectrum (Artemev et al. 2022). The results showed the gas temperature as below 5500 K at beam power densities lower than 0.1 GW/m^2. Measurements using continuous spectra as well as band spectra of molecules might yield more useful information in the future.

Density and Temperature Measurements in Other Ambient Gases

Plasma jets allow many applications such as chemical decomposition for pollutant processing, surface treatment, bacterial inactivation, and biomedical treatment (Sintsov et al. 2020). Microwave discharges particularly are useful because of the non-equilibrium characteristics between electrons and heavy particles. One earlier report (Sintsov et al. 2020) described a non-equilibrium argon plasma torch sustained using a 24 GHz microwave beam in the external atmosphere of air. It was investigated using optical emission spectroscopy for temperature measurements. Results obtained from the measurement of electron excitation temperature of argon and vibrational and rotational temperatures of molecular nitrogen indicated that the electrons and heavy particles are in a non-equilibrium state.

Shimamura investigated microwave argon discharge plasma applying a collisional radiative (CR) model in optical emission spectroscopy at beam power densities of 0.1–0.2 GW/m^2 and atmospheric pressures of 40–100 kPa (Shimamura et al. 2021). This method solves the rate equations for collisions of various types in plasma such as electronic excitation, atomic excitation, electron impact ionization, atomic impact ionization, photon-excitation, and photon-ionization. The population density of each excited level and the resulting luminescence intensity are calculable. The optimal electron number density and electron temperatures can be determined by comparing the calculated spectrum with experimentally obtained spectra. The results of the literature (Shimamura et al. 2021) showed that the measured electron number density was almost at the cutoff density. Furthermore, the electron temperature was higher than argon gas temperature (approx. 2 eV) though it depends on the atmospheric pressure and beam power densities. Optical emission spectroscopy using a CR model is an informative tool for diagnosing non-equilibrium microwave or millimeter-wave discharges, for which contact measurements such as probe measurements are difficult.

References

Akatsuka H (2012) In: Patel V (ed) Chemical kinetics. IntechOpen, Rijeka. https://doi.org/10.5772/36512

Arcese E, Rogier F, Boeuf JP (2017) Plasma fluid modeling of microwave streamers: approximations and accuracy. Physics Plasma 24(11):113517. https://doi.org/10.1063/1.5006651

Artem'ev KV, Batanov GM, Berezhetskaya NK, Borzosekov VD, Davydov AM, Kozhevnikova NA, Konchekov EM, Kossyi IA, Sarksyan KA, Stepakhin VD, Sysoev SO, Temchin SM, Kharchev NK (2019) Discharge in a subthreshold microwave beam as an effective means for mercaptan decomposition. Plasma Phys Rep 45:523. https://doi.org/10.1007/s11141017-9784-0

Artemev KV, Batanov GM, Berezhetskaya NK, Davydov AM, Kossyi IA, Nefedov VI, Sarksyan KA, Kharchev NK (2018) A subthreshold high-pressure discharge excited by a microwave beam: physical basics and applications. Plasma Phys Rep 44(6):615. https://doi.org/10.1134/S1063780X18050021

Artemev KV, Batanov GM, Berezhetskaya NK, Borzosekov VD, Davydov AM, Kolik LV, Konchekov EM, Kossyi IA, Malakhov DV, Moryakov IV, Petrov AE, Sarksyan KA, Stepakhin VD, Kharchev NK (2022) Changes in structure of subthreshold discharge in air occurring with decreasing microwave radiation intensity. Plasma Phys Rep 48(2):170. https://doi.org/10.1134/S1063780X22020027

Balcon N, Aanesland A, Boswell R (2007) Pulsed RF discharges, glow and filamentary mode at atmospheric pressure in argon. Plasma Sources Sci Technol 16(2):217. https://doi.org/10.1088/0963-0252/16/2/002

Barmashova T, Sidorov AV, Vodopyanov AV, Luchinin A, Murzanev A, Razin S et al (2023) Study of THz gas discharge spatial dynamic in Argon. IEEE Trans Terahertz Sci Technol 13 (1):3–9. https://doi.org/10.1109/TTHZ.2022.3164546

Boeuf JP, Chaudhury B, Zhu GQ (2010) Theory and modeling of self-organization and propagation of filamentary plasma arrays in microwave breakdown at atmospheric pressure. Phys Rev Lett 104:015002. https://doi.org/10.1103/PhysRevLett.104.015002
Bogatov NA, Golubev SV, Zorin VG (1983) Ionizing radiation from a microwave discharge. Sov Tech Phys Lett 9(7):382
Borodacheva TV, Semenov VE, Sov N (1985) Electrodynamics of the photoionization mechanism for gasdynamic propagation of a microwave discharge. Tech Phys Lett 30(9):1019
Bournot P, Pincosy P, Inglesakis G, Autric M, Dufresne D, Caressa JP (1979) Propagation of a laser-supported detonation wave. Acta Astronaut 6(3):257. https://doi.org/10.1016/0094-5765(79)90097-3. https://www.sciencedirect.com/science/article/pii/0094576579900973
Brodskii YY, Venediktov IP, Colubev SV, Zorin VG, Kossyi IA (1984) Nonequilibrium microwave discharge in air at atmospheric pressure. Sov Tech Phys Lett 10(2):77
Bufetov IA, Prokhorov AM, Fedorov VB, Fomin VK (1987) Optical discharge accompanying a restriction imposed on lateral expansion of gas and a reduction in the threshold of light-induced detonation. JETP Lett 39(5):258. http://jetpletters.ru/ps/1288/article19461.html
Djurović S, Konjević N (2009) On the use of non-hydrogenic spectral lines for low electron density and high pressure plasma diagnostics. Plasma Sources Sci Technol 18(3):035011. https://doi.org/10.1088/0963-0252/18/3/035011
Fukunari M, Arnault A, Yamaguchi T, Komurasaki K (2014) Replacement of chemical rocket launchers by beamed energy propulsion. Appl Opt 53(31):I16. https://doi.org/10.1364/AO.53.000I16
Fukunari M, Komurasaki K, Nakamura Y, Oda Y, Sakamoto K (2017) Rocket propulsion powered using a gyrotron. J Energy Power Eng 11:363. https://doi.org/10.17265/1934-8975/2017.06.001
Fukunari M, Tanaka S, Shinbayashi R, Yamaguchi Y, Tatematsu Y, Saito T (2019) Observation of a comb-shaped filamentary plasma array under subcritical condition in 303-GHz millimetre-wave air discharge. Sci Rep 9:17972. https://doi.org/10.1038/s41598-019-54333-5
Gigosos MA, González MÁ, Cardeñoso V (2003) Computer simulated Balmer-alpha, -beta and -gamma Stark line profiles for non-equilibrium plasmas diagnostics. Spectrochimica Acta Part B Atom Spectrosc 58(8):1489. https://doi.org/10.1016/S0584-8547(03)00097-1. https://www.sciencedirect.com/science/article/pii/S0584854703000971. 5th European Furnace Symposium and 10th International Solid Sampling Colloquium with Atomic Spectroscopy
Gilmore FR, Laher RR, Espy PJ (1992) Franck–Condon Factors, r-Centroids, electronic transition moments, and Einstein coefficients for many nitrogen and oxygen band systems. J Phys Chem Ref Data 21(5):1005. https://doi.org/10.1063/1.555910
TM Hamasaki, KN Ohnishi (2018) HPLA: International high power laser ablation symposium
Hidaka Y, Choi EM, Mastovsky I, Shapiro MA, Sirigiri JR, Temkin RJ, Edmiston GF, Neuber AA, Oda Y (2009) Plasma structures observed in gas breakdown using a 1.5 MW, 110 GHz pulsed gyrotron. Physics Plasma 16(5):055702. https://doi.org/10.1063/1.3083218
Hummelt JS, Shapiro MA, Temkin RJ (2012) Spectroscopic temperature measurements of air breakdown plasma using a 110 GHz megawatt gyrotron beam. Physics Plasma 19(12):123509. https://doi.org/10.1063/1.4773037
Kakinuma K, Fukunari M, Yamaguchi T, Nakamura Y, Koizumi H, Komurasaki K, Parkin K (2016) Two-stage-to-orbit transporting system combining microwave rocket and microwave thermal rocket for small satellite launch. Aerospace Technol Japan 14(ists30):Pb99. https://doi.org/10.2322/tastj.14.Pb99
Kariya T, Mitsunaka Y, Imai T, Saito T, Tatematsu Y, Sakamoto K, Minami R, Watanabe O, Numakura T, Endo Y (2007) Optimization of 28 GHz gyrotron output performance for ECRH experiment of the GAMMA 10. Fusion Sci Technol 51(2T):397. https://doi.org/10.13182/FST07-A1414
Komurasaki K, Tabata K (2018) Development of a novel launch system microwave rocket powered by millimeter-wave discharge. Int J Aerospace Engin 2018:9247429. https://doi.org/10.1155/2018/9247429

Komurasaki K, Shimamura K, Fukunari M, Oda Y, Imai T, Takahashi M, Katsurayama H, Ohnishi N (2017) Microwave rocket propelled by millimeter wave. J Plasma Fusion Res 93(10):465

Konjević N, Ivković M, Sakan N (2012) Hydrogen Balmer lines for low electron number density plasma diagnostics. Spectrochim Acta B At Spectrosc 76:16. https://doi.org/10.1016/j.sab.2012.06.026. https://www.sciencedirect.com/science/article/pii/S0584854712001589. Honoring issue: a collection of papers on plasma and laser Spectrochemistry. Dedicated to Gary M. Hieftje on the occasion of his 70th birthday

Kossyi I (n.d.) Self-/non-self sustained microwave discharge as a new type of large-size plasma source. https://doi.org/10.2514/6.2006-1457

Kourtzanidis K, Boeuf JP, Rogier F (2014) Three dimensional simulations of pattern formation during high-pressure, freely localized microwave breakdown in air. Physics Plasma 21 (12):123513. https://doi.org/10.1063/1.4905071

Litbak A (1991) Strong microwaves and terahertz waves: sources and applications, pp 267–286

Matsui K, Shimano T, Ofosu JA, Komurasaki K, Schoenherr T, Koizumi H (2017) Accurate propagation velocity measurement of laser supported detonation waves. Vacuum 136:171. https://doi.org/10.1016/j.vacuum.2016.07.011. https://www.sciencedirect.com/science/article/pii/S0042207X16302287

Nakagawa T, Mihara Y, Komurasaki K, Takahashi K, Sakamoto K, Imai T (2004) Propulsive impulse measurement of a microwave-boosted vehicle in the atmosphere. J Spacecr Rocket 41(1):151. https://doi.org/10.2514/1.2540

Nakamura Y, Komurasaki K (2020) Theory and modeling of under-critical millimeter-wave discharge in atmospheric air induced by high-energy excited neutral-particles carried via photons. Plasma Sources Sci Technol 29(10):105017. https://doi.org/10.1088/1361-6595/ab8e4c

Nakamura Y, Komurasaki K, Fukunari M, Koizumi H (2018) Numerical analysis of plasma structure observed in atmospheric millimeter-wave discharge at under-critical intensity. J Appl Phys 124(3):033303. https://doi.org/10.1063/1.5023269

Oda Y, Komurasaki K, Takahashi K, Kasugai A, Sakamoto K (2006) Plasma generation using high-power millimeter-wave beam and its application for thrust generation. J Appl Phys 100(11): 113307. https://doi.org/10.1063/1.2399899

Oda Y, Shibata T, Komurasaki K, Takahashi K, Kasugai A, Sakamoto K (2009a) Thrust performance of microwave rocket under repetitive-pulse operation. J Propuls Power 25(1):118. https://doi.org/10.2514/1.37623

Oda Y, Kajiwara K, Takahashi K, Kasugai A, Sakamoto K, Komurasaki K (2009b) In-tube shock wave driven by atmospheric millimeter-wave plasma. Jpn J Appl Phys 48(11):116001. https://doi.org/10.1143/jjap.48.116001

Oda Y, Takahashi M, Tabata K, Ohnishi N, Komurasak K, Sakamoto K (2018) 43rd International conference on infrared, millimeter, and terahertz waves (IRMMW-THz), pp 1–2. https://doi.org/10.1109/IRMMW-THz.2018.8510051

Park C, Griffith W (1991) Nonequilibrium Hypersonic Aerothermodynamics. Phys Today 44(2):98. https://doi.org/10.1063/1.2809999

Perrin MY, Rivière P, Soufiani A (2009) RTO-EN-AVT-162 non-equilibrium gas dynamics-from physical models to hypersonic flights, ed. by P. presented during the AVT-162 RTO AVT/VKI lecture series held at the von Karman institute, RS Genèse, Belgium, VKI lecture series—ISBN 978-92-837-0091-3. http://www.rta.nato.int/Pubs/RDP.asp?RDP=RTO-EN-AVT-162 (NATO research and Technology organisation. pp 8.1,8.40. https://hal.archives-ouvertes.fr/hal00492190

Phillips MGR (1975) A new technique for the study of test time and rarefaction wave effects in hypersonic shock tubes. J Phys D Appl Phys 8(12):1359. https://doi.org/10.1088/0022-3727/8/12/007

Plavcan J, Grolmusova Z, Rakovsky J, Cermak P, Veis P (2010) WDS'10 proceedings of contributed papers, Part II pp. 101–104

Prevosto L, Kelly H, Mancinelli B (2016) Modelling of an atmospheric pressure nitrogen glow discharge operating in high-gas temperature regimes. Plasma Chem Plasma Process 36:973. https://doi.org/10.1007/s11090-016-9716-3

Schaub SC, Hummelt JS, Guss WC, Shapiro MA, Temkin RJ (2016) Electron density and gas density measurements in a millimeter-wave discharge. Physics Plasma 23(8):083512. https://doi.org/10.1063/1.4959171

Shimamura K, Matsukura M, Ozaki N, Miyawaki K, Yokota S, Minami R, Kariya T, Imai T (2020) Wireless power transmission efficiency for microwave rocket using 28 ghz gyrotron. J Spacecr Rocket 57(3):632. https://doi.org/10.2514/1.A34726

Shimamura K, Yamasaki J, Miyawaki K, Minami R, Kariya T, Yang J, Yokota S (2021) Propagation of microwave breakdown in argon induced by a 28 GHz gyrotron beam. Physics Plasmas 28(3):033505. https://doi.org/10.1063/5.0045350

Sidorov AV, Razin SV, Golubev SV, Safronova MI, Fokin AP, Luchinin AG, Vodopyanov AV, Glyavin MY (2016) Measurement of plasma density in the discharge maintained in a nonuniform gas flow by a high-power terahertz-wave gyrotron. Physics Plasma 23(4):043511. https://doi.org/10.1063/1.4947219

Sintsov S, Tabata K, Mansfeld D, Vodopyanov A, Komurasaki K (2020) Optical emission spectroscopy of non-equilibrium microwave plasma torch sustained by focused radiation of gyrotron at 24 GHz. J Phys D Appl Phys 53(30):305203. https://doi.org/10.1088/1361-6463/ab8999

Staack D, Farouk B, Gutsol AF, Fridman AA (2006) Spectroscopic studies and rotational and vibrational temperature measurements of atmospheric pressure normal glow plasma discharges in air. Plasma Sources Sci Technol 15(4):818. https://doi.org/10.1088/0963-0252/15/4/027

Tabata K (2022) Study on atmospheric millimeter-wave discharge using gyrotron, Ph.D. dissertation submitted to the University of Tokyo

Tabata K, Harada Y, Nakamura Y, Komurasaki K, Koizumi H, Kariya T, Minami R (2020a) Experimental investigation of ionization front propagating in a 28 GHz gyrotron beam: observation of plasma structure and spectroscopic measurement of gas temperature. J Appl Phys 127 (6):063301. https://doi.org/10.1063/1.5144157

Tabata K, Komurasaki K, Kariya T, Minami R (2020b) Optical emission spectroscopy of millimeter-wave discharge plasma for the measurement of vibrational and rotational temperatures. J Phys D Appl Phys 28(1):19

Tabata K, Oda Y, Komurasaki K, Manabe A, Kawashima R (2023) Non-equilibrium aerodynamics between ionization-wave and shock-wave fronts in millimetre-wave supported detonation Abstract. Jpn J Appl Phys. https://doi.org/10.35848/1347-4065/ad00a1

Takahashi M, Komurasaki K (2018) Discharge from a high-intensity millimeter wave beam and its application to propulsion. Adv Phys: X **3**(1):1417744. https://doi.org/10.1080/23746149.2017.1417744

Vodopyanov AV, Glyavin MY, Golubev SV, Luchinin AG, Razin SV, Safronova MI, Sidorov AV, Fokin AP (2017) Plasma density in discharge sustained in inhomogeneous gas flow by high-power radiation in the terahertz frequency range. Tech Phys Lett 43:186. https://doi.org/10.1134/S1063785017020286

Zhou Q, Dong Z (2011) Modeling study on pressure dependence of plasma structure and formation in 110 GHz microwave air breakdown. Appl Phys Lett 98(16):161504. https://doi.org/10.1063/1.3583452

Chapter 5
Experimental Studies of Microwave Discharge Induced by Gyrotron

Masafumi Fukunari

Abstract This chapter comprehensively reviews experimental investigations on gyrotron-induced microwave discharges and associated theories, comparing over-critical and under-critical conditions. Simple calculation models of the ionization threshold that separates over-critical and under-critical conditions are explained, and specific calculation examples are given. Filamentary structures of the induced plasma in over-critical and under-critical conditions are discussed with high-speed camera images. The study underscores the need for experiments on the ionization front's propagation velocity in various incident beam frequencies for assessing microwave rocket's thrust performance. Investigations also reveal limitations imposed by the transition from nonequilibrium to equilibrium discharge in the focusing optics of the incident beams and abnormal ignitions due to residual plasma.

Keywords Gyrotron-induced microwave discharges · Over-critical condition · Under-critical condition · Ionization threshold · Structure formation mechanism · Microwave rocket propulsion · Air discharges · Ionization front propagation velocity · Nonequilibrium to equilibrium discharge · Residual plasma

Introduction

This chapter will review experimental studies on gyrotron-induced microwave discharges and related theories. The microwave is usually defined as frequencies between 3 and 30 GHz. Frequencies from 30–300 GHz and 300–3000 GHz correspond to the millimeter and submillimeter waves, respectively. However, here, we refer to the gas discharge in these frequencies as microwave discharge in contrast to the light region, such as laser discharge.

The region between radio waves and infrared light is the so-called terahertz gap, an unexplored area with little research due to the lack of high-power oscillation

M. Fukunari (✉)
Research Center for Development of Far-Infrared Region, University of Fukui, Fukui, Japan
e-mail: fukunari@fir.u-fukui.ac.jp

K. Mori et al. (eds.), *Beamed-mobility Engineering*, Springer Tracts in Electrical and Electronics Engineering, https://doi.org/10.1007/978-981-99-4618-1_5

sources. Gyrotrons are vacuum-tube oscillators based on electron cyclotron maser instability and the only radiation source capable of high-power oscillation on a laboratory scale in the millimeter and submillimeter bands (Nusinovich 2004; Kartikeyan et al. 2004; Edgcombe 1993) (see Chap. 8). High-power gyrotrons have been developed as plasma heating devices in the field of nuclear fusion. Continuous-wave (CW) operation in 1 h at 1 MW, pulse modulation at 5 kHz, and oscillation efficiency of over 50% have already been achieved (Sakamoto et al. 2007).

The development of high-power power sources has driven research on microwave discharges since before the 1960s. Microwave discharges have been studied for engineering applications as well as from the standpoint of preventing discharges on transmission lines of high-power systems such as radar systems. MacDonald (1966) wrote a famous book leading research on microwave discharge in the waveguide in 1966. Raizer (1972) described in a 1971 paper that "A discharge appears from time to time in the waveguide, and the resulting plasma formation travels against the direction of propagation of the microwaves." This feature is the same as the discharges observed in the Microwave Rocket studies described in Chap. 2. Regarding the microwave discharge induced in the free space, Khodataev observed in 1957 a freely located microwave discharge at a wavelength of 30 mm in the focus of radiation for the first time (Khodataev 2008).

In radio communication, the creation of an artificial ionized region in the atmosphere by microwave discharge was studied to increase the reflection frequency of the radio waves (Gurevich et al. 1997). According to a review by Litvak (1991), there were exciting science fiction ideas in Tekhnika-Molodezhi popular science magazine (1957), about igniting an atmospheric discharge in the region of crossing focused beams over cities to light them. Recent progress has been made in the research of a point-like source of extreme ultraviolet (EUV) radiation using the discharge induced in a nonuniform gas flow injected into a vacuum vessel by powerful gyrotron radiation in the terahertz band (Glyavin et al. 2014; Bratman et al. 2011; Sidorov et al. 2020; Sidorov 2022).

A review of engineering applications of microwave discharges, such as an engine, was summarized by Btanov et al. (1985). The title of the second chapter of Batanov et al.'s paper, "High-pressure microwave discharges," published in 1985, is "Non-Self-Sustained Discharge and Microwave Rocket Concept." The idea of using electromagnetic waves to supply the energy necessary for propulsion to aircraft or spacecraft from remote locations is natural, considering the weight of the propellant used in rockets, as mentioned in Chap. 1. According to this paper, the first study on the "Microwave Rocket Concept" was published by J. L. Shad in 1964. The year 1964 was in the midst of the Apollo program.

In addition to these engineering applications, the evolution of microwave discharges is a nonlinear process involving instabilities, which has attracted purely physical interest. Microwave discharge forms beautiful self-organized filamentary discharge structures depending on frequency, gas, and pressure. With the development of high-speed imaging technology, these time-evolved images can now be observed, and studies have been conducted to clarify the physical mechanism.

In the section "Ionization Threshold Models," simple models of the ionization thresholds are described. Microwave discharges can be classified into over-critical condition, where the local electric field strength E_{lo} is higher than the ionization threshold E_{cr} of the gas, and under-critical condition, where the E_{lo} is lower than E_{cr}. The value of E_{cr} has been derived theoretically and experimentally with certain assumptions or limitations.

The section "Microwave Discharge in the Over-Critical Condition" reviews the experimental results of the discharge at the over-critical condition. The microwave discharge induced in gas is a key to the energy conversion process in the Microwave Rocket reviewed in Chap. 2. The over-critical condition corresponds to the parabolic mirror's focal point mounted on the thruster head of the Microwave Rocket. In experimental studies, focusing lenses or mirrors are utilized to ignite the discharge. The discharge plasma forms filamentary arrays. The ionization front (the head of the main plasma density growth) of the discharge plasma propagates backward to the incident beam. The propagation velocity is typically much higher than the sonic velocity.

The experimental results of the discharge at the under-critical condition are reviewed in the section "Microwave Discharge in the Under-Critical Condition." The discharge at under-critical condition is developed from the initial plasma generated by an external source or generated in the over-critical condition. The ionization front propagates, absorbing the incident beam energy. The propagation velocity of the ionization front is in the same order as the sound speed, and the shock wave is driven in this condition. Although the discharge plasma has filamentary structures, the spatial structure is entirely different from the over-critical condition. This condition corresponds to the discharge in the cylindrical thruster tube of the Microwave Rocket. The high pressure behind the shock wave imports the thrust impulse. When the ionization front and the shock wave propagate at the same velocity, this phenomenon is called microwave supported detonation (MSD) due to its analogy with the detonation in chemical combustion. The same system of fluid equations can be used for the analysis of the pressure, temperature, density, flow velocity, and energy absorption rate with measured or estimated propagation velocity. Currently, however, no reasonable model can capture the ionization mechanism at the under-critical condition where the local electric field is roughly one order of magnitude lower than the ionization threshold and reproduce the propagation velocity measured in the experiments.

Ionization Threshold Models

The over-critical and under-critical conditions are characterized by the ionization threshold in the local electric field or power density. Let us review simple models of the ionization threshold in air and not-too-light noble gases. The breakdown is interpreted as an avalanche-like increase in the number density of the initial seed electron n_0 in the gas to a specific value n_c by the intense electric field. Nevertheless, it is generally difficult to strictly determine the specific value of n_c. The ionization threshold of the electric field E_{cr} is here, the minimum electric field strength that can cause this breakdown.

Taylor et al. (1971) developed a model for the ionization threshold of air based on the empirical relationship between the ionization frequency ν_i and the root-mean-square (rms) electric field E_{rms} [V/cm].

$$\frac{\nu_i}{p} = 4 \times 10^7 \left[\frac{E_{eff}}{p \times 10^2}\right]^{5.33} - 6.4 \times 10^4, \tag{5.1}$$

where p (here it is for room temperature work) is pressure in Torr. E_{eff} indicates the effective electric field defined as

$$E_{eff} = \frac{E_{rms}}{\left(1 + \frac{\omega^2}{\nu_m^2}\right)^{1/2}} \tag{5.2}$$

where $\nu_m = 5.3 \times 10^9\, p$ [Hz] is the average electron collision frequency with neutral particles as a function of the ambient pressure in Torr and ω[Hz] is the angular frequency of the incident electromagnetic wave. E_{cr} [V/cm] in rms field is represented as

$$E_{cr} = 3.75p\left(1 + \frac{\omega^2}{\nu_m^2}\right)^{1/2}\left[\frac{pD}{p^2\Lambda^2} + \frac{\ln\left(n_c/n_0\right)}{p\tau} + 6.4 \times 10^4\right]^{3/16} \tag{5.3}$$

where D [cm^2/s] is the diffusion coefficient, Λ [cm] is the characteristic diffusion length, and τ[s] is the pulse width of the incident wave. Because D depends on the electron density and electron temperature, D varies during the breakdown. They employed an approximated free diffusion coefficient $D = 1.3 \times 10^6/p$ as a function of the pressure in this model. Therefore, pD becomes a constant. For spatial scales smaller than the Debye length λ_D , the assumption of free electron diffusion is sufficiently reasonable. Assuming initial electron density $n_0 \approx 10^9$ m^{-3} incidentally present in the gas due to cosmic rays, etc., and its temperature of 0.03 eV, the Debye length becomes $\lambda_D \approx 4$ cm, which is larger than the typical focal spot diameter. For other models of microwave discharges induced in air, see also Gurevich et al. (1997), and Batanov et al. (1985).

Vyskrebentsev and Raizer (1973) analytically derived a kinetic model of the ionization threshold for monatomic not-too-light gases. The threshold field is determined by

$$E_{cr}^2 = 5.7 \times 10^{-16} I_1(\omega^2 + \nu_m^2)\frac{\nu^*}{\nu_m}\Phi(\frac{\nu_i}{\nu^*}) \tag{5.4}$$

$$\frac{\nu_{\mathrm{i}}}{\nu^*} = \frac{(\nu_d + \nu_t)}{\nu^*}, \nu_{\mathrm{d}} = \frac{D}{\Lambda^2} \approx 5.8 \times 10^{14} \frac{I^*}{\nu_{\mathrm{m}} \Lambda^2}, \nu_{\mathrm{t}} = \frac{\ln(n_{\mathrm{c}}/n_0)}{\tau} \tag{5.5}$$

Here, E_{cr} is also rms field in V/cm, I_1 and I^* are the ionization and excitation energy in eV, ν_i is the ionization frequency, ν_d is the characteristic frequency of the diffusion drift, and ν^* is the frequency of exciting collisions. D and ν_{m} are different from Taylor's model. The function is defined as

$$\Phi\left(\frac{\nu_{\mathrm{i}}}{\nu^*}\right) = \frac{\nu_{\mathrm{E}}}{\nu^*}. \tag{5.6}$$

where ν_E represents the "frequency of energy set," the frequency of electrons in the field acquire energy I_1, which is derived from the elementary theory. The dimensionless function $\Phi(\eta)$ is determined by

$$e^{(a-1)y}\left(\cosh\frac{y}{z} + z\sinh\frac{y}{z}\right) - e^{-(a-1)y}\left(\cosh\frac{y}{z} - z\sinh\frac{y}{z}\right) = 2a(1+\beta)y$$
$$+ 2\left(1 - z^{-2}\right)\{y\cosh[(a-1)y] + \sinh[(a-1)y] - ay\} \tag{5.7}$$

$$y = a^{-1}\sqrt{\frac{6(\nu_{\mathrm{i}} + \nu^*)}{\nu_{\mathrm{E}}}}, z = \sqrt{1 + \frac{\nu^*}{\nu_{\mathrm{i}}}}, a = \sqrt{\frac{I_1}{I^*}}. \tag{5.8}$$

In this model, it is assumed that the electrons having the energy I_1 instantaneously experience an inelastic collision, ionizing an atom with probability β. Based on the excitation and ionization cross-sections of the atoms, they assumed that β could be set to 0.2 for almost all atoms except in the case of rapid ionization due to the photoelectric effect. In this case, the excitation of the upper states leads to fast breeding, and they take $\beta = 1$. The function Φ is deduced from the relationship between y and z in Eq. (5.8), which is easily calculated numerically using, for example, the Newton-Raphson method. Vyskrebentsev and Raizer take $\nu_{\mathrm{m}} = 7 \times 10^9 p$, $\nu^* = 2.6 \times 10^8 p$, $I_1 = 16.8$ eV, and $I_1 = 11.5$ eV for the argon, and $\nu_{\mathrm{m}} = 1.5 \times 10^{10} p$, $\nu^* = 4 \times 10^8 p$, $I_1 = 13.1$ eV, and $I_1 = 8.4$ eV for the xenon; here, p is in Torr.

Figures 5.1 and 5.2 show the calculated ionization thresholds in air and argon at various conditions, respectively. The ionization threshold is similar to the well-known Paschen curve for discharges in an electrostatic field. At the minimum point of the ionization threshold, $\nu_{\mathrm{m}} \approx \omega$.

The dependency of Λ, τ, and ω on E_{cr} is similar for both gases. If the pulse width is insufficient, the overall E_{cr} increases due to the term $\ln(n_c/n_0)/\tau$. A large Λ suppresses the effect of diffusion, and thus, E_{cr} in the low-pressure region decreases. E_{cr} increases with the increase in frequency and the pressure at the minimum E_{cr} moves with the frequency. Note that the minimum focusing diameter determined by the diffraction limit decreases with frequency.

Cook et al. (2010) conducted experiments using a 110 GHz gyrotron and compared the experimental results with the models. Cook et al. confirmed that the resulting ionization threshold agreed well with the model when $\Lambda \approx w/\pi$ [cm] is

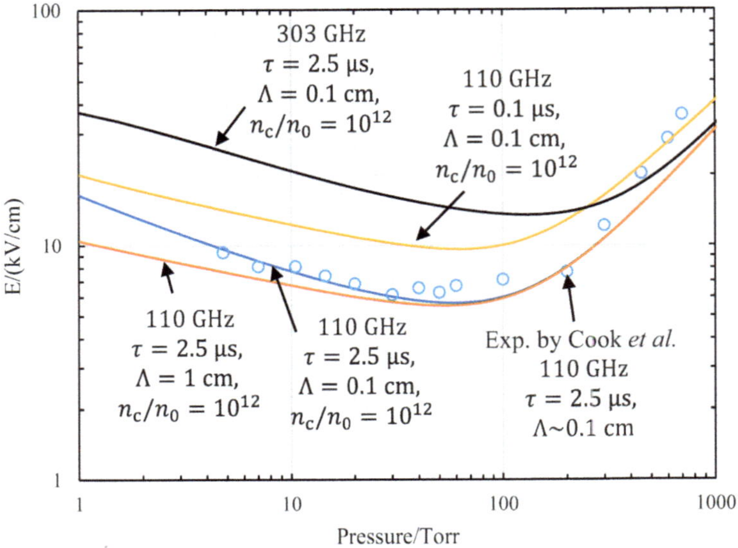

Fig. 5.1 Ionization threshold electric field (rms) of air estimated by the Taylor model (Taylor et al. 1971). The circle plots the experimental results by Cook et al (2010)

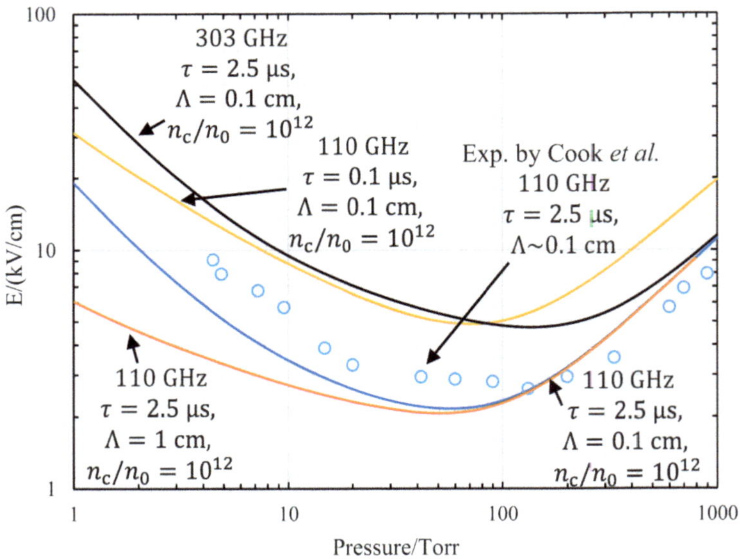

Fig. 5.2 Ionization threshold electric field(rms) of argon estimated by the Vyskrebentsev and Raizer model (1973). The circle plots the experimental results by Cook et al. (2010)

used, where w is the incident beam radius. $n_c/n_0 \approx 10^8 \sim 10^{12}$ was assumed. In the experiments, the output power of the gyrotrons ranged from 10 kW to 1.3 MW, and the pulse width ranged from 2 to 3 μs. The pressure range is from 5 Torr to 700 Torr. Cook et al. also compared the experimental data of MacDonald et al. (1963) at 24.1 GHz and 9.4 GHz with Taylor's model. The formation time of the electron avalanche ionization in 110 GHz air discharge was investigated as a function of pressure (Cook et al. 2011a). The ionization threshold of argon, krypton, and xenon at 670 GHz, and argon and krypton at 263 GHz and 250 GHz were summarized in a review paper (Sidorov 2022).

There is no such simple model for E_{cr} of other general molecular gases and light monatomic gases. In these cases, a Boltzmann equation solver, BOLSIG+ developed by the LAPLACE laboratory in France (BOLSIG+ n.d.), could be utilized to calculate E_{cr}. Some numerical studies of the discharge at over-critical and under-critical conditions have used the BOLSIG+ to determine the local ionization frequency as a function of the local electric field (Boeuf et al. 2010a; Takahashi et al. 2017) (see also Chap. 6).

Next, let us consider the tendency of the ionization threshold qualitatively with the equation of motion (see ref. (Raizer 1977) in Chap. 1). We do not have to consider the elementary effects in microwave bands. The equation of motion of the electron fluid in the electric field is written as

$$m_e n \frac{d\boldsymbol{v}_e}{dt} = -en\boldsymbol{E} - \nabla \boldsymbol{p} - m_e n \nu_m \boldsymbol{v}_e. \tag{5.9}$$

Hereby, $\boldsymbol{E}$ denotes the electric field vector of the incident beam, n represents the number density, and $\boldsymbol{v}_e$ is the velocity vector of the electron. Consider a scalar wave with an angular frequency of ω, $E_0 \exp(-i\omega t)$ and the electron velocity of $v_{eo} \exp(-i\omega t)$, and ignore the pressure gradient $\nabla \boldsymbol{p}$,the equation becomes

$$\frac{d}{dt}[v_{eo} \exp(-i\omega t)] = -\frac{eE_0}{m_e} \exp(-i\omega t) - \nu_m v_{eo} \exp(-i\omega t). \tag{5.10}$$

From the above equation, the amplitude of the electron velocity v_{eo} is deduced as

$$v_{eo} = -\frac{eE_0}{m_e} \frac{\nu_m + i\omega}{\nu_m^2 + \omega^2}. \tag{5.11}$$

The average work gained by an electron from the electric field per unit time is then

$$\left(\frac{d\varepsilon}{dt}\right)_E = \frac{1}{2} \mathrm{Re}(-eE_0 v_{eo}) = \frac{e^2 E_{rms}^2}{m_e \nu_m \left(1 + \omega^2/\nu_m^2\right)}. \tag{5.12}$$

In the case of high-pressure $\omega^2/\nu_m^2 \ll 1$, the electron energy lost through elastic and inelastic collision with neutral particle affect E_{cr}. In the elastic collision, the electron loses a small portion of the energy on the order of the ratio of the electron to the atomic mass. Assuming that electron energy $\varepsilon_{\text{loss}}$ is lost ν_{loss} times per unit time on average, the energy balance required to achieve the breakdown becomes

$$\left(\frac{d\varepsilon}{dt}\right)_E \frac{e^2\tilde{E}_{\text{rms}}^2}{m_e\nu_m} > \nu_{\text{loss}}\varepsilon_{\text{loss}}. \tag{5.13}$$

Consequently, the threshold field is determined by

$$E_{\text{cr}}^2 \frac{m_e\nu_m\tilde{\nu}_{\text{loss}}\varepsilon_{\text{loss}}}{e^2}. \tag{5.14}$$

Since both ν_m and ν_{loss} are proportional to the pressure, we obtain $E_{\text{cr}} \propto p$.

In contrast, the energy loss due to the collision is slightly at low pressure, and the primary loss mechanism is diffusion. In this case, from Eq. (5.5), the ionization frequency is required to be higher than the characteristic frequency of the diffusion, $\nu_i > \nu_d$ assuming that the pulse width of the incident beam is sufficiently long, ν_t~0. For simplicity, we set the ν_i by using the ionization energy I_1 with $\omega^2/\nu_m^2 \gg 1$ as

$$\nu_i \frac{\tilde{1}}{I_1}\left(\frac{d\varepsilon}{dt}\right)_E \frac{e^2\tilde{E}_{\text{rms}}^2\nu_m}{I_1 m_e\omega^2}. \tag{5.15}$$

Using the ν_d in Eq. (5.5), we can derive the threshold field from $\nu_i \sim \nu_d$.

$$E_{\text{cr}}^2 \propto \frac{\omega^2}{\nu_m^2\Lambda^2} \tag{5.16}$$

Therefore in the low-pressure condition, we obtain $E_{\text{cr}} \propto \omega/(p\Lambda)$. These features derived here are consistent with those in Figs. 5.1 and 5.2.

In various papers, the ionization threshold is expressed in terms of the local electric field strength or power density. The power density is determined by

$$S\,[\text{W/m}] = \frac{E_0^2}{2Z} = \frac{E_{\text{rms}}^2}{Z} \tag{5.17}$$

Here, Z [Ω]~120 π is the characteristic impedance of the vacuum. E_0 represents the peak electric field strength. For instance, the ionization threshold of the atmospheric air in the peak field is roughly 40 kV/cm or 4 MV/m, corresponding to 21 GW/m^2 or 2.1 MW/cm^2.

The E_{cr} estimated in this chapter provides only a reference value of the order. Practically, the ionization threshold depends on the laboratory environment, such as aerosols, humidity, and metals near the discharge area. If sharp metal spikes are present in the incident beam path, the ignition of the discharge would be enhanced by the electric field concentration. The parabolic mirror attached to the top of the

Microwave Rocket has a threaded hole in the center so that a sharp thread can be placed at the focal point to ensure the ignition.

Measurements of the incident beam power, radiation profile, and polarization direction with high accuracy are also tricky tasks. The radiation pattern of such a high-power beam is generally measured thermally by observing the temperature rise of a thin plastic plate or paper target using an infrared camera (Tatematsu et al. 2014, 2015; Oda et al. 2013). Temperature-sensitive liquid crystal sheets (e.g., ref. (Edmundoptics n.d.)) can also be used to measure the radiation pattern – the color changes as the temperature rises. The total incident beam power is measured with water dummy loads. A typical water load consists of a metal container and a water tube made from a material transparent to electromagnetic waves, such as Teflon. The water flowing through the tube absorbs the electromagnetic waves incident in the metal container, and the beam power is estimated from the temperature rise of the water. The polarization direction of radiation from the gyrotron is well-defined. However, the direction of polarization would change depending on the optics as well as cross-polarization components would be generated. The oscillation frequency is measured by a heterodyne detector or a Fabry-Perot resonator.

The incident beam power density is estimated from the measured radiation pattern and total beam power. However, when the spot size at the focal is small, considering the spatial resolution of the measurement, or when, as in the case of the parabolic mirror, the focusing and the incident beams overlap, the direct measurements of the radiation profile are difficult. In most cases, the incident beam in the free space has a Gaussian or Gaussian-like beam. Therefore, the Gaussian beam approximation is often used. This approximation makes it easy to handle power density and its variation with beam propagation. In this approximation, the beam power density is characterized by the beam radius (spot size) w (generally, 1/e radius of the electric field intensity) and peak power density S_{pk}. The beam power density distribution is described as

$$S(x, y) = S_{pk} \exp\left[-2\frac{(x - x_0)^2 + (y - y_0)^2}{w^2}\right] \tag{5.18}$$

If the radiation pattern is elliptical, we can approximate it to the following bi-Gaussian beam:

$$S(x, y) = S_{pk} \exp\left[-2\left\{\frac{(x - x_0)^2}{w_x^2} + \frac{(y - y_0)^2}{w_y^2}\right\}\right] \tag{5.19}$$

Hereby, the x and y coordinates are defined on the cross-section of the beam perpendicular to the propagation axis z. The x_0 and y_0 represent the beam center. The w_x and w_y are beam radii in the x- and y-directions, respectively. S_{pk} is determined from the total power P_{tot} as

$$S_{pk} = \begin{cases} \dfrac{2P_{tot}}{\pi w^2} & \text{(axisymmetry)} \\ \dfrac{2P_{tot}}{\pi w_x w_y} & \text{(elliptical)} \end{cases} \tag{5.20}$$

The average power density is just half of S_{pk}.

The first-order and second-order moments of the Gaussian beam provide the incident beam center and radius (Shimozuma et al. 2005). The nth moment in the x-direction is estimated from beam power density $S(x, y)$

$$\langle x^n \rangle = \int x^n S(x,y)dxdy / \int S(x,y)dxdy \tag{5.21}$$

The first-order moment directly represents the beam center. The beam radius in x-direction is deduced as

$$w_x = 2\sqrt{\langle x^2 \rangle - \langle x^1 \rangle^2} \tag{5.22}$$

The temperature rise measured using the infrared camera $\Delta T(x, y)$ is proportional to $S(x, y)$. Therefore, the beam center and radius are deduced from $\Delta T(x, y)$. However, simple Gaussian fitting would give more accurate results if the infrared camera image does not contain the entire beam, has sidelobes, or has biased background noise. Variation of the beam radius is described as

$$w = w_0\sqrt{1 + [\frac{\lambda(z - z_0)}{\pi w_0^2}]} \tag{5.23}$$

where w_0 represents minimum beam radius (beam waist), z_0 denotes the beam waist position, and λ is the wavelength. The w_0 and z_0 can be estimated from a fitting between measured beam radii and Eq. (5.23).

The actual ionization threshold depends on these values. The experimental setup should be designed with a sufficient margin over the predicted values.

Microwave Discharge in the Over-Critical Condition

In the over-critical condition, the local electric field E_{lo} generated by the incident beam is higher than E_{cr}, and a self-sustaining plasma is generated through the breakdown. This condition is satisfied at around the focal point of the lenses or mirrors. The ionization front jumps and propagates back to the incident beam. The spatial structure of the discharge plasm is different in the ***E*-*k*** plane, which is parallel to the wavenumber vector ***k*** and electric field vector ***E*** of the incident millimeter-

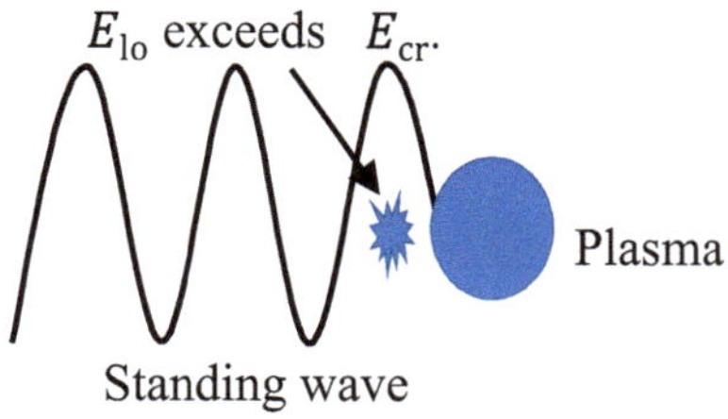

Fig. 5.3 Schematic image of new plasma ignition in the over-critical condition

wave, and in the ***B***-***k*** plane, which is parallel to ***k*** and magnetic field vector ***B***. Two-dimensional arrays of plasma filaments elongated along ***E*** in the ***E***-***k*** plane have been reported by Vikharev et al. at 37.5 GHz (Litvak 1991; Vikharev et al. 1988, 1992), a group at Massachusetts Institute of Technology (MIT) at 110 GHz (Cook et al. 2010, 2011a,b; Hidaka et al. 2008, 2009; Hummelt et al. 2012; Schaub et al. 2016) and 124.5 GHz (Schaub et al. 2016), and Fukunari et al. (2019) at 303 GHz.

Figure 5.3 shows the schematic image of the standing wave and new plasma ignition. When the number density n of the election in the discharge reaches the cutoff frequency determined by

$$n_{\mathrm{cut}} = \frac{m_{\mathrm{e}}\varepsilon_0}{e^2}\omega^2 \tag{5.24}$$

where e, m_{e}, and ε_0 denote the elementary charge, the electron mass, and permittivity in the vacuum, respectively; the discharge plasma reflects the incident beam like a conductor. Note that here, we employ the MKSA unit. At the cutoff density, the angular frequency ω equals the plasma frequency $\omega_{\mathrm{p}} = e^2n/(m_{\mathrm{e}}\varepsilon_0)$ and the wave number k of the incident beam becomes zero from the dispersion relation $\omega^2 = c^2k^2 + \omega_{\mathrm{p}}^2$ of the collisionless plasma, where c is the speed of light. Subsequently, a standing wave forms in front of the discharge plasma. The electric field distribution has nodes and antinodes (hotspots) where the E_{lo} is higher than E_{cr}. New discharge grows in the hotspot, and the ionization front jumps. The boundary of the plasma becomes a node of the standing wave, and the hot spot is formed at the distance of a quarter wavelength. Although multiple hot spots are generated, new discharges are initiated in hot spots around the plasma where background seed electrons reach a certain number density due to diffusion from the bulk plasma. In the ***E***-***k*** plane, the electric field intensity in the polar regions of the discharge plasma is amplified, and thus, the plasma stretches along with the ***E*** vector, resulting in the plasma filament formation. In the ***B***-***k*** plane, plasmas with a grain-like shape (plasmoid) appear. This structure is called λ/4 structure because filaments are aligned every quarter wavelength, or the fish-bone structure because the plasma filaments look like middle bones protruding from the backbone of a fish.

Figure 5.4 shows the spatial structures of the 303 GHz discharge induced in the atmospheric air observed with a high-speed shutter camera, pco.pixelfly usb (Fukunari 2019). The minimum exposure time of the camera is 1 μs. The elapsed

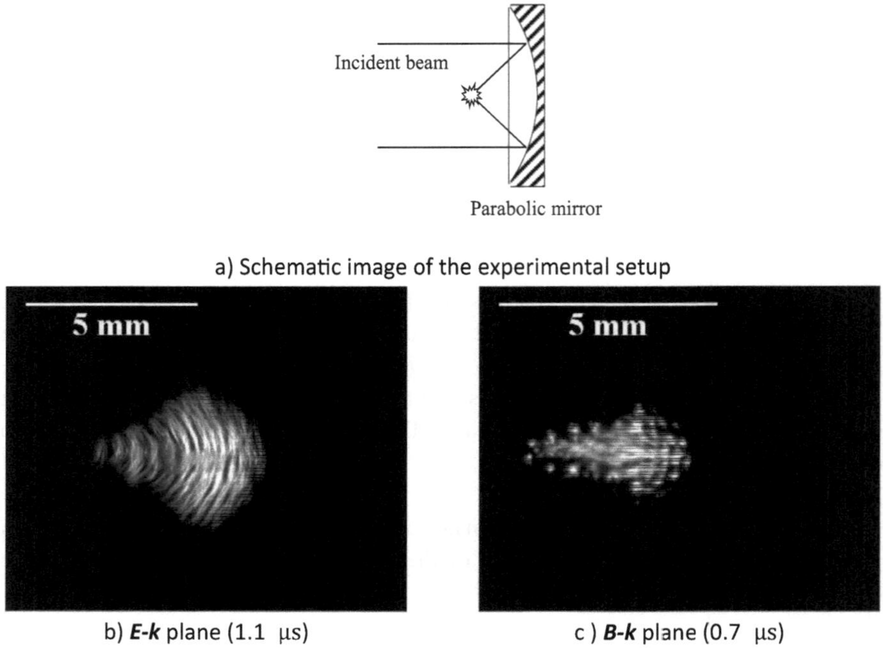

Fig. 5.4 Spatial structure of the 303 GHz microwave discharge at the over-critical condition

time from the ignition was estimated as 1.1 μs for the image in ***E***-***k*** plane and 0.7 μs for ***B***-***k*** plane. The ignition timing was detected from the self-emission of the plasma measured by a photosensor, H10723. A 303 GHz high-power gyrotron developed at the Research Center for Development of Far-Infrared Region, University of Fukui, was used as a beam source (Yamaguchi et al. 2015). The incident beam was collimated by quasi-optical mirrors and had a quasi-axisymmetric Gaussian beam profile (non-focused beam). The total power of the incident beam was measured as 104 kW with a water dummy load. The beam radius was estimated as 14.2 mm by radiation pattern measurements using an infrared camera. The peak power density and the electric field at the peak of the incident beam were estimated to be 33 kW/cm^2 and 3.5 kV/cm in rms, respectively, which is lower than E_{cr}. A parabolic mirror with a diameter and focal length of 50 mm and 18 mm, located on the right-hand side of the figure, focused the incident beam. Assuming that the incident beam radius is ideally focused to the same as the wavelength, the average electric field reaches 36 kV/cm in rms, which exceeds E_{cr} (see Fig. 5.1). Note that in ref. (Fukunari et al. 2019), the peak electric field strength was used. In this experimental setup with the parabolic mirror, the ionization front, the head of the breakdown wave, propagates mainly toward the parabolic mirror. Because the formation process of the λ /4 structure proceeded less than 1 μs, the process cannot be observed using

Table 5.1 Structure transition pressures(Vikharev et al. 1988)

Gas	p_1[Torr]	p_2[Torr]
Nitrogen	3–5	40–50
Helium	10–20	100–150
Argon	–	30–40

the high-speed shutter camera. 303 GHz is (as far as the author knows) the highest frequency at which the λ/4 structure was observed.

The discharge structure depends on the incident beam frequency, gas species, and pressure. Vikharev et al. (Litvak 1991; Vikharev et al. 1988, 1992) reported three types of discharge structures (diffuse, multiplasmoid, and filamentary) that succeeded one another as the pressure increased. In these experiments, a gyrotron at 37.5 GHz (wavelength of 8 mm) was used as a beam source. The incident beam was transferred into a linearly polarized Gaussian beam. The pulse width and the incident beam power were τ~100 μs and P~100 kW, respectively. The beam radius at the focal point was a~1.5 cm. An echo-free chamber was filled with technically pure nitrogen, helium, or argon. The measurements were performed in the focal region of the beam at pressures in the range $p = 0.1 - 600$ Torr in helium and argon and $p = 0.3 - 70$ Torr in nitrogen. The dynamics and structure of the discharge were investigated by observing the plasma emission using a high-speed camera. The system can photograph the discharge during the pulse in nine frames at an interval of 2–32 μs with an exposure time of 0.2 μs. The emission image showed (1) the continuous (diffuse) structure at low pressures, (2) bright bands parallel to the beam axis (multiplasmoid) at intermediate pressures, and (3) bright filaments parallel to $\boldsymbol{E}$ (filamentary) at high pressures. The filamentary structure of (3) is the λ/4 structure described above. Transition pressures of type (1) to type (2), p_1, and type (2) to type (3), p_2, for each gas are summarized in Table 5.1. In argon, type (2) structure was not observed, and a transition from type (1) to type (3) structure was observed at p_2 in the table. The plasma structure in $\boldsymbol{B}$-$\boldsymbol{k}$ plane was reported in Ref. (Vikharev et al. 1992). The estimated propagation velocity of 10^2–10^3 m/s was quite slow for the discharge in the over-critical condition.

The mechanism by which a discharging plasma breaks up and forms a spatial structure can be interpreted in terms of instability (Gurevich et al. 1997; Vikharev et al. 1988; Gil'denburg and Kim 1980, 1978). The primary instabilities are ionization-field instability and ionization-thermal instability. The ionization-field instability results from the enhancement of electric field strength and ionization in regions of high electron number density. In the ionization-thermal instability, the gas temperature increase due to the heating contributes to the increase in the ionization. The growth rate of the instability has been investigated from linear instability analysis using the wave equation of the electric field and the rate equation of the electron density.

According to Vikharev et al., the ionization-field instability predominates in the case of $\nu_m/\omega \ll 1$ until the pressure is higher than p_1. If the ambient pressure is lower than p_1, the instability is dumped by diffusion. In high-pressure ($\nu_m/\omega \gg 1$), the ionization-thermal instability is then developed. The elongation of the filaments parallel to the ***E*** vector is attributed to the enhancement of the electric field at the polar of plasma and the ionization-thermal instability. However, they also mentioned that the linear theory does not fully describe the evolution of the discharge with well-developed perturbations associated with the nonlinear stage.

Since 2007, a group at the Massachusetts Institute of Technology has conducted intensive experiments on microwave discharge at 110 GHz using a 1.5 MW gyrotron with a pulse width of 2–3 μs. The output beam of the gyrotron is a linearly polarized Gaussian beam. Hidaka et al. (2008) investigated volume (free-space) and surface breakdown. A high-density polyethylene lens (HDPE) was used to focus the beam. The focal length and the beam radius at the focal point are 14 cm and roughly 4 mm, respectively. The plasma emission was detected with two axially displaced photodiodes, and the radiation transmitted to the focal spot was measured with an rf diode. The time-integrated plasma image of the volume breakdown shows a typical $\lambda/4$ structure, filamentary plasma array parallel to the ***E*** vector. A standard radius of the plasma filament was estimated as 0.1–0.2 mm. The average distance between the filament was 0.76 mm (~0.28 λ). The waveform of the rf diode showed that the discharge plasma cut the incident radiation. The signal from the photodiode at downstream decayed before the pulse off. This implied that the new filaments ignited at the upstream absorbed or cut off the incident beam completely. For the surface breakdown, a dielectric sample, polycarbonate, HDPE, or Teflon was placed at the beam waist with a tilt angle. The time-integral image showed a highly periodic array of plasma filaments. There were two distinct discharge areas, one near the surface of the sample and the other far from it. Near the sample surface, filaments were arranged parallel to the surface. The filaments were aligned with the beam axis in an area far from the sample surface. There was no clear dependence on materials and the tilt angle.

Hidaka et al. then described the filamentary array formation by the electric field concentration (hot spot formation) due to diffracted waves from the plasma. They simulated the electric field based on the High-Frequency Structure Simulator code (High Frequency Structure Simulator n.d.), assuming the plasma filament as a perfect conductor post, and reproduced the observed structure qualitatively.

Nam and Verboncoeur (2009) explained these experimental results at 110 GHz, using a one-dimensional plasma fluid model coupled with the plane wave equation of the electric field. Subsequently, a group at the University of Toulouse, Laboratoire Plasma et Conversion d'Energie (LAPLACE), developed simulation models coupling Maxwell equations with plasma fluid equations to describe the two-dimensional structure formation (Boeuf et al. 2010b; Chaudhury et al. 2010; Chaudhury and Boeuf 2010; Kourtzanidis et al. 2015). A group at Tohoku University has extended this model by coupling it with a fluid model to reproduce the discharge at the under-critical condition at 170 GHz (see Chap. 6 for the numerical modeling for detail).

Discharge structure in various gases using the 110 GHz gyrotron was also reported (Hidaka et al. 2009). In the experiments, a polymethylpentene lens with a focal length of 9.2 cm and a beam waist of 4–5 mm was used to focus the beam. The maximum incident power density was estimated as 5.6 MV/m from peak electric field strength. The open-shutter images and fast-gated images of plasma emission were measured in ambient air, nitrogen, SF_6, and helium. In air discharge, the movement of the ionization front toward the beam source was clearly observed. The filament at the upstream absorbed and reflected the incident beam, resulting in the downstream plasma being extinguished. The discharge structure in nitrogen and in air was almost identical. The discharge structure in helium is qualitatively different from that in the air and nitrogen. The plasma has a discrete structure at downstream (after the beam spot) while that is diffusive at upstream.

Cook et al. (2010) have investigated the pressure dependence of the discharge structure at 110 GHz. An HDPE lens focuses the linearly polarized Gaussian beam to a focal point inside a chamber filled with air or argon. The beam radius is estimated as w~4 mm. They reported that distinct plasma filamentary arrays appear only at high pressure above the minimum of the ionization threshold ($\omega \sim \nu_m$). The discharge structure is diffusive at low pressures of less than 10 Torr. In the air, the $\lambda/4$ structure first appears at around 40 Torr. Note that assuming $\nu_m = 5.3 \times 10^9 p$,the condition $\nu_m(p_2)/\omega \approx 1$ is satisfied when $p_2 \approx 130$ Torr. Although they did not mention in the paper, the discharge structure in the ***E*-*k*** plane at 125 Torr and 180 Torr seems like the multiplasmoid structure. The plasma filaments are well-defined at a pressure above 400 Torr. At high pressure, a dark center is observed. In argon, $\lambda/4$ structure is observed from around 40 Torr and becomes distinct as the pressure approaches atmospheric levels. Cook et al. (2011b) have also measured the discharge structure induced in atmospheric air by a circularly polarized beam generated by a metal grating polarizer. In this case, the plasma filament was expanded in the cross section of the incident beam, and an azimuthally symmetric disklike structure formed.

The propagation velocity of the ionization front U_i measured at 110 GHz was much faster than the sonic speed of the gas. U_i in air and nitrogen at atmospheric pressure with the incident power density of 3 MW/cm^2 were measured from the high-speed camera images as 14 km/s and 12 km/s, respectively. The propagation velocity of 14 km/s for air discharge decreased to 5 km/s with a decrease of the incident beam power density to 1.5–2 MW/cm^2. U_i also depends on the pressure. U_i increased from 5 km/s at atmospheric pressure to 20 km/s at 420 Torr (Hidaka et al. 2009).

Cook et al. (2011b) measured U_i and the expansion velocity of filaments transverse to the beam axis with a linearly polarized or circularly polarized Gaussian beam. The peak power density of the incident beam was estimated as 5 MW/cm^2. The U_i drastically decreased from >100 km/s to 10 km/s due to the decrease of the local intensity of the incident beam as the ionization front moves. The tendency was similar for both polarizations. The transverse expansion velocity of the individual plasma filament was ~5–10 km/s. Cook et al. showed that the transverse expansion agreed qualitatively with the one-dimensional ionization-diffusion theory (Ebert and Saarloos 2000). In theory, the motion of the ionization front in an external electric field is described. The propagation velocity is deduced as

$$U_{\mathrm{i}} = 2\sqrt{D_{\mathrm{e}}\nu_{\mathrm{i}}} \tag{5.25}$$

where D_{e} is the free-electron diffusion, and the ionization frequency is estimated by

$$\nu_{\mathrm{i}} = \nu_a \left[\left(\frac{E_{\mathrm{lo}}}{E_{\mathrm{cr}}} \right)^{\beta} - 1 \right] \tag{5.26}$$

Here, $\nu_a = 5 \times 10^4 p$ [Hz] is the electron-neutral attachment frequency in the air at p [Torr], which represents an electron loss by the attachment. They employed in the paper, $E_{\mathrm{cr}} = 3200p$ [V/m], which corresponds only to the high-pressure region in Fig. 5.1. $\beta \approx 5.33$ is an empirical constant for air.

The microwave discharge at the over-critical condition is regarded as thermal nonequilibrium. Here, the term "thermal nonequilibrium" means just that the electron temperature T_{e} and gas temperature T_{g} are different and do not mention whether the velocity distribution function is Maxwellian or not. In air discharge, the vibrational temperature T_{v} and rotational temperature T_{r} of the nitrogen molecules are also different. If the pressure is sufficiently high, T_{r} is approximately the same as the translational temperature of a neutral particle, which is T_{g}. The plasma emission of the microwave discharge in the air is primarily the second positive system in neutral molecular nitrogen. Hummelt et al. (2012) estimated the T_{r} and T_{v} in air discharge from emission spectroscopy measurements of the second positive system over a pressure range of 1–100 Torr. The 110 GHz gyrotron was used to ignite the discharge. The emission spectra were analyzed by the SPECAIR code. T_{v} decreased from 6200 K to 4200 K with the increase of pressure from 1 Torr to 100 Torr. T_{r} was in the range of 300–500 K and also decreased with the increase of the pressure. Both T_{r} and T_{v} increased slowly with the increase of the incident beam power density.

Schaub et al. (2016) measured the electron density of the microwave discharge in the over-critical condition at 110 GHz and 124.5 GHz in various pressures by emission spectroscopy. The Stark broadening of the H_{α} line from dissociated water vapor introduced bubbling in the dry air. In both frequencies, the electron density was close to the critical density of the collisional plasma defined as

$$n_{\mathrm{cr}} = \frac{m_{\mathrm{e}}\varepsilon_0}{e^2}\left(\omega^2 + \nu_{\mathrm{m}}^2\right). \tag{5.27}$$

The following equation expresses the relative permittivity ε_p of the collisional plasma:

$$\varepsilon_p = 1 - \frac{\omega_p^2}{\omega^2 + \nu_{\mathrm{m}}^2} - i\frac{\nu_{\mathrm{m}}\omega_p^2}{\omega(\omega^2 + \nu_{\mathrm{m}}^2)} \tag{5.28}$$

Therefore, the real part of Eq. (5.28) becomes zero at the critical density. As the collision frequency decreases, Eq. (5.28) approaches the cutoff density determined by Eq. (5.24), asymptotically. Here, the MKSA unit is employed. They assumed

$\nu_m = 5.3 \times 10^9\, p$ as a function of the pressure in Torr in the analysis. Note that in the case of 37.5 GHz, the electron density measured from the frequency shift of the open resonator was 20–30% of the critical density.

Kourtzanidis et al. (2015) investigated the effects of gas heating (thermal-ionization instability) in microwave plasma streamers by numerical simulations. Schaub et al. (2016) measured the gas density around the discharge using a Mach-Zehnder interferometer with two probe light wavelengths of 532 nm and 635 nm. The change in gas density appeared in hundreds of nanoseconds after the plasma emission forms. The gas density along a single plasma filament was then compared with the numerical calculations. The gas density was reduced by more than 80% of the initial condition, where the electron density peaked by heating. Around the filament tip, compression waves were generated as predicted by the simulation. In the simulation, the elongation of the microwave streamer was limited by the changes in gas density distribution due to the gas heating. However, the experimentally observed filament length was longer than the calculated value.

Experimental results in discharges under over-critical condition were discussed here. The formation process of $\lambda/4$ structures, propagation velocities, electron number densities, temperature, and their pressure dependence have been studied at limited frequencies. Numerical simulations have already successfully reproduced the structure formation at the frequency of 110 GHz. The formation mechanism has been qualitatively understood at the over-critical condition. Nevertheless, there are some issues with validating the numerical model at different frequencies and gas species as well as the interaction with the fluid.

Microwave Discharge in the Under-Critical Condition

In order to ignite the discharge, the local electric field induced by the incident beam needs to be higher than E_{cr}. For example, let us consider Eq. (5.26), which is used to calculate the ionization frequency in the numerical model. If the local electric field is lower than E_{cr}, the electron number density decreases due to attachment, recombination, and diffusion. However, in practice, even if the electric field is below E_{cr}, if there is initial plasma with substantial electron density, the discharge can be maintained, and the ionization front can propagate toward the beam source. The microwave discharge at the under-critical condition has been observed over a wide frequency range in 28 GHz in air (Tabata et al. 2020) and in argon (Shimamura et al. 2021), 75 GHz in air (Artem'ev et al. 2018a,b) and in argon (Artem'ev et al. 2020), 85.7 GHz in air (Brodskii et al. 1983, 1984; Bogatov et al. 1986), 170 GHz in air (Oda et al. 2006, 2009, 2020), 303 GHz in air (Fukunari et al. 2019), and 550 GHz and 670 GHz in noble gases (Glyavin et al. 2014; Bratman et al. 2011; Sidorov et al. 2020; Sidorov 2022). Beautiful photo images of the discharge structure in the under-critical condition were obtained in the atmospheric air and noble gases using a high-speed camera. These images show that plasma filaments are also formed even in the under-critical condition. However, the plasma filaments are not oriented in the

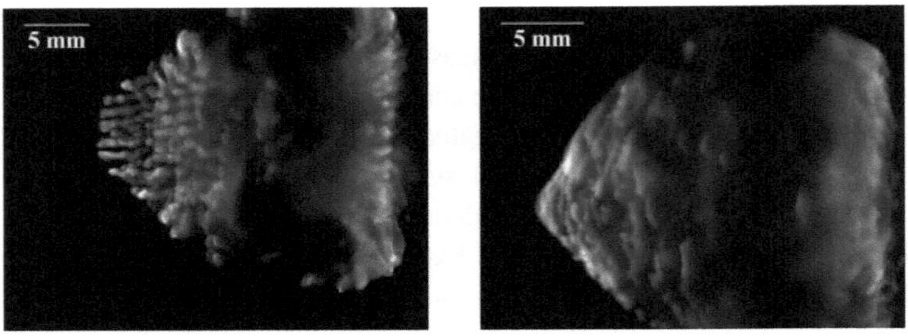

a) ***E-k*** plane (61 μs) b) ***B-k*** plane (60 μs)

Fig. 5.5 Fast-camera images of the 303 GHz microwave discharge in the under-critical condition

direction of ***E*** of the incident beam but are either parallel or slightly oblique to the wavenumber vector ***k*** of the incident beam. The plasma filaments at the ionization front were almost parallel to the ***k***. The λ/4 structure does not appear in this condition.

In order to initiate the discharge in the under-critical condition, the initial plasma needs to be provided. A simple method is to use parabolic mirrors or lenses to focus the incident beam and ignite the discharge in the over-critical condition. In other words, with the typical experimental setup, if the pulse width is sufficiently long, the discharge in the over-critical condition develops into a discharge in the under-critical condition. In addition, screws, metal brushes, and jagged metals can be used to ignite the discharge through the electric field concentration. Even if the electric field intensity at the focal point does not reach E_{cr}, discharge generated by an external high-voltage power supply can provide the initial plasma. When the ionization front propagates away from the ignition point, the difference in the initiator does not affect the discharge, as Batanov et al. (1985) mentions that the discharge "forgets" the initiator.

Microwave discharges in the under-critical condition have been known for longer than those in the over-critical condition because this discharge often occurs in waveguides for high-power microwave transmission lines at a power level lower than the ionization threshold. According to Raizer (1972), the discharge initiated some impurity or foreign metal in a waveguide, and the resulting plasma traveling toward the radiation source was familiar to those working with high-power microwave devices. In addition, the power threshold of the discharge was lower by three orders of magnitude than that of the ordinal breakdown threshold in air.

Figure 5.5 shows fast-camera images of the 303 GHz microwave discharge in the under-critical condition (Fukunari et al. 2019) in the ***E***-***k*** plane and ***B***-***k*** plane. These images are successive from Fig. 5.4. The incident beam propagates from left to right, and the parabolic mirror is located on the right-hand side in this figure. The exposure time of the camera was set at 1 μs. The elapsed time from the ignition was estimated at 61 μs for the image in ***E***-***k*** plane and 60 μs for ***B***-***k*** plane. Plasma filaments parallel to the incident beam(***k*** vector) grew. The pitch size (the distance between the filaments) was estimated as about that of a wavelength. The formation of the comb-shaped filamentary plasma array was predicted by a numerical study (Nakamura et al. 2018) at 170 GHz with a nonphysical tuning parameter in the ionization model to simulate the discharge in the under-critical condition (see Chap. 6). The maximum radius of the plasma in the direction perpendicular to the incident beam was estimated as about 11 mm. Consequently, about 30% of the incident beam power was not irradiated to the discharge plasma and was focused by the parabolic mirror. This fraction of the incident beam probably supported the plasm on the parabolic mirror side. The direction of plasma filaments formed on the parabolic mirror side in the ***E-k*** plane reflects the ***k*** vector of the focused beam from the parabolic mirror. Because the plasma cut off the incident beam and the reflected beam from the parabolic mirror, there is a dark spot at the plasma center. Therefore, the formation of the filamentary structure on the left-hand side was determined by the incident beam independently of the focused beam from the parabolic mirror. Note that the ionization front on the parabolic mirror side was far from the focal point in this figure and penetrated the region in the under-critical condition.

In contrast, in the ***B-k*** plane, the plasma formed a diffuse structure, and the head became pointed. The difference in spatial structure between the ***E-k*** and ***B-k*** planes in the under-critical condition was reported first in ref. (Fukunari et al. 2019). Oda et al. have reported photographs of the atmospheric air discharge at 170 GHz. In the top view of the 170 GHz discharge, there seems to be no such difference between the ***E-k*** and ***B-k*** planes (Oda et al. 2020). The spatial structure in the 670 GHz discharges in argon and krypton also shows filament formation in the ***B-k*** plane (Sidorov et al. 2020). The mechanism of this structural difference in 303 GHz discharges is still unknown. The discharge structures in the ***E-k*** and ***B-k*** planes suggest that the plasma filaments observed in the ***E-k*** plane are layer structures.

Figure 5.6 plots the ionization front position propagating backward to the incident beam in the 303 GHz air discharge. The propagation velocity decreased from 280 m/s to 170 m/s abruptly at 20 μs when the plasma filaments parallel to the ***k*** vector began to grow. A review (Sidorov 2022) mentions that the abrupt decrease in the propagation velocity may be associated with the transition of the discharge to the region of pre-breakdown fields (over-critical condition to under-critical). However, the incident beam was an almost parallel (Gaussian) beam, and the reflected beam from the parabolic mirror was cut off by the plasma. Therefore, the electric field intensity at the ionization front should be almost constant at 3.5 kV/cm in rms, which is one order of magnitude lower than E_{cr}. In addition to this, the propagation velocity of the ionization front in the over-critical condition is much faster than 10^2 m/s. The ionization front in the over-critical condition in the experimental setup propagated toward the parabolic mirror. The plasma filaments formed at 20 μs might change the electric field distribution by the diffraction. The variation of the electric field distribution may affect the propagation velocity.

In the under-critical condition, gas heating due to the Joule heating from plasma plays an important role in maintaining the discharge. Raizer first developed an analytical model of the discharge by analogy with combustion. The model is based on the one-dimensional heat conduction equation. In this model, the heat transfer

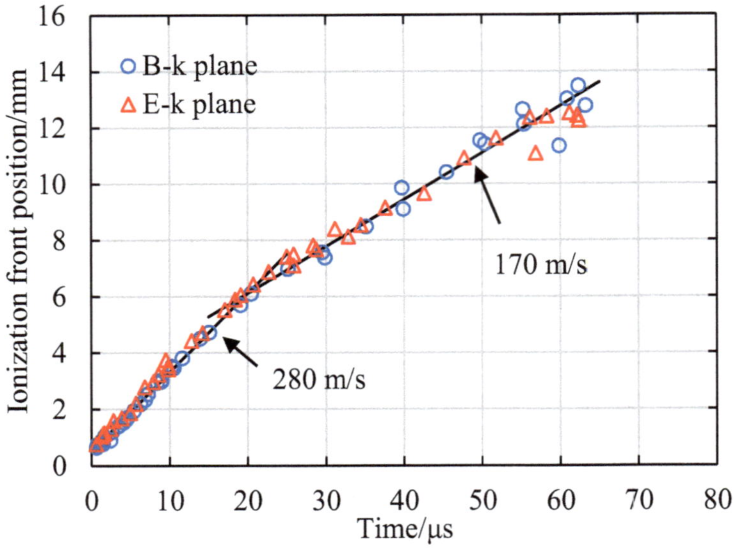

Fig. 5.6 Ionization front position of 303 GHz atmospheric air discharge in the under-critical condition

from plasma to cold gas is the main mechanism of ionization front propagation. The plasma absorbs the incident radiation energy. The cold gas at the further layer is then heated by heat transfer and ionized.

Brodskii et al. (1983) introduced a new mechanism including ultraviolet radiation (UVR) from the bulk plasma ionizes the cold gas by photoionization. The pre-plasma is heated, absorbing the microwave beam energy to the thermal-ionization temperature, resulting in the ionization front propagation. Brodskii et al. experimentally investigated the discharge dynamics in an oversized waveguide system: a Gaussian beam with a beam radius of 2.5 cm and a divergence angle of 4° incident on a waveguide with a radius of 4 cm. They measured the electron density distribution with two double probes and the electron density averaged over the discharge diameter. The beam source was a gyrotron at a frequency of 85.7 GHz and a pulse duration from 0.1 ms to 20 ms. The maximum output power was 250 kW. The incident beam causes an initial corona-like discharge at sharp points or small inhomogeneities of the edge in the waveguide. When the incident beam power is high enough, the ionization front detached and propagated. The velocity increased from 10 m/s to 100 m/s with the increase of the incident power from 10 kW to 250 kW. In this experiment, the main growth of the electron density in the region with a thickness of less than 1 mm was observed. The electron number density increased from 10^{13} cm^{-3} to more than 10^{14} cm^{-3}. The most important finding is a pre-plasma region (so-called plasma halo or precursor) about 1 cm thick in front of the ionization front. This plasma halo is evidence of the photoionization by UVR from the plasma. The electron density in this region decreased slowly with distance from the ionization front. They estimated that the plasma halo absorbed up to 20% of the incident beam power, and the gas was effectively heated.

Bogatov et al. (1986) and Brodskii et al. (1984) further extended the work of Brodskii et al. in ref. (Brodskii et al. 1983) to the nonequilibrium microwave discharge induced by high-power irradiation. As incident microwave power increases, the ionization mechanism changes from thermal equilibrium (thermal ionization dominant) to nonequilibrium in the under-critical condition. The gas temperature decreases in the nonequilibrium state, and the difference from the electron temperature increases (see also Fig. 5 in ref. (Litvak 1991)). They reported experimental results of microwave discharge in air induced by the 85.7 GHz gyrotron. The microwave discharge is initiated using a "whiskbroom," fine wires inserted in the beam pass.

They concluded that the experimental results support the photoionization mechanism for the ionization front propagation rather than the heat conduction model. In the one-dimensional steady-state heat conduction model, U_i is determined by a balance of the increase of the specific enthalpy q of the gas absorbing the incident beam and the power density incident on the plasma. The absorption length and the absorption rate depend on the heating mechanism, whether heat conduction or photoionization. Assuming isobaric heating, $q=\int_0^{T_\mathrm{g}} c_p dT$ with specific heat at constant pressure c_p when the gas is heated to a temperature T_g. Consequently,

$$U_\mathrm{i}=\frac{\eta S}{\rho q}. \tag{5.29}$$

where S is the incident power density, η is the absorption rate, and ρ is the gas density. According to Bogatov et al., if the thermal conductivity transports the energy absorbed by the plasma,

$$\eta S \tilde{k}_\mathrm{g}\frac{T_\mathrm{g}}{l_\mathrm{g}} \tag{5.30}$$

is satisfied. Here, k_g and l_g denote the thermal conductivity of the gas and the characteristic length of the gas temperature change, respectively. Then, η in the dense plasma is proportional to the characteristic length (skin depth) $\eta \propto l_\mathrm{g}$. Eliminating l_g from the Eq. (5.30) yields

$$\eta \propto \left(k_\mathrm{g}\frac{T_\mathrm{g}}{S}\right)^{\frac{1}{2}} \tag{5.31}$$

Thus, η should decrease with $S^{-1/2}$ in the heat conduction model. If the incident beam power is sufficiently high and the discharge becomes the nonequilibrium state, η decreases more rapidly because T_g decreases with S. Subsequently, the propagation velocity decreases with S since a function of U_i~ηS. In the experiments, the transition threshold from the equilibrium to the nonequilibrium state was estimated as S^*~

3 kW/cm^2. The tendency of U_i was measured as U_i[m/s] = 3.5 × S[kW/cm^2] in the equilibrium state 0.6 kW/cm^2 $< S < S^*$. In the nonequilibrium state, U_i is proportional to the square of the beam power density U_i~S^2. These results contradict the behavior predicted by the discussion above. However, the discussion above raises some questions because q also depends on T_g. For instance, assuming q~c_pT_g, if T_g~S^1, η and U_i become constant, and if T_g~S^{-1}, U_i increases linearly with S.

Anyway, Bogatov et al., directly measured the electron density distribution along the incident beam direction with double electrostatic probes and confirmed the plasma halo. The electron density in the plasma halo was estimated as 10^{13} cm^{-3} and increased with the increase of incident beam power. They concluded that the ohmic heating of the gas in the plasma halo has an important role in the maintenance of the nonequilibrium microwave discharge.

Recently, numerical modeling of the microwave discharge in the under-critical condition with the precursor generated by the UV radiation from the bulk plasma, including excitation of the molecular, has proceeded (see Chap. 6).

Why does gas heating contribute to ionization? The average energy gained by an electron between collisions is deduced from Eq. (5.12) as

$$(\varepsilon_{\mathrm{av}})_E = \frac{1}{\nu_\mathrm{m}}\left(\frac{d\varepsilon}{dt}\right)_E = \frac{e^2E_{\mathrm{rms}}^2}{m_\mathrm{e}\nu_\mathrm{m}^2(1+\omega^2/\nu_\mathrm{m}^2)} \propto \left(\frac{E_{\mathrm{eff}}}{N_\mathrm{m}}\right)^2. \tag{5.32}$$

assuming that $\nu_\mathrm{m} \propto p \propto N_\mathrm{m}$, where N_m is the density of the neutral particle (gas). From the effective reduced electric field $E_\mathrm{eff}/N_\mathrm{m}$, one can see that the energy gained by the electrons in each collision cycle increases with the decrease in gas density due to the gas heating in the region of $\omega^2/\nu_\mathrm{m}^2 \ll 1$. When N_m becomes extremely small and/or ω is sufficiently high, $\omega^2/\nu_m^2 \gg 1$; eventually E_eff/N_m asymptotically approaches E_rms/ω. Thus, in this case, the gas heating effects would be limited. In addition to this, at the under-critical condition, the energy loss due to the diffusion would not be dominant because the diffusion becomes ambipolar and the characteristic length is sufficiently large.

In numerical studies (Takahashi 2017), the Maxwell equations, the diffusion equation for electrons, and the Euler equations for fluids were coupled to capture the effects of the gas heating. The ionization frequency was calculated as a function of E_eff/N_m using the Boltzmann solver BOLSIG+. As a result, the ionization front propagated following an expansion wave generated by the heating in the under-critical condition. However, U_i obtained in the calculations does not agree with the experiments. Note that this model did not include photoionization. It also should be noted that in order to calculate the ionization frequency from E_eff/N_m, the variation of the electromagnetic field, which depends on the electron density gradient, should be smaller than the electron mean free path. In addition to this, ν_m should be high enough so that the relaxation time of the electron energy becomes much shorter than the period of the incident microwave. It is unclear whether these assumptions are valid for the ionization front of plasma filaments, where the electron and gas density

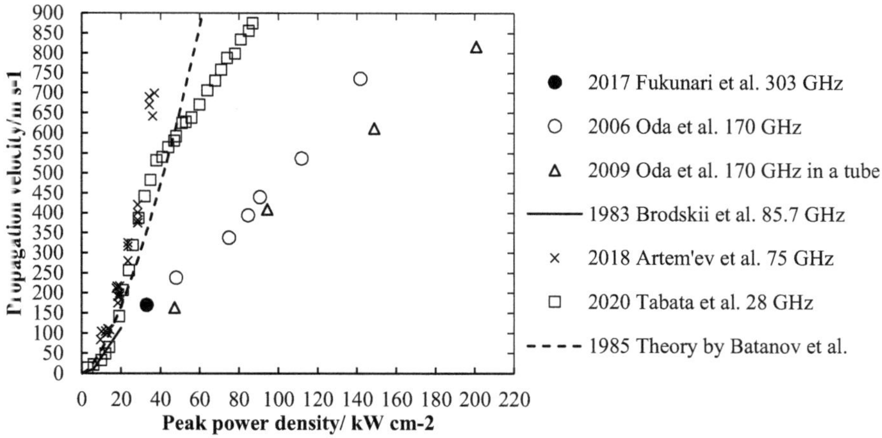

Fig. 5.7 The propagation velocity of the ionization front in the air discharge: 303 GHz (Fukunari et al. 2019), 170 GHz (Oda et al. 2009), and 170 GHz in a tube (Oda et al. 2020) and 85.7 GHz (Brodskii et al. 1984; Bogatov et al. 1986), 75 GHz (Artem'ev et al. 2018b), and 28 GHz (Tabata et al. 2020) in theory (Batanov et al. 1985)

gradients should be large (Takahashi et al. 2017; Kourtzanidis et al. 2015) (see Chap. 6).

Next, let us review the experimental results on the propagation velocity U_i of the ionization front in air discharge. Figure 5.7 shows the propagation velocity in air discharge at various frequencies in atmospheric air. Bogatov et al. (1986) and Brodskii et al. (1984) reported that in atmospheric discharge at 85.7 Hz, the propagation velocity increased linearly with the incident power density in the region of $S < S^*$, and the propagation velocity increased proportionally to the square of S n the region of $S > S^*$. They showed the measured energy density distribution of the incident beam. The beam radius is roughly estimated to be ~1.5 cm from Fig. 1 in ref. (Bogatov et al. 1986). In Brodskii's paper, which contains the same data of U_i versus power density, they mentioned that the gyrotron power reaches 200 kW and the power density up to 20 kW/cm^2. Thus, the power density shown in the series of papers is probably the average power density. The power densities shown in Brodskii et al.'s paper are doubled in the figure for comparison in the peak power density.

Artem'ev et al. (2018b) measured U_i using a 75 GHz gyrotron with a pulse duration of 4 ms or 8 ms. A quasi-optical coupler was used to measure the radiation reflected from the plasma. U_i was estimated by the phase change of the reflected signal. The experimental results show good agreement with a theoretical model proposed by Batanov et al. (1985). They proposed the model of U_i taking into account the photoionization.

$$U_{\mathrm{i}}\frac{\tilde{l_{\mathrm{f}}}}{\tau^*}, \tau^*\frac{c_p\tilde{\rho}\Delta T}{\sigma_0\left|E_{\mathrm{rms}}\right|^2}$$

$$\sigma_0 = \frac{e^2 n\nu_m}{m_e\left(\omega^2+\nu_m^2\right)} \tag{5.33}$$

Hereby, n is the electron density, l_{f} is the mean free path of the ionization radiation from the plasma, ΔT is the temperature increase in the absorption layer, and σ_0 is the electric conductivity. Assuming a high-pressure region $\nu_{\mathrm{m}} \gg \omega$, l_{f}~0.5 cm,and critical reduced electric field $\gamma^* = E_0/N$~10^{-15} Vcm^2. They proposed that as

$$U_{\mathrm{i}}\tilde{1}0\frac{nE_0^3}{p^4} \propto S^{\frac{3}{2}} \tag{5.34}$$

E_{rms} is in kV/cm, p is in Torr, and n is in cm^{-3}.

Since the experimental results of Artem'ev et al. (2018a) are presented for the incident power P_{tot}, we estimated the peak power density from the beam profile, an ellipse with semiaxes of w_{a}= 1.9 cm and w_{b}= 4 cm long by $S_{\mathrm{pk}} = 2P_{\mathrm{tot}}/(\pi w_{\mathrm{a}} w_{\mathrm{b}})$. Eq. (5.34) is also shown in this figure. According to ref. (Batanov et al. 1985), the electron density is set at n~10^{13} cm^{-3}. Note that in 28 GHz case (Tabata et al. 2020), the electron density was measured at about n~10^{14} cm^{-3}. n~10^{13} cm^{-3} corresponds to the plasma halo measured in the case of 85.7 GHz. Nevertheless, here, n and p are just tuning parameters, which are not very important.

Oda et al. (2006, 2009) observed the ionization front propagation of the 170 GHz discharge induced in the atmospheric air. U_{i} in the 170 GHz discharge was linear with S_{pk}, which is clearly different from the experiment of Artem'ev et al. In the 303 GHz discharge experiment, the tendency of U_{i} on S_{pk} was not measured (Fukunari et al. 2019). However, the measured velocity was close to the values obtained in the experiments of Oda et al. At 28 GHz discharge, streamer-like plasmas elongated from the filament. These structural differences at low frequencies may affect the tendency of the propagation velocity. Nevertheless, the measured velocity at low power density is close to Artem'ev et al.'s results.

Apparently, U_{i} is divided into two regions, fast and slow, depending on the incident beam frequency. At high frequencies, the tendency of U_{i} is roughly linear with respect to S_{pk} while U_{i} seems to be $U_{\mathrm{i}} \propto S_{\mathrm{pk}}^{3/2}$ at low frequencies. However, the experimental data is still insufficient to conclude these tendencies. In addition to the Fig. 5.7, Oda et al. (2018) reported that the tendency of U_{i} at 137 GHz is almost the same as that at 170 GHz (slow) in a conference. According to ref. (Batanov et al. 1985), Brodskii et al. first reported that U_{i} exceeded the sonic velocity. The 85.7 GHz discharge would correspond to the fast region considering the incident beam power. Therefore, the transition frequency would be between 85.7 GHz and 137 GHz in the case of atmospheric air discharge. It should be noted that in the case of air discharge induced by a CO_2 laser, the ratio of U_{i} on S_{pk} is much smaller than

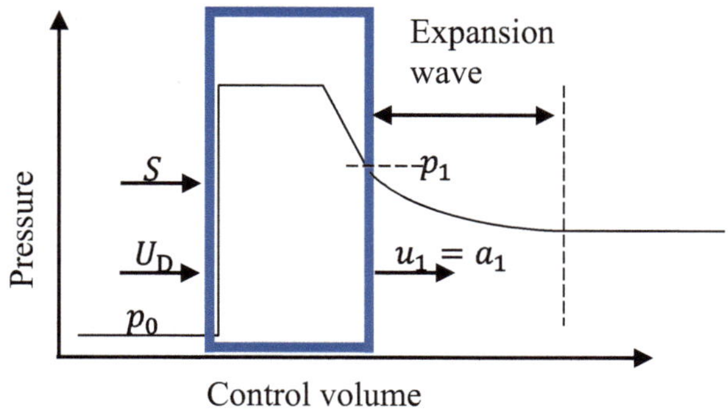

Fig. 5.8 Schematic image of the pressure distribution in an optical detonation in C-J state

the microwave discharge (Bufetov et al. 1984; Matsui et al. 2016). There might be another transition frequency in terahertz bands.

The hydrodynamic mechanism has to be considered for the discharge in the under-critical condition because U_i is the same order as the sonic velocity, and a shock wave driven by the discharge may affect the discharge through the acceleration and compression of the ambient gas. Oda et al. 2009 reported U_i and U_s as a function of S_{pk} measured in 170 GHz discharge induced in atmospheric air. U_i and U_s increases with the incident beam power density.

In particular, in microwave discharge induced in a tube or induced by a parabolic mirror, the reflected shock wave from the tube wall or mirror surface catches up with the ionization front because the flow behind the ionization front is accelerated. In addition, when the shock wave is exhausted from an open end of the tube, an expansion wave begins to propagate in the opposite direction. If the discharge is ignited in the tube with a closed end as in the Microwave Rocket, the flow at the closed end is at rest, and thus, an expansion wave behind the ionization front exists.

If the ionization front and the shock wave propagate with the same velocity backward to the incident beam, we can consider the same system of fluid equations for the MSD. Raizer (1977, 1965) proposed the optical detonation model with an analogy of chemical detonation. In the case of laser discharge, the phenomenon is called laser-supported detonation (LSD).

A detonation wave consisting of a shock wave and an ionization front propagates with a velocity U_D. The ionization front has a certain width and works as an absorbing medium for the incident beam. We assume that the incident beam is an ideal plane wave and the absorption length of the plasma is much shorter than its radius so that MSD becomes a one-dimensional phenomenon. Then, we define a control volume fixed with the detonation wave. Figure 5.8 shows the schematic image of the pressure distribution in the optical detonation. The area in the blue square is the control volume. Note that the rear surfaces of the control volume coincide with that of the detonation wave because the flow becomes unsteady in some cases behind the detonation wave.

In the detonation wave fixed coordinate, the fluid flows into the control volume at the velocity of U_D. Assume that the conservation laws of mass, momentum, and energy hold at the exit and entrance of the control volume.

$$\rho_1 u_1 = \rho_0 U_D > 0 \tag{5.35}$$

$$p_1 + \rho_1 u_1^2 = p_0 + \rho_0 U_D^2 \tag{5.36}$$

$$\frac{1}{2}u_1^2 + C_p T_1 = \frac{1}{2}U_D^2 + C_p T_0 + q \tag{5.37}$$

where C_p p, T, u, and ρ represent the specific heat at constant pressure, pressure, temperature, flow velocity, and density, respectively. The subscripts 0 and 1 indicate the quantities at the front and rear of the control volume. The contributions of viscosity and heat conduction are assumed to be negligible. The chemical composition is assumed to be fixed. C_p and the specific heat ratio γ are assumed to be constant. q represents the amount of heat provided per unit mass from the incident wave.

$$q = \frac{\eta S}{\rho U_D}. \tag{5.38}$$

The gas is assumed to be an ideal gas and satisfies

$$p = \rho R T. \tag{5.39}$$

Here, R is the gas constant. The parameters, T_0, p_0, ρ_0, S, and η are given. The quantities of the rear surface of the detonation p_1, T_1, u_1, ρ_1, and U_D are set to be unknown. Therefore, we must have one more equation to determine the five unknown quantities. Therefore, we add the assumption of the Chapman-Jouguet (C-J) state. The flow velocity behind the shock wave is subsonic, and the flow velocity increases due to heating. In the C-J state, the flow velocity behind the detonation is equal to the sonic velocity $u_1 = a_1 = \sqrt{\gamma R T_1}$, and the flow is thermally choked due to heating. The C-J state is a reasonable assumption because if $u_1 < a_1$, the expansion wave penetrates into the detonation wave, and the system becomes nonstationary. Subsequently, the detonation wave is weakened by the expansion wave until the C-J state is satisfied.

Let us derive the relationship between U_D and S under the C-J state. The Mach number of detonation is decided as $M_D = U_D/a_0$. From Eqs. (5.35) and (5.36), we obtain

$$\frac{\rho_0}{\rho_1} = \frac{\left(1 + \gamma M_D^2\right)}{\gamma M_D^2} - \frac{1}{\gamma M_D^2}\left(\frac{p_1}{p_0}\right). \tag{5.40}$$

Dividing Eq. (5.37) by $C_p T_0$ and rearranging it yields

$$\frac{T_1}{T_0} = 1 + \frac{\gamma - 1}{2} M_D^2 [1 - (\frac{\rho_0}{\rho_1})^2] + \frac{q}{C_p T_0}. \quad (5.41)$$

Note that $C_p = \gamma R/(\gamma - 1)$, and thus, $U_D^2/\left(2C_pT_0\right) = (\gamma - 1)M_D^2/2$. Substituting Eq. (5.40) into Eq. (5.41) yields

$$\frac{T_1}{T_0} = 1 + \frac{\gamma - 1}{2} M_D^2 - \frac{(\gamma - 1)}{2\gamma^2 M_D^2} \left[1 + \gamma M_D^2 - \left(\frac{p_1}{p_0}\right)\right]^2 + \frac{q}{C_p T_0}. \quad (5.42)$$

From Eq. (5.39),

$$\frac{p_1}{p_0} = \left(\frac{\rho_1}{\rho_0}\right)\left(\frac{T_1}{T_0}\right). \quad (5.43)$$

Substituting Eqs. (5.40) and (5.41), we obtain a quadratic equation for p_1/p_0

$$\left(\frac{p_1}{p_0}\right)^2 - \frac{2\left(1 + \gamma M_D^2\right)}{\gamma + 1}\left(\frac{p_1}{p_0}\right) + \frac{2\gamma M_D^2 - (\gamma - 1)}{\gamma + 1} + \frac{2\gamma^2 M_D^2}{\gamma + 1}\frac{q}{C_p T_0} = 0. \quad (5.44)$$

Solving Eq. (5.44) for p_1/p_0 yields

$$\frac{p_1}{p_0} = \frac{1 + \gamma M_D^2}{\gamma + 1} \pm \left[\frac{\gamma\left(M_D^2 - 1\right)}{\gamma + 1}\right]\sqrt{1 - \frac{2(\gamma + 1)M_D^2}{\left(M_D^2 - 1\right)^2}\frac{q}{C_p T_0}} \quad (5.45)$$

This is a pressure ratio that is valid not only for the C-J state but also for normal detonation and deflagration. Deflagration is a combustion wave whose wavefront propagates slower than the sonic velocity. If we take the + sign and set $q = 0$, Eq. (5.45) is consistent with the ratio of the pressure before and after the shock wave. On the other hand, if we take the − sign and set $q = 0$, p_1/p_0 becomes 1. Consequently, the difference in sign indicates the presence or absence of the shock wave in the control volume. The solution with a positive sign is called strong detonation, where a shock wave is present. The solution with a negative sign is called weak detonation.

In the C-J state, q becomes its maximum value, and the inside of the root of Eq. (5.45) is zero. Therefore, we obtain

$$M_{CJ}^4 - 2\left(1 + \frac{\gamma^2 - 1}{a_0^2} q\right) M_{CJ}^2 + 1 = 0 \quad (5.46)$$

Solving Eq. (5.46) for M_{CJ}, and substituting Eq. (5.38), we will have

$$U_{\mathrm{CJ}} = a_0 M_{\mathrm{CJ}} = \sqrt{\frac{1}{2}(\gamma^2 - 1)\frac{\eta S}{\rho_0 U_{\mathrm{CJ}}} + a_0^2} \pm \sqrt{\frac{1}{2}(\gamma^2 - 1)\frac{\eta S}{\rho_0 U_{\mathrm{CJ}}}} \quad (5.47)$$

The ± of the compound sign indicates deflagration with the − sign and detonation with the + sign. Assuming that $\frac{1}{2}(\gamma^2 - 1)q \gg a_0^2$, and taking the + sign,

$$U_{\mathrm{CJ}} = \left[2(\gamma^2 - 1)\frac{\eta S}{\rho_0}\right]^{\frac{1}{3}} \quad (5.48)$$

Raizer deduced this equation from the conservation rows, assuming that the pressure and energy in front of the detonation are small. The ratio of U_i/S of the microwave discharge is higher than that of laser discharge. Thus, the assumption $\frac{1}{2}(\gamma^2 - 1)q \gg a_0^2$ is not valid for microwave discharge. In this case, we have to solve Eq. (5.47) directly with a numerical method such as the Newton-Raphson method.

Raizer's famous book "Laser-Induced Discharge Phenomena" (1977), in Chap. 6 at 20.5, explains that the attainability of the optical detonation in the microwave regime is not simple. First of all, the equations derived from the hydrodynamic theory are independent of the wave frequency. We can utilize the same system of equations for microwave discharge. The problems are the following: The beam power density needed to maintain the detonation is almost independent of the frequency. On the other hand, E_{cr} decreases with the frequency, with approximately ω^2. Thus, the breakdown may occur in microwave bands before a detonation wave is generated in the atmospheric air. Raizer estimated the detonation threshold from the temperature behind the detonation. The assumption that the absorption length in the detonation is sufficiently smaller than the transverse dimensions is necessary to maintain the detonation. Otherwise, the lateral expansion loses a large amount of energy, and the shock wave cannot be maintained. Assuming that the ionization is in the equilibrium state and the threshold temperature of a few thousand Kelvin, Raizer calculated the detonation threshold at the order of 10 MW/cm^2 that is much higher than E_{cr} of microwave discharge in air.

Are these concerns that Raizer describes in the book correct? When this book was published, very little experimental data existed on microwave discharges induced by gyrotrons. The key is the difference of the U_i in the microwave discharge and the laser discharge. Raizer assumed that the detonation threshold is independent of frequency, but in fact, the detonation threshold should change because U_i varies with frequency. For example, substituting $U_{\mathrm{D}} = \alpha S_{\mathrm{pk}}$ to Eq. (5.47), the detonation threshold is deduced as

$$S_{\mathrm{pk,CJ}} = \frac{1}{\sqrt{\alpha^3 \rho_0}}\left[\sqrt{\frac{1}{2}(\gamma^2 - 1)\eta + \alpha\gamma p_0} + \sqrt{\frac{1}{2}(\gamma^2 - 1)\eta}\right] \quad (5.49)$$

As mentioned above, the coefficient α of the microwave discharge is much higher than that of laser discharge. For example, in the laser discharge, the propagation

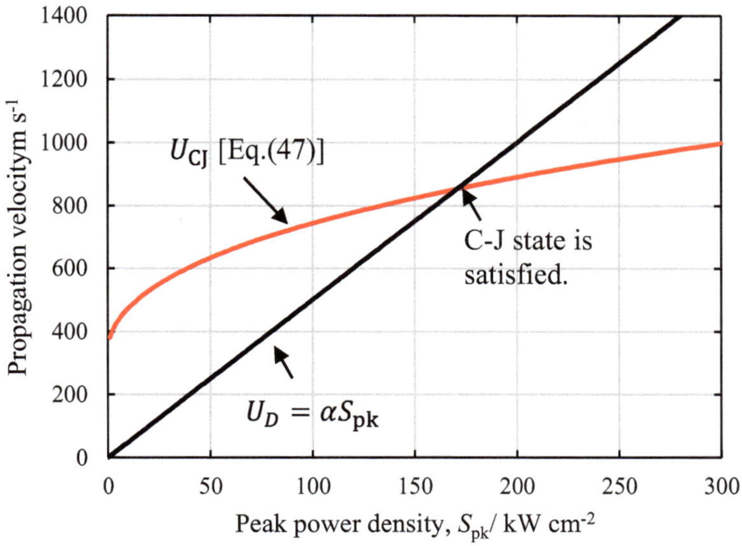

Fig. 5.9 U_{CJ} [Eq. (5.47)] and $U_D = \alpha S_{pk}$ as a function of S_{pk}. $\alpha = 5 \times 10^{-7}$ m^3/J, $\gamma = 1.33$, $\eta = 0.2$ (Brodskii 1983), and $\rho_0 = 1.18$ kg/m^3

velocity is $U_D \sim 1.1 \times 10^5$ m/s at the incident beam power density of 2×10^{15} W/m^2 (Raizer 1965). Note that the propagation velocity is not linear with the beam power density in the laser discharge.

Considering microwave discharge in the atmospheric air and substituting $p_0 = 101325$ P a, $\alpha \sim 5 \times 10^{-7}$ m^3/J, $\gamma = 1.33$ (for air discharge (Raizer 1977)), $\eta = 0.2$ (Brodskii 1983), and $\rho_0 = 1.18$ kg/m^3, the detonation threshold at the C-J state in the microwave discharge is calculated as 171 kW/cm^2. This value is one order of magnitude lower than the ionization threshold of 2 MW/cm^2. Figure 5.9 shows the U_{CJ} calculated by Eq. (5.47) with the Newton-Raphson method and $U_D = \alpha S_{pk}$. The intersection point of the two lines indicates the detonation threshold.

Regarding the temperature limit, because the discharge is in the nonequilibrium state when U_i reaches the sonic velocity, the electron number density would increase to about the critical density, and the absorption length can be sufficiently thin. Further research is needed on this topic. The total power required to achieve a peak power density of 171 kW/cm^2 is ~1.1 MW, assuming a Gaussian beam with a beam radius of 20 mm. This is feasible with the current gyrotrons.

From Eq. (5.49), the detonation threshold decreases with a decrease of η and an increase of α. This can be easily understood by drawing Rayleigh and Hugoniot curves on a $p - v$ diagram, where $v = 1/\rho$. However, M_{CJ} also decrease with the detonation threshold. Because the pressure behind the detonation wave at the C-J state is determined by

$$\frac{p_1}{p_0} = \frac{1 + \gamma M_{CJ}^2}{\gamma + 1}, \tag{5.50}$$

the pressure behind the detonation decreases in this case.

If the beam power density is lower than the detonation threshold, the ionization front and the shock wave propagate separately. This condition is called microwave-supported combustion (MSC) with the analogy of the combustion wave. Nevertheless, MSC is not the combustion process.

Note that Eqs. (5.47 and 5.48) represent a relationship when the C-J state is satisfied. If the ionization front propagation is not a shock wave-induced phenomenon (e.g., seed electrons induced by the shock wave absorb the incident beam), Eq. (5.47) does not necessarily hold even if the incident beam power density exceeds the detonation threshold. From the order with respect to S of Eq. (5.47) and the experimental results, the ionization front could overtake the shock wave as the incident beam power increases. At this time, the system becomes a weak detonation where the shock wave is absent, and the absorption region alone propagates. In this case, the flow behind the detonation is supersonic, and the expansion wave cannot influence the detonation wave. Therefore, the assumption of the C-J state is not guaranteed.

Brodskii et al. (1983, 1984) and Bogatov et al. (1986) found that when $S < S^{*} \sim 3$ kW/cm^2, the 85.7 GHz discharge induced in the atmospheric air in the undercritical condition becomes in the thermal equilibrium where the $T_e \sim T_g$. Bogatov et al. (1986) showed the time-integrated images of the equilibrium and nonequilibrium discharge in free space. The plasma filaments propagating parallel to the incident beam were observed in the nonequilibrium discharge, $S > S^{*}$. On the other hand, the plasma structure was diffusive at the low-power density, $S < S^{*}$. Bogatov et al. explained that this is because T_g reached the thermal ionization temperature in the equilibrium discharge, and the gas temperature distribution determined the spatial structure. Besides, the plasma decay time after the pulse off is significantly different in the equilibrium and nonequilibrium states. In the equilibrium discharge, the decay time is determined by the time needed to cool the gas by heat conduction. In contrast, the decay time is dominated by the inelastic collision between the electron and neutral particle and the recombination process in the nonequilibrium plasma. In this experiment, the decay time of the plasma drastically varied from 800 μs to several tens of microseconds at $S \sim S^{*}$.

This transition of the thermal nonequilibrium to equilibrium discharge was observed in long-distance transmission experiments on the Microwave Rocket. In order to transmit the microwave beam over long distances, the beam diameter should be enlarged to prevent diffraction loss. Yamaguchi et al. (2011) developed a pair of mirror systems: a transmission mirror system and a receiving mirror system. The transmission mirror system expands the beam diameter from 40.8 to 240 mm. The receiving mirror system is mounted on the thruster. The receiving mirror system focuses the beam diameter to 40.8 mm and collimates the beam to guide it into the thruster of the Microwave Rocket.

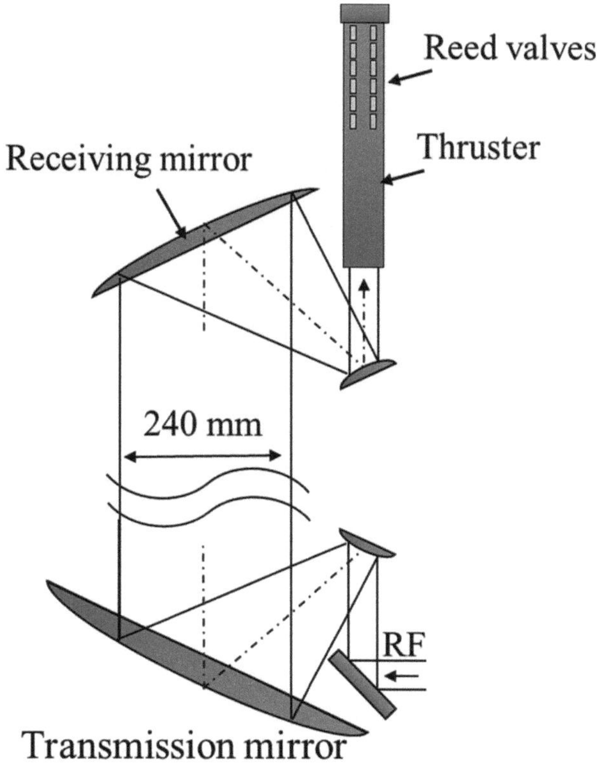

Fig. 5.10 The experimental setup of the long-distance transmission experiments

Figure 5.10 shows the experimental setup of the long-distance transmission experiments. A 1 MW class, 170 GHz gyrotron was used as a beam source (Sakamoto 2007). The thruster was a cylindrical tube with a closed end at the top. The tube's inner diameter and length were 56 and 500 mm, respectively. A parabolic mirror was attached to the closed end. Reed valves were mounted on the thruster for air-breathing during the pulse off. The incident beam power on the transmission mirror system was estimated at 650 kW. The transmission efficiency of the mirror system was not measured in this experiment. The peak power density of the incident beam on the receiving mirror was estimated from the total power of 650 kW and the beam radius of 120 mm to be less than 3 kW/cm^2, neglecting the transmission loss. In the same way, the peak power density in the thruster was deduced from the ideal beam radius of 20.4 mm as 100 kW/cm^2.

Figure 5.11 shows high-speed camera images of the discharge in a multi-pulse experiment at third pulse in 10 pulses. The frame rate was set at 0.1 ms. The pulse width and the repetition frequency were 0.8 ms and 200 Hz, respectively. The propagation velocity U_{i} of the ionization front was increased at the later pulse in this experimental condition, maybe because the air-breathing performance was insufficient and the gas inside the thruster became hot and low density due to heating. Therefore, the ionization front was exhausted from the thruster and approached the beam-focusing part in the receiving mirror system. In this region,

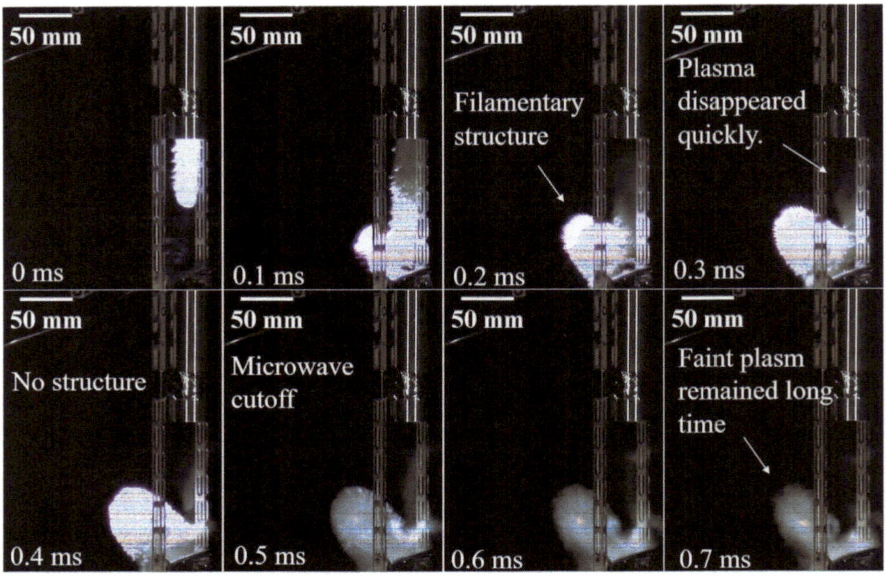

Fig. 5.11 High-speed camera images of the discharge in a multi-pulse experiment at the third pulse in 10 pulses

U_{i} decreased with the propagation as the local electric field decreased. As shown in the figure, the filamentary structure disappeared in this region, and the plasma remained for a long time between the mirror. This feature is consistent with the equilibrium discharge reported by Bogatov et al.

Figure 5.12 shows high-speed camera images from the fourth pulse to the sixth pulse. After the fourth pulse, ignition occurred in the cloudlike drifting plasma. The abnormal ignition disturbed the energy supply to the thruster completely. The plasma in the receiving mirror system grew by absorbing the beam energy with each pulse.

The abnormal ignition strongly limits the duty cycle, a product of pulse repetition frequency and width of the incident beam. Once the abnormal ignition occurs and the discharge turns into the equilibrium discharge, the plasma remains several tens of milliseconds after the pulse. In multi-pulse operation, U_{i} would vary depending on the air-breathing performance and duty cycle. In addition, if ignition occurs at an intermediate position of the thruster due to inhomogeneities of metal or its heating, the ionization front would propagate outside of the thruster. The abnormal ignition due to the equilibrium discharge was also observed in the long-distance transmission experiments with a tapered-tube receiver system (Fukunari 2018). Further studies of the propagation velocity and equilibrium discharge in the under-critical condition are necessary.

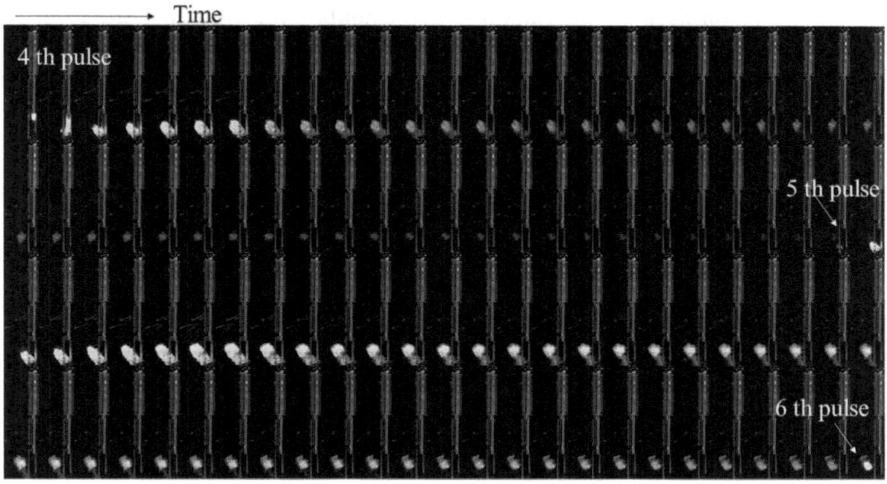

Fig. 5.12 High-speed camera images from the fourth pulse to the sixth pulse

Summary

The microwave discharges in the over-critical and under-critical conditions classified by the ionization threshold were reviewed. Since Microwave Rockets utilize the ambient air as a propellant, we reviewed mainly air discharges. In the over-critical condition, the $\lambda/4$ structure observed at 110 GHz was reproduced by numerical studies. Although some problems remain, such as gas heating effects, the principle of the structure formation mechanism becomes clear qualitatively. On the other hand, the mechanisms of the structure formation and reduction of the ionization threshold in the under-critical condition remain unclear.

In addition to these, few experiments have been conducted on shock waves generated in the under-critical condition. Oda et al. (2011) visualized the one-dimensional shock wave supported by a microwave discharge induced in a shock tube filled with air. In order to clarify the detonation threshold and energy absorption rate, the propagation of the shock wave and ionization front at various conditions should be measured experimentally. The relationship between ionization front propagation velocity and peak power density measured in previous experiments in air discharge reveals that the tendency is divided depending on frequency. This tendency would naturally vary with pressure and gas species. For example, the ionization front propagation velocity of the microwave discharge induced in argon is much faster than that of air discharge (Sidorov 2022). The detonation threshold strongly depends on the ionization front propagation velocity. The detonation threshold of the microwave discharge would be drastically lower than that of laser discharge. In order to evaluate the thrust performance of the Microwave Rocket, it is necessary to understand the mechanism to determine the ionization front propagation velocity.

The transition from the nonequilibrium to equilibrium discharge in the focusing optics of the incident beams and the abnormal ignitions due to the remaining plasma strongly limit the duty cycle of the Microwave Rocket. Further studies are required to prevent the transition.

References

Artem'ev KV, Batanov GM, Berezhetskaya NK, Borzosekov VD, Kolika LV, Konchekov EM, Kossyi IA, Malakhov DV, Petrov AE, Sarksyan KA, Stepakhin VD, Kharchev NK (2018a) Discharge in the atmosphere in a Gaussian beam of subthreshold millimeter waves. JETP Lett 107:219–222. https://doi.org/10.1134/S0021364018040045

Artem'ev KV, Batanov GM, Berezhetskaya NK, Borzosekov VD, Davydov AM, Kolik LV, Konchekov EM, Kossyi IA, Petrov AE, Sarksyan KA, Stepakhin VD, Kharchev NK (2018b) Discharge in a subthreshold microwave beam as an unusual type of ionization wave. Plasma Phys Rep 44:1146–1153. https://doi.org/10.1134/S1063780X18120012

Artem'ev KV, Batanova GM, Berezhetskaya NK, Borzosekov VD, Davydov AM, Kolik LV, Konchekov EM, Kossyi IA, Moryakov IV, Petrov AE, Sarksyan KA, Stepakhin VD, Kharchev NK (2020) Features of a supersonic ionization wave in argon at atmospheric pressure in a sub-threshold microwave field. Plasma Phys Rep 46:1220–1226. https://doi.org/10.1134/S1063780X20120016

Batanov GM, Grisinin SI, Kossyy IA, Magunov AN, Silakov VP, Tarasova NM (1985) High-pressure microwave discharges. In: Kovrizhnykh LM (ed) Plasma physics and plasma electronics. Nova Science Publishers, Commack, pp 241–282

Boeuf JP, Chaudhury B, Zhu GQ (2010a) Theory and modeling of self-organization and propagation of filamentary plasma arrays in microwave breakdown at atmospheric pressure. Phys Rev Lett 104(1):015002. https://doi.org/10.1103/PhysRevLett.104.015002

Boeuf JP, Chaudhury B, Qiang Zhu G (2010b) Theory and modeling of self-organization and propagation of filamentary plasma arrays in microwave breakdown at atmospheric pressure. Phys Rev Lett 104(1):015002. https://doi.org/10.1103/PhysRevLett.104.015002

Bogatov NA, Bykov YV, Venediktov NP, Golubev SV, Zorin VG, Eremeev AG, Semenov VE (1986) Gasdynamic propagation of a nonequilibrium microwave discharge. Sov J Plasma Phys 12(6):416–420

BOLSIG+ n.d. BOLSIG+ Electron Boltzmann equation solver http://www.bolsig.laplace.univ-tlse.fr/

Bratman VL, Zorin VG, Kalynov YK, Koldanov VA, Litvak AG, Razin SV, Sidorov AV, Skalyga VA (2011) Plasma creation by terahertz electromagnetic radiation. Phys Plasma 18:083507. https://doi.org/10.1063/1.3622202

Brodskii YY, Golubev SV, Zorin VG, Luchinin AG, Semenov VE (1983) New mechanism of gasdynamic propagation of a discharge. Sov Phys JETP 57(5):989–993

Brodskii YY, Venediktov IP, Golubev SV, Zorin VG, Kossyi IA (1984) Nonequilibrium microwave discharge in air at atmospheric pressure. Sov Tech Phys Lett 10(2):77–79

Bufetov IA, Prokhorov AM, Fedorov VB, Fomin VK (1984) Optical discharge accompanying a restriction imposed on lateral expansion of gas and a reduction in the threshold of light-induced detonation. J Exp Theor Phys Lett 39:258–261

Chaudhury B, Boeuf JP (2010) Computational studies of filamentary pattern formation in a high power microwave breakdown generated air plasma. IEEE Transac Plasma Sci 38(9):2281. https://doi.org/10.1109/TPS.2010.2055893

Chaudhury B, Boeuf JP, Zhu GQ (2010) Pattern formation and propagation during microwave breakdown. Phys Plasma 17:123505. https://doi.org/10.1063/1.3517177

Cook A, Shapiro M, Temkin R (2010) Pressure dependence of plasma structure in microwave gas breakdown at 110 GHz. Appl Phys Lett 97:011504. https://doi.org/10.1063/1.3462320

Cook AM, Hummelt JS, Shapiro MA, Temkin RJ (2011a) Measurements of electron avalanche formation time in W-band microwave air breakdown. Phys Plasma 18:080707. https://doi.org/10.1063/1.3626383

Cook AM, Hummelt JS, Shapiro MA, Temkin RJ (2011b) Observation of plasma array dynamics in 110 GHz millimeter-wave air breakdown. Phys Plasma 18:100704. https://doi.org/10.1063/1.3656980

Ebert U, Saarloos WV (2000) Front propagation into unstable states: universal algebraic convergence towards uniformly translating pulled fronts. Physica D 146(1–4):1–99. https://doi.org/10.1016/S0167-2789(00)00068-3

Edgcombe CJ (1993) Gyrotron oscillators. Taylor & Francis, Routledge

Edmundoptics n.d. Temperature sensitive liquid crystal sheets, edmund optics. https://www.edmundoptics.jp/f/temperature-sensitive-liquid-crystal-sheets/11878/

Fukunari M, Yamaguchi T, Nakamura Y, Komurasaki K, Oda Y, Kajiwara K, Takahashi K, Sakamoto K (2018) Thrust generation experiments on microwave rocket with a beam concentrator for long distance wireless power feeding. Acta Astronaut 145:263–267. https://doi.org/10.1016/j.actaastro.2018.01.057

Fukunari M, Tanaka S, Shinbayashi R, Yamaguchi Y, Tatematsu Y, Saito T (2019) Observation of a comb-shaped filamentary plasma array under subcritical condition in 303-GHz millimetre-wave air discharge. Sci Rep 9:17972. https://doi.org/10.1038/s41598-019-54333-5

Gil'denburg VB, Kim A (1980) Ionization-thermal instability of an rf discharge in an electromagnetic wave. Sov J Plasma Phys 6(4):496–499

Gil'denburg VB, Kim AV (1978) Ionization instabilities of an electromagnetic wave. Sov Phys JETP 47(1):72–75

Glyavin MY, Golubev SV, Izotov IV, Litvak AG, Luchinin AG, Razin SV, Sidorov AV, Skalyga VA, Vodopyanov AV (2014) A point-like source of extreme ultraviolet radiation based on a discharge in a non-uniform gas flow, sustained by powerful gyrotron radiation of terahertz frequency band. Appl Phys Lett 105:174101. https://doi.org/10.1063/1.4900751

Gurevich AV, Brisove ND, Milikh GM (1997) Physics of microwave discharges:artificially ionized regions in the atmosphere. Gordon & Breach Science Publishers, Philadelphia

Hidaka Y, Choi EM, Mastovsky I, Shapiro MA, Sirigiri JR, Temkin RJ (2008) Observation of large arrays of plasma filaments in air breakdown by 1.5 MW, 110 GHz Gyrotron pulses. Phys Rev Lett 100(3):035003. https://doi.org/10.1103/PhysRevLett.100.035003

Hidaka Y, Choi EM, Mastovsky I, Shapiro MA, Sirigiri JR, Temkin RJ, Edmiston GF, Neuber AA, Oda Y (2009) Plasma structures observed in gas breakdown using a 1.5 MW, 110 GHz pulsed gyrotron. Phys Plasma 16:055702. https://doi.org/10.1063/1.3083218

High Frequency Structure Simulator n.d.. www.hfss.com Ansoft corporation, Pittsburgh, PA

Hummelt JS, Shapiro MA, Temkin RJ (2012) Spectroscopic temperature measurements of air breakdown plasma using a 110 GHz megawatt gyrotron beam. Phys Plasma 19:123509. https://doi.org/10.1063/1.4773037

Kartikeyan MV, Borie E, Thumm M (2004) Gyrotrons: high-power microwave and millimeter wave technology. Springer, Cham

Khodataev KV (2008) Microwave discharges and possible applications. J Propuls Power 24(5). https://doi.org/10.2514/1.24409

Kourtzanidis K, Rogier F, Boeuf JP (2015) Gas heating effects on the formation and propagation of a microwave streamer in air. J Appl Phys 118:103301. https://doi.org/10.1063/1.4930163

Litvak A (1991) Nonlinear dynamics of a freely localized gas discharge in microwave beams. Strong Microwave Plasma 1:267–286

MacDonald D (1966) Microwave breakdown in gases. Wiley, New York

MacDonald AD, Gaskel DU, Gitterman HN (1963) Microwave breakdown in air, oxygen, and nitrogen. Phys Rev 130(5):1841–1850. https://doi.org/10.1103/PhysRev.130.1841

Matsui K, Shimano T, Ofosu JA, Komurasaki K, Schoenherr T, Koizumi H (2016) Accurate propagation velocity measurement of laser supported detonation waves. Vacuum 136:171–176. https://doi.org/10.1016/j.vacuum.2016.07.011

Nakamura Y, Komurasaki K, Fukunari M, Koizumi H (2018) Numerical analysis of plasma structure observed in atmospheric millimeter-wave discharge at under-critical intensity. J Appl Phys 124:033303. https://doi.org/10.1063/1.5023269

Nam SK, Verboncoeur JP (2009) Theory of filamentary plasma Array formation in microwave breakdown at near-atmospheric pressure. Phys Rev Lett 103(5):055004. https://doi.org/10.1103/PhysRevLett.103.055004

Nusinovich GS (2004) Introduction to the physics of Gyrotrons. Johns Hopkins University Press, Baltimore

Oda Y, Komurasaki K, Takahashi K, Kasugai A, Sakamoto K (2006) Plasma generation using high-power millimeter-wave beam and its application for thrust generation. J Appl Phys 100:113307. https://doi.org/10.1063/1.2399899

Oda Y, Kajiwara K, Takahashi K, Kasugai A, Sakamoto K, Komurasaki K (2009) In-tube shock wave driven by atmospheric millimeter-wave plasma. Jpn J Appl Phys 48(11R):116001. https://doi.org/10.1143/JJAP.48.116001

Oda Y, Yamaguchi T, Shiraishi Y, Komurasaki K, Kajiwara K, Takahashi K, Kasugai A, Sakamoto K (2011) A one-dimensional propagation of shock wave supported by atmospheric millimeter-wave plasma. J Infrared Milli Terahz Waves 32:877–882. https://doi.org/10.1007/s10762-011-9798-1

Oda Y, Kajiwara K, Takahashi K, Mitsunaka Y, Sakamoto K (2013) High efficiency coupling of radio frequency beams from the dual frequency gyrotron with a corrugated waveguide transmission system. Rev Sci Instrum 84:013501. https://doi.org/10.1063/1.4772572

Oda Y, Takahashi M, Tabata K, Ohnishi N, Komurasaki K (2018) Frequency dependence of atmospheric millimeter-wave breakdown plasma. 43rd International conference on infrared, millimeter, and terahertz waves (IRMMW-THz), Nagoya. doi: https://doi.org/10.1109/IRMMW-THz.2018.8510051

Oda Y, Takahashi M, Ohnishi N, Komurasaki K, Sakamoto K, Imai T (2020) A study on the macroscopic self-organized structure of high-power millimeter-wave breakdown plasma. Plasma Sources Sci Technol 29(7):075010. https://doi.org/10.1088/1361-6595/ab9d67

Raizer YP (1965) Heating of a gas by a powerful light pulse. Sov Phys JETP 21(5):1009–1017

Raizer YP (1972) Propagation of a high-pressure microwave discharge. Sov Phys JETP 34(1):114–120

Raizer YP (1977) Laser-induced discharge phenomena. Plenum Pub Corp, New York

Sakamoto K, Kasugai A, Takahashi K, Minami R, Kobayashi N, Kajiwara K (2007) Achievement of robust high-efficiency 1 MW oscillation in the hard-self-excitation region by a 170 GHz continuous-wave gyrotron. Nat Phys 3:411–414. https://doi.org/10.1038/nphys599

Schaub SC, Hummelt JS, Guss WC, Shapiro MA, Temkin RJ (2016) Electron density and gas density measurements in a millimeter-wave discharge. Phys Plasma 23:083512. https://doi.org/10.1063/1.4959171

Shimamura K, Yamasaki J, Miyawaki K, Minami R, Kariya T, Yang J, Yokota S (2021) Propagation of microwave breakdown in argon induced by a 28GHz gyrotron beam. Phys Plasma 28:033505. https://doi.org/10.1063/5.0045350

Shimozuma T, Idei H, Shapiro M, Temkin R, Ito S, Notake T, Kubo S, Yoshimura Y, Kobayashi S, Mizuno Y, Takita Y, Ohkubo K (2005) Alignment method of ech transmission lines based on the moment and phase retrieval method using ir images. J Plasma Fusion Res 81(3):191–196. https://doi.org/10.1585/jspf.81.191

Sidorov AV (2022) Terahertz gas discharge: current progress and possible applications. J Phys D Appl Phys 55:293001. https://doi.org/10.1088/1361-6463/ac5556

Sidorov A, Razin S, Veselov A, Viktorov M, Vodopyanov A, Luchinin A, Glyavin M (2020) Dynamics of the gas discharge in noble gases sustained by the powerful radiation of 0.67 THz gyrotron. Phys Plasma 27:093509. https://doi.org/10.1063/5.0012583

Tabata K, Harada Y, Nakamura Y, Komurasaki K, Koizumi H, Kariya T, Minami R (2020) Experimental investigation of ionization front propagating in a 28 GHz gyrotron beam: observation of plasma structure and spectroscopic measurement of gas temperature. J Appl Phys 127: 063301. https://doi.org/10.1063/1.5144157

Takahashi M, Kageyama Y, Ohnishi N (2017) Joule-heating-supported plasma filamentation and branching during subcritical. AIP Adv 7:055206. https://doi.org/10.1063/1.4983569

Tatematsu Y, Yamaguchi Y, Idehara T, Kawase T, Ogawa I, Saito T, Fujiwara T (2014) Characteristics of the mode converter of Gyrotron FU CW GII radiating Gaussian beams in both the fundamental and second harmonic frequency bands. J Infrared Milli Terahz Waves 35:517–524. https://doi.org/10.1007/s10762-014-0072-1

Tatematsu Y, Yamaguchi Y, Ichioka R, Kotera M, Saito T, Idehara T (2015) Development of the multifrequency Gyrotron FU CW GV with Gaussian beam output. J Infrared Milli Terahz Waves 36:697–708. https://doi.org/10.1007/s10762-015-0173-5

Taylor WC, Scharfman WE, Morita T (1971) Voltage breakdown of microwave antennas. In: Advances in microwaves, vol 7. Academic, New York, pp 59–130

Vikharev AL, Gil'denburg VB, Golubev SV, Eremin BG, Ivanov OA, Litvak AG, Stepanov AN, Yunakovskii AD (1988) Nonlinear dynamics of a freely localized microwave discharge in an electromagnetic wave beam. Sov Phys JETP 67(4):724–728

Vikharev AL, Gorbachev AM, Kim AV, Kolysko AL (1992) Formation of the small-scale structure in a microwave discharge in high-pressure gas. Sov J Plasma Phys 18(8):554–560

Vyskrebentsev A, Raizer Y (1973) A simple theory of breakdown of monatomic nonlight gases in fields of frequency from low to optical. J Appl Mech Tech Phys 14(1):32–38. https://doi.org/10.1007/BF00850574

Yamaguchi T, Komurasaki K, Oda Y, Kajiwara K, Takahashi K, Sakamoto K (2011) Millimeter-wave beam conversion with quasi-optical mirrors for microwave rocket launch demonstration. AIP Conference Proceedings 1402:467. https://doi.org/10.1063/1.3657055

Yamaguchi Y, Kasa J, Saito T, Tatematsu Y, Kotera M, Kubo S, Shimozuma T, Tanaka K, Nishiura M (2015) High power 303 GHz gyrotron for CTS in LHD. J Instrum 10:C10002. https://doi.org/10.1088/1748-0221/10/10/C10002

Chapter 6
Modeling and Theoretical Studies on Beamed-Induced Plasma

Masayuki Takahashi and Yusuke Nakamura

Abstract Modeling and theory for millimeter-wave discharge were described to understand the plasma-formation physics at over- and under-critical intensities. On overcritical discharge, plasma fluid or particle model was combined with electromagnetic wave propagations via a current-density feedback. Simulation results and theory indicated that electron-impact ionization, electron diffusion, and wave reflection caused the propagation of the ionization front. On the other hand, at under-critical conditions, modeling studies revealed that neutral gas dynamics, excitation reaction, and photon transport played important roles on the propagation of the ionization front.

Keywords Millimeter-wave discharge · Overcritical discharge · Under-critical discharge · Ionization-front propagation · Plasma fluid model · Particle model · Finite-difference time-domain method · Electron-impact ionization · Neutral gas expansion · Photon transport

Introduction to Theoretical Studies

Beam-induced plasma has fascinating characteristics and has received attention in the fields of plasma physics and engineering. An interesting feature of beam-induced plasma is that the plasma pattern drastically changes depending on the parameters of the incident beam (power, wavelength, and pulse width) and the background gas. Since the 1950s, the technologies of Maser and laser technologies have been developed, and using these beam sources, many breakdown experiments have been conducted to capture the behavior of dense plasma. Moreover, high-power

M. Takahashi (✉)
Department of Aerospace Engineering, Tohoku University, Sendai, Japan
e-mail: masayuki.takahashi.c8@tohoku.ac.jp

Y. Nakamura
Department of Aerospace Engineering, Nagoya University, Nagoya, Japan
e-mail: yusuke.nakamura@mae.nagoya-u.ac.jp

K. Mori et al. (eds.), *Beamed-mobility Engineering*, Springer Tracts in Electrical and Electronics Engineering, https://doi.org/10.1007/978-981-99-4618-1_6

gyrotron technology was developed in the 1960s, opening up a new field for plasma physics and devices using a beam with a millimeter wavelength. In recent years, with an increase in the output power of the gyrotron, millimeter-wave beam-induced plasma has been utilized as a rocketing technology to launch and transport vehicles to the outer space. Such a launch system that uses a millimeter-wave beam is called a microwave rocket. A detailed discussion of the rocket launching technology was conducted in the above chapters, and in this section, we mention modeling studies of the millimeter-wave plasma behavior because understanding the millimeter-wave discharge is important in atmospheric plasma physics and engineering application.

Millimeter-wave breakdown is classified into overcritical and subcritical conditions depending on the incident beam intensity. The balance between electron-impact ionization and electron attachment frequencies determines the threshold between them. When the intensity of the incident millimeter-wave beam is higher than the threshold intensity, the electron-impact ionization frequency becomes larger than the attachment frequency, and thus, plasma production is constantly maintained. However, when the intensity of the incident beam is lower than the threshold intensity, some processes are necessary to defeat electron disappearance by the attachment reaction and maintain plasma production. Because the plasma production process is simpler, modeling for millimeter-wave plasma was first proposed to describe the overcritical breakdown. Subsequently, the overcritical breakdown model was updated to describe the subcritical breakdown. Therefore, to understand the modeling studies of millimeter-wave breakdown, we first discuss modeling studies for overcritical discharge.

Fluid Modeling for Overcritical Millimeter-Wave Discharge

The first paper on modeling overcritical breakdown was written by Nam and Verboncoeur in 2009 (Nam and Verboncoeur 2009). They developed a one-dimensional (1D) plasma fluid model with a module to integrate the following continuity and energy equations:

$$\frac{\partial n_e}{\partial t} = -\nabla \cdot J_e + K_{ion} n_e n_{gas}, \tag{6.1}$$

$$\frac{\partial}{\partial t}\left(\frac{3}{2} n_e T_e\right) = -\nabla \cdot q_e + P_{abs} - \left(\epsilon_{ion} K_{ion} n_e n_{gas} + \epsilon_{exc} K_{exc} n_e n_{gas} + \tilde{K}_{mom} n_e n_{gas}\right), \tag{6.2}$$

where t is the time, n_e and n_{gas} are the electron and neutral number densities, J_e is the electron flux, K_{ion} and K_{exc} are the ionization and excitation reaction coefficients, T_e is the electron temperature, q_e is the energy flux, P_{abs} is the energy absorption rate, ϵ_{ion} and ϵ_{exc} are the ionization and excitation threshold energies, and $\tilde{K}_{mom}$ is the momentum energy transfer rate coefficient. The free diffusion and ambipolar drift terms were considered in the electron number density and energy fluxes. The energy absorption rate was determined using the electric field in the perpendicular direction $E_{\perp}$, which was calculated by solving the 1D wave equation. Reflection from the dense plasma was introduced by the complex wave number in the plasma medium.

The 1D plasma fluid model was combined with the 1D wave equation, which revealed that a standing wave was induced in front of the bulk plasma because the incident microwave was reflected by the plasma when the plasma density exceeded the overcritical density. At the antinode of the standing wave, the ionization reaction rate increased owing to an increase in the electron temperature; therefore, the electron number density increased rapidly. The next plasma spot is then generated at the antinode of the standing wave. The plasma front propagating toward the incident beam source was reproduced because the processes of wave reflection, ionization, and standing wave formation were repeated. However, because the interaction between the plasma and the incident microwave could not be captured correctly because of insufficient modeling of the complex wave number in the plasma medium, the number density of electrons was much higher than the critical density.

To capture the interaction dynamics between the plasma and the wave precisely, Boeuf et al. proposed a plasma-wave coupling model via feedback of the current density (Beouf et al. 2010). In their model, a simple electron diffusion equation was numerically integrated as follows:

$$\frac{\partial n_{\mathrm{e}}}{\partial t} - D_{\mathrm{eff}} \Delta n_{\mathrm{e}} = n_{\mathrm{e}}(\nu_{\mathrm{i}} - \nu_{\mathrm{a}}) - r_{\mathrm{ei}} n_{\mathrm{e}}^2, \tag{6.3}$$

where D_{eff} is the effective diffusion coefficient; ν_{i} and ν_{a} are the electron-impact ionization and electron attachment frequencies, respectively; and r_{ei} is the recombination coefficient. Here, the effective diffusion coefficient was introduced to capture the transition between ambipolar and free diffusion. If the plasma dimension becomes much longer than the Debye length at the plasma front, the free-electron diffusion coefficient should be used instead of the ambipolar diffusion coefficient. To consider this transition of the diffusion coefficient, the transition parameter α was calculated at the local point as follows:

$$\alpha = \nu_{\mathrm{i}} \tau_{\mathrm{M}} = \lambda_{\mathrm{D}}^2 / L^2, \tag{6.4}$$

where τ_{M} is the Maxwell relaxation time, λ_{D} is the Debye length, and L is the plasma scale length. By using α, the effective diffusion coefficient can be written as

$$D_{\mathrm{eff}} = \frac{\alpha D_{\mathrm{e}} + D_{\mathrm{a}}}{\alpha + 1}, \tag{6.5}$$

where D_{e} and D_{a} are the free and ambipolar diffusion coefficients, respectively. The transport coefficients, such as the ionization frequency, were evaluated as functions that were obtained from Bolsig+ (Hagelaar and Pitchford 2005) of the local effective field E_{eff}. Here, the local effective field is given by

$$E_{\text{eff}} = \frac{E_{\text{rms}}}{\sqrt{1+\omega^2/\nu_{\text{m}}^2}}, \tag{6.6}$$

where E_{rms} is the root-mean-square (RMS) electric field, ω is the angular frequency of the incident wave, and ν_{m} is the elastic collision frequency between the electron and neutral particles. The RMS electric field E_{rms} was calculated by integrating Maxwell's equations as follows:

$$\nabla \times \mathbf{E} = -\frac{\partial \mathbf{B}}{\partial t}, \tag{6.7}$$

$$\nabla \times \mathbf{H} = \mathbf{J} + \epsilon_0 \frac{\partial \mathbf{E}}{\partial t}, \tag{6.8}$$

where $\mathbf{E}$ is the electric field vector, $\mathbf{B}$ is the magnetic flux density vector, $\mathbf{H}$ is the magnetic field vector, $\mathbf{J}$ is the current density vector, and ϵ_0 is the permittivity in a vacuum. Maxwell's equations were solved using the finite-difference time-domain (FDTD) method. In the FDTD method, the electric and magnetic fields are defined on the Yee mesh, and these fields are updated alternately over time. In the case of 2D FDTD, the set of E_z, H_x, and H_y (TM wave) and the set of E_x, E_y, and H_z (TE wave) can be discretized, where E_x, E_y, and E_z are the electric field components in the x-, y-, and z-directions, and H_x, H_y, and H_z are the magnetic field components in the x-, y-, and z-directions, respectively. For the TM wave, H_x is defined at the grid point of $(i, j + 1/2)$ and the time step of $n - 1/2$, and H_y is defined at the grid point $(i + 1/2, j)$ and the time step of $n - 1/2$ when E_z is defined at the grid point of (i, j) and the time step of n. Note that the spatial and temporal definitions of the magnetic and electric fields are shifted. Using these definitions, with the z-directing current density component J_z, the discretized form of the electromagnetic field is written as

$$\begin{aligned} E_z^n(i,j) = E_z^{n-1}(i,j) &+ \frac{\Delta t}{\epsilon_0 \Delta x}\left\{H_y^{n-\frac{1}{2}}\left(i+\frac{1}{2},j\right) - H_y^{n-\frac{1}{2}}\left(i-\frac{1}{2},j\right)\right\} - \frac{\Delta t}{\epsilon_0 \Delta y} \\ &\times \left\{H_x^{n-1/2}(i,j+1/2) - H_x^{n-1/2}(i,j-1/2)\right\} - \frac{\Delta t}{\epsilon_0} J_z, \end{aligned} \tag{6.9}$$

$$H_x^{n+1/2}(i,j+1/2) = H_x^{n-1/2}(i,j+1/2) - \frac{\Delta t}{\mu \Delta y}\{E_z^n(i,j+1) - E_z^n(i,j)\}, \tag{6.10}$$

$$H_y^{n+1/2}(i+1/2,j) = H_y^{n-1/2}(i+1/2,j) + \frac{\Delta t}{\mu \Delta x}\{E_z^n(i+1,j) - E_z^n(i,j)\}. \tag{6.11}$$

For the TE wave, similar to the TM wave, the electromagnetic field component was defined at different grid points and time steps. If the magnetic field component H_z is defined at the grid point of $(i + 1/2, j + 1/2)$ and the time step of $n - 1/2$, the electric field component E_x is defined at the grid point of $(i + 1/2, j)$ and the time step of n, and the electric field component E_y is defined at the grid point of $(i, j + 1/2)$ and the time step of n. Using the current density components J_x and J_y in the x- and y-directions, the discretization form for the TM wave is obtained as follows:

$$E_x^n(i+1/2,j) = E_x^{n-1}(i+1/2,j) + \frac{\Delta t}{\epsilon_0 \Delta y} \times \left\{ H_z^{n-\frac{1}{2}}\left(i+\frac{1}{2},j+\frac{1}{2}\right) - H_z^{n-\frac{1}{2}}\left(i+\frac{1}{2},j-\frac{1}{2}\right) \right\} - \frac{\Delta t}{\epsilon_0} J_x, \tag{6.12}$$

$$E_y^n(i,j+1/2) = E_y^{n-1}(i,j+1/2) + \frac{\Delta t}{\epsilon_0 \Delta x} \times \left\{ H_z^{n-\frac{1}{2}}\left(i+\frac{1}{2},j+\frac{1}{2}\right) - H_z^{n-\frac{1}{2}}(i-1/2,j+1/2) \right\} - \frac{\Delta t}{\epsilon_0} J_y, \tag{6.13}$$

$$\begin{aligned} H_z^{(n+1/2)}(i+1/2,j+1/2) = {} & H_z^{(n-1/2)}(i+1/2,j+1/2) \\ & - \frac{\Delta t}{\mu \Delta x} \left\{ E_y^n(i+1,j+1/2) - E_y^n(i,j+1/2) \right\} \\ & + \frac{\Delta t}{\mu \Delta y} \left\{ E_x^n(i+1/2,j+1) - E_x^n(i+1/2,j) \right\}. \end{aligned} \tag{6.14}$$

Moreover, the electron equation of motion is numerically integrated to evaluate the current density vector $\mathbf{J}$ at the local position and to consider the interaction between the plasma and electromagnetic wave as follows:

$$\frac{\partial \mathbf{v}_e}{\partial t} = -\frac{e\mathbf{E}}{m_e} - \nu_m \mathbf{v}_e, \tag{6.15}$$

where $\mathbf{v}_e$ is the electron velocity vector, e is the elementary charge, and m_e is the electron mass. Here, the ion current is ignored because the timescale of ion motion is much slower than that of electron motion. The current density vector was evaluated as $\mathbf{J} = -en_e\mathbf{v}_e$, which was fed back to Maxwell's equations to reproduce the interaction between the plasma dynamics and the electromagnetic waves.

Note that in the plasma-electromagnetic wave coupling model, the timescale of the electromagnetic wave propagation is much shorter than that of the plasma drift-diffusion. If the integration time steps for every module are uniformly determined based on the electromagnetic wave propagation, the computational cost increases. Therefore, in Boeuf's model, the time step for the FDTD was determined by the Courant-Friedrichs-Lewy (CFL) condition using the speed of light, but the time step for the plasma diffusion equation was set as one period of the incident microwave to save computational cost. The RMS electric field was obtained using one-period integration for the FDTD model. Subsequently, using the RMS electric field, the transport coefficients were evaluated by Bolsig+ to integrate the plasma diffusion equation. If the time step of the plasma diffusion equation, which is determined by the CFL condition, is smaller than one period of the incident microwave, it is necessary to use the time step based on the CFL condition on a framework of explicit integration; however, the time step determined by the CFL condition tends to be large in the atmospheric plasma because the electron diffusion coefficient is not very large.

The 2D spatial distributions of the plasma were numerically obtained in the *H*- and *E*-planes using the FDTD-plasma coupling model with a feedback term of the electron current density. In previous breakdown experiments using a high-speed camera (Hidaka et al. 2008, 2009), a discrete plasma structure, such as stepping stones, was captured in the *H*-plane, which was successfully reproduced by the coupling simulation. Moreover, their model numerically reproduced a fish-bone structure, which was experimentally obtained in the *E*-plane, because a wave reflection was induced at the plasma front and the standing wave was formed when the electron density reached the overcritical value. Electron-impact ionization rapidly occurred at the antinode of the standing wave, which discretely increased the electron number density. By introducing the current feedback term, the interaction between the electromagnetic wave and the plasma could be precisely captured; therefore, an overlarge plasma density was not induced in their model. In addition, the propagation speed of the plasma front reproduced by the simulation was approximately 30 km/s, which is in agreement with the theoretical prediction of the front velocity $2\sqrt{D_\mathrm{e}\nu_\mathrm{i}}$ (Ebert et al. 1996, 1997a). The 3D structure for the microwave breakdown was also captured using the same model (Kourtzanidis et al. 2014). Instead of the explicit FDTD method, an alternating direction implicit (ADI) FDTD method was used to speed up the electromagnetic wave calculation. The 3D modeling indicated that the electric field at the edge of the filaments was enhanced, and the elongation length of the single filament increased.

Particle Modeling for Overcritical Millimeter-Wave Discharge

In addition to fluid modeling, particle modeling was conducted while coupling with electromagnetic wave propagation. Takahashi and Ohnishi conducted a 1D particle-in-cell (PIC) simulation to capture the dynamics of microwave plasma under an external magnetic field (Takahashi and Ohnishi 2014, 2016, 2018). The computational cost of the PIC model was higher than that of the fluid model; however, the particle model could precisely capture the resonance dynamics between the electromagnetic wave and the cyclotron motion of the electron. Moreover, the mobility and diffusivity of the charged particles under a magnetic filed were correctly obtained using the particle model. In the PIC model, the equations of motion for the electrons and ions are numerically integrated as follows:

$$m\frac{d\mathbf{v}}{dt} = q(\mathbf{E} + \mathbf{v} \times \mathbf{B}), \tag{6.16}$$

where m is the mass of the electron and ion, $\mathbf{v}$ is the velocity vector of the electron and ion, and q is the electric charge of the electron and the ion. The effect of the magnetic field can be introduced in the Lorentz force term of the equation of motion, which induces the rotational motion of charged particles. The velocity vector $\mathbf{v}$ had three components in the x-, y-, and z-directions to capture the Larmor motion of the charged particles under a magnetic field. If the spatial dimension is only in the x-direction and the velocity vector has three components, it is called the 1D3V PIC. We call the 2D3V PIC if the spatial dimension has two components and the velocity vector has three components. Because the particle simulation tends to have an extremely high computational cost, the 1D3V or 2D3V PIC is frequently conducted instead of the 3D3V PIC. Moreover, to reduce the computational cost, several particles are regarded as a single particle (super-particle), and the equations of motion of the super-particle can be integrated. The weight of the super-particle is a controllable parameter, and depending on the plasma density and spatial length, we have to determine the suitable weight of the super-particle while noting that there is a trade-off relationship between the numerical noise and the super-particle weight. In general, more than 100 super-particles are necessary for one cell to reduce the numerical noise and obtain a meaningful solution in the discharge region.

In the PIC, the equations of motion for the super-particle were numerically integrated using the Buneman-Boris method (Boris 1970). In the Buneman-Boris method, the equations of motion are discretized as follows:

$$\frac{\mathbf{v}^{n+1/2} - \mathbf{v}^{n-1/2}}{\Delta t} = \frac{q}{m}\left(\mathbf{E}^n + \frac{\mathbf{v}^{n+1/2} + \mathbf{v}^{n-1/2}}{\Delta t} \times \mathbf{B}^n\right), \tag{6.17}$$

where n is the time step and Δt is the time interval for integration. In the above equation, the following velocities are introduced to obtain $\mathbf{v}^{n-1/2}$:

$$\mathbf{v}^{-} = \mathbf{v}^{n-1/2} + \frac{q}{m}\mathbf{E}^n\frac{\Delta t}{2}, \tag{6.18}$$

$$\mathbf{v}^{+} = \mathbf{v}^{n+1/2} - \frac{q}{m}\mathbf{E}^n\frac{\Delta t}{2}. \tag{6.19}$$

By substituting these equations into the equations for $\mathbf{v}^{n+1/2}$ and $\mathbf{v}^{n-1/2}$, we obtain the following form:

$$\frac{\mathbf{v}^{+} - \mathbf{v}^{-}}{\Delta t} = \frac{1}{2}\frac{q}{m}(\mathbf{v}^{+} + \mathbf{v}^{-}) \times \mathbf{B}^{n}. \tag{6.20}$$

This equation implies that only the Lorentz force acting on the particles during the velocity changes from $\mathbf{v}^{-}$ to $\mathbf{v}^{+}$. Here, we transform the above equation as follows:

$$\mathbf{v}^{+} = \mathbf{v}^{-} + \frac{\mathbf{2}}{1 + \mathbf{T}^{2}}(\mathbf{v}^{-} + \mathbf{v}^{-} \times \mathbf{T}) \times \mathbf{T}, \tag{6.21}$$

where $\mathbf{T} = (q/m)\Delta t\mathbf{B}^{n}/\mathbf{2}$. The following equations were obtained when $\mathbf{S} = \mathbf{2T}/(\mathbf{1} + \mathbf{T}^{2})$ was introduced into the above equation:

$$\mathbf{v}^{0} = \mathbf{v}^{-} + \mathbf{v}^{-} \times \mathbf{T}, \tag{6.22}$$

$$\mathbf{v}^{+} = \mathbf{v}^{-} + \mathbf{v}^{0} \times \mathbf{S}. \tag{6.23}$$

Consequently, the velocity vector of the charged particles was updated using the following four steps:

$$\mathbf{v}^{-} = \mathbf{v}^{n-1/2} + \frac{q}{m}\mathbf{E}^{n}\frac{\Delta t}{\mathbf{2}}, \tag{6.24}$$

$$\mathbf{v}^{0} = \mathbf{v}^{-} + \mathbf{v}^{-} \times \mathbf{T}, \tag{6.25}$$

$$\mathbf{v}^{+} = \mathbf{v}^{-} + \mathbf{v}^{0} \times \mathbf{S}. \tag{6.26}$$

$$\mathbf{v}^{n+1/2} = \mathbf{v}^{+} + \frac{q}{m}\mathbf{E}^{n}\frac{\Delta t}{\mathbf{2}}, \tag{6.27}$$

When the velocity vector was calculated, the position vector of the particles was updated as follows:

$$\mathbf{x}^{n+1} = \mathbf{x}^{n} + \mathbf{v}^{n+1/2}\Delta t, \tag{6.28}$$

When the particle velocity is updated, the current density is calculated and fed back into Maxwell's equations to capture the interaction between the wave and the plasma, similar to Boeuf's coupling model. However, contrary to the plasma fluid model, the current density is evaluated through a weighting procedure between the particles and field data. Note that weighting is necessary for the PIC simulation because the particle information is defined as Lagrangian values, and the information of the electromagnetic field is defined as Eulerian values on the grid points. In the 1D

PIC, the current density $\mathbf{J} = qn\mathbf{v}$, which is defined on a grid point X_i, is obtained using the particle velocity and electric charge as follows:

$$\mathbf{J}(X_i) = \frac{1}{\Delta x}\sum_{p=1}^{N} q_p \mathbf{v}_p S_i(x_p), \tag{6.29}$$

where Δx is the grid interval, p is the particle index, N is the maximum number of particles, q_p is the electric charge of particle p, $\mathbf{v}_p$ is the velocity vector of particle p, and $S_i(x_p)$ is the shape function for the particle position x_p. The shape function determines the area of influence of the charged particles whose location is x_p, which affects the amplitude of the numerical noise. In the 1D PIC, the following linear function was proposed as the shape function for first-order could-in-cell (CIC) weighting:

$$S_i(x_p) = \begin{cases} 1 - \dfrac{|x_p - X_i|}{\Delta x}, \mathit{if} \ \dfrac{|x_p - X_i|}{\Delta x} \leq 1 \\ 0, \text{else} \end{cases} \tag{6.30}$$

However, a more than second-order shape function is frequently used to reduce numerical noise. In the second-order weighting, the second-order spline function is given by the following functions:

$$S_i(x_p) = \begin{cases} \dfrac{3}{4} - \left(\dfrac{x_p - X_i}{\Delta x}\right)^2, \text{if} \ \dfrac{|x_p - X_i|}{\Delta x} \leq \dfrac{1}{2} \\ 0, \text{else} \end{cases} \tag{6.31}$$

$$S_{i\pm1}(x_p) = \begin{cases} \dfrac{1}{2}\left(\dfrac{1}{2} \pm \dfrac{x_p - X_i}{\Delta x}\right)^2, \text{if} \ \dfrac{|x_p - X_i|}{\Delta x} \leq \dfrac{1}{2} \\ 0, \text{else} \end{cases} \tag{6.32}$$

When the current density was obtained, Maxwell's equations were numerically solved using the FDTD method to reproduce the interaction between the plasma and the electromagnetic wave propagation. After obtaining the electromagnetic fields, the motion and spatial distribution of the charged particles were updated by solving the equations of motion.

However, when the current density and charge distribution are calculated separately, as discussed above, the following charge conservation law is not satisfied, and the numerical error can accumulate.

$$\frac{\partial \rho_e}{\partial t} + \nabla \cdot \mathbf{J} = 0. \tag{6.33}$$

Therefore, a charge conservation method was proposed to maintain the charge conservation law. In the charge conservation method, the density decomposition scheme, which divides one charge into $q/2$, is known for its low computational cost (Zh 2001). In the 3D case, each component of the current density vector was calculated as follows:

$$J^1_{i+1,j,k} - J^1_{i,j,k} = -Q\frac{dx}{dt}W^1_{i,j,k}, \tag{6.34}$$

$$J^2_{i,j+1,k} - J^2_{i,j,k} = -Q\frac{dy}{dt}W^2_{i,j,k}, \tag{6.35}$$

$$J^3_{i,j,k+1} - J^3_{i,j,k} = -Q\frac{dz}{dt}W^3_{i,j,k}, \tag{6.36}$$

where i, j, and k are the grid indexes in the x-, y-, and z-directions and J^1, J^2, and J^3 are the x-, y-, and z-directing components of the current density vector, respectively; Q is the charge of the super-particle. W^1, W^2, and W^3 satisfy the following equations, which consist of arbitrary shape functions S:

$$\begin{aligned} W^1 = &\frac{1}{3}S(x+\Delta x, y+\Delta y, z+\Delta z) - \frac{1}{3}S(x, y+\Delta y, z+\Delta z) + \frac{1}{6}S(x+\Delta x, y, z+\Delta z) \\ &- \frac{1}{6}S(x, y, z+\Delta z) + \frac{1}{6}S(x+\Delta x, y+\Delta y, z) - \frac{1}{6}S(x, y+\Delta y, z) \\ &+ \frac{1}{3}S(x+\Delta x, y, z) - \frac{1}{3}S(x, y, z), \end{aligned} \tag{6.37}$$

$$\begin{aligned} W^2 = &\frac{1}{3}S(x+\Delta x, y+\Delta y, z+\Delta z) - \frac{1}{3}S(x+\Delta x, y, z+\Delta z) + \frac{1}{6}S(x, y+\Delta y, z+\Delta z) \\ &- \frac{1}{6}S(x, y, z+\Delta z) + \frac{1}{6}S(x+\Delta x, y+\Delta y, z) - \frac{1}{6}S(x+\Delta x, y, z) \\ &+ \frac{1}{3}S(x, y+\Delta y, z) - \frac{1}{3}S(x, y, z), \end{aligned} \tag{6.38}$$

$$\begin{aligned} W^3 = &\frac{1}{3}S(x+\Delta x, y+\Delta y, z+\Delta z) - \frac{1}{3}S(x+\Delta x, y+\Delta y, z) + \frac{1}{6}S(x, y+\Delta y, z+\Delta z) \\ &- \frac{1}{6}S(x, y+\Delta y, z) + \frac{1}{6}S(x+\Delta x, y, z+\Delta z) - \frac{1}{6}S(x+\Delta x, y, z) \\ &+ \frac{1}{3}S(x, y, z+\Delta z) - \frac{1}{3}S(x, y, z). \end{aligned} \tag{6.39}$$

The computational cost of density decomposition tends to be low in the charge conservation method because this scheme uses only simple polynomes of the shape function. The amount of computation on the simple weighting method is smaller than that of the density decomposition method; however, the density decomposition method is selected when capturing the detailed wave and plasma structure is important.

Additionally, to model the plasma generation process under intense microwaves, it is necessary to consider a collisional reaction because the electron number density can be determined through production by an electron-impact ionization reaction. To introduce the collisional source term, the Monte Carlo collision (MCC) model, which uses a random number to determine the collision event, is coupled with PIC. One of the commonly used MCC methods is the Nanbu method (Nanbu 1980), which can determine the collision type and the presence or absence of a collision in one trial. Let us consider the case in which particle i collides with particle k. The collision probability $P_e(m)$ is given by

$$P_e(m) = N_k \sigma_m(\epsilon) g_{ik} \Delta t, \tag{6.40}$$

where m is the type of collisional reaction; N_k is the number density of the particle k; $\sigma_m(\epsilon)$ is the collisional cross section, which is a function of the collision energy ϵ; and g_{ik} is the relative velocity of the particles i and k. Using the above equation, the flowchart for an evaluation of the collisional reaction is explained as follows: There are many methods to evaluate collisional reactions; however, here is one example. When the collision form has two types (e.g., elastic and ionization collisions), a line segment with a length of unity is divided into 1/2 and 1/2. The collision probabilities of $P_e(1)$ and $P_e(2)$ were then calculated using the relative velocity and collision cross section. The collision probabilities of $P_e(1)$ and $P_e(2)$ were then arranged inside each line segment of length 1/2. Therefore, in the line segment of unity, the collision type of $m = 1$ is arranged from 0 to $P_e(1)$, no collision is arranged from $P_e(1)$ to 1/2, the collision type of $m = 2$ is arranged from 1/2 to $P_e(2)$, and no collision is arranged from $P_e(2)$ to 1. It is necessary to select a sufficiently small Δt to maintain the collision probabilities of $P_e(1)$ and $P_e(2)$ less than 1/2. Pseudorandom numbers were used in the range from 0 to 1. If the random number is within the range from 0 to $P_e(1)$, it is considered that a collision of $m = 1$ occurs. If the random number is within the range of 1/2 to $P_e(2)$, a collision of $m = 2$ is considered to occur. If a random number has another value, it is considered that collisional reactions do not occur.

If an elastic collision reaction occurs between particles i and k, the following equation is established because the momentum is conserved before and after the collisional reaction:

$$m_i \mathbf{c}_i + m_k \mathbf{c}_k = m_i \mathbf{c}'_i + m_k \mathbf{c}'_k, \tag{6.41}$$

where m_i and m_k are the masses of the particles i and k, $\mathbf{c}_i$ and $\mathbf{c}_k$ are the velocity vectors of particles i and k before the collision, and $\mathbf{c}'_i$ and $\mathbf{c}'_k$ are the velocity vectors of particles i and k after the collision. Moreover, in an elastic collision, the following equation is established from the energy conservation:

$$\frac{1}{2}\mu \mathbf{g}^2 = \frac{1}{2}\mu \mathbf{g}'^2, \tag{6.42}$$

where $\mu = m_i m_k/(m_i + m_k)$ is the conversion mass and $\mathbf{g}$ and $\mathbf{g}'$ are the relative velocity vectors before and after the collision, respectively. Therefore, we obtain $\mathbf{g} = \mathbf{g}'$ from energy conservation. Here, if the hard-sphere (HS) model is introduced to determine the scattering angle, the unit vector for the scattering direction $\mathbf{R}$ is given by

$$R_x = \sin\theta\cos\phi, R_y = \sin\theta\sin\phi, R_z = \cos\theta, \tag{6.43}$$

where R_x, R_y, and R_z are the x-, y-, and z-directing components of vector $\mathbf{R}$. Here, the scattering angles of θ and ϕ are determined as

$$\cos\theta = 1 - 2R_1, \phi = 2\pi R_2, \tag{6.44}$$

where R_1 and R_2 are the independent random numbers. By combining the momentum and energy conservations with the scattering angle vector, the velocity vector after the collision is calculated by

$$\mathbf{c}'_i = \frac{m_i \mathbf{c}_i + m_k \mathbf{c}_k + m_k \mid \mathbf{g} \mid \mathbf{R}}{m_i + m_k}. \tag{6.45}$$

If an inelastic collision occurs, the ionization or excitation potential energy is subtracted from the system energy before the collisional reaction. Here, we consider the ionization reaction. In the singly ionized process, in addition to subtraction of the ionization energy, it is necessary to produce one electron. Here, the electron colliding with the heavy particle is called the parent electron, and the electron produced by the ionization reaction is called the children electron. The total energy of the parent and children's electrons $\epsilon_p + \epsilon_c$ is determined by

$$\epsilon_p + \epsilon_c = \epsilon_b - E_i, \tag{6.46}$$

where ϵ_p is the parent electron energy, ϵ_c is the children electron energy, ϵ_b is the system energy before the collision, and E_i is the ionization potential. Here, a proportion of energies of ϵ_p and ϵ_c has arbitrariness, but it is often the case that $\epsilon_p : \epsilon_c = 9 : 1$ is chosen. After ϵ_p and ϵ_c are calculated by the above equation, by using the scattering angle vector, the velocity vectors of the parent and child electrons are randomly determined, while maintaining the energies of ϵ_p and ϵ_c.

Past Simulations by PIC-MCC Model

Takahashi and Ohnishi conducted the 1D3V PIC in a super-particle manner to reduce the computational cost of numerical modeling of microwave discharge (Takahashi and Ohnishi 2014, 2016, 2018). Simulations for the 5-MV/m and 110-GHz microwave were conducted to follow the previous breakdown experiment. To consider the electron production and energy loss through the inelastic collision reaction, ionization and excitation reactions were introduced in the PIC model by utilizing the Nanbu method. The current density was evaluated using the weighting method, which was fed back into Maxwell's equations. Maxwell's equation for the TE wave was numerically solved by the FDTD method and coupled with a 1D3V PIC-MCC simulation, which indicated that the microwave plasma was successfully reproduced in a manner similar to the plasma fluid model because the interaction physics between the electromagnetic wave and the plasma was captured. A discrete plasma structure with intervals of one quarter of the incident beam wavelength was obtained at higher pressures because the electron-impact ionization rapidly occurred at the antinode of the standing wave; however, the microwave plasma became diffusive with a decrease in the background gas pressure because each peak of the discrete structure was smeared out by rapid electron diffusion. Moreover, the plasma-front propagation speed increased with a decrease in the background gas pressure because the elastic collision frequency decreased, and the electron rapidly diffused from the plasma front. This increase in the plasma-front propagation speed can be understood by theoretical modeling of $2\sqrt{D_e \nu_i}$ as similar to the discussion of the plasma fluid model because a higher D_e was also obtained in the PIC simulation.

In terms of application to microwave rockets, a speeding up of the plasma-front propagation can decrease the thrust performance because the gas heating is concentrated only at the plasma front, and the gas-heating region rapidly escapes from the rocket nozzle. Moreover, at a lower pressure, the energy absorption rate of the electron drastically decreased because the elastic collision frequency was smaller, and the phase difference between the wave and electron oscillations could not be obtained. This implies that the shock wave driven inside the rocket nozzle weakened, and the thrust performance of the microwave rocket decreased. To avoid the high-speed propagation of the plasma front at a lower pressure, an external magnetic field was applied to the discharge domain. When an external magnetic field of 1 T is applied to the discharge domain, the Hall parameter becomes sufficiently high at a pressure of less than 0.01 atm. At a high Hall parameter, the electron diffusion coefficient can be described as

$$D_{\perp} = \frac{D_e}{\sqrt{1 + \omega_c^2 / \nu_e^2}}, \tag{6.47}$$

where $D_{\perp}$ is the electron diffusion coefficient under a magnetic field, ω_c is the electron cyclotron frequency, and ν_e is the elastic collision frequency between electrons and heavy particles. The PIC-MCC simulation with a magnetic field indicated that the propagation speed of the plasma front decreased at a lower pressure. A theoretical model of the plasma-front propagation speed was also modified into $2\sqrt{D_\perp \nu_i}$ using the electron diffusion coefficient under a magnetic field, which was in agreement with the PIC-MCC simulation result. This result indicates that the gas-heating region at the plasma front can be confined inside the microwave rocket nozzle. Moreover, in terms of the enhancement of the energy absorption rate, electron cyclotron resonance (ECR), which occurs when the electron cyclotron frequency is in agreement with the incident microwave frequency, can be utilized at a lower pressure. Therefore, the magnetic field required for the ECR is changed depending on the incident microwave frequency and is 3.93 T for the breakdown induced by the 110-GHz microwave. The second-harmonic ECR condition was also satisfied by using a magnetic field of 1.97 T. The PIC-MCC simulation with these magnetic fields revealed that the energy absorption rates drastically increased at lower pressures.

By combining the enhancement of the energy absorption rate with the suppression of the plasma-front propagation, it is considered that the thrust performance of the microwave rocket is improved. The fluid dynamics induced by the microwave discharge were examined using computational fluid dynamics (CFD) simulations. The energy absorption rate at the plasma front was evaluated using the PIC-MCC, which was introduced in the heating source term in the CFD model. In the CFD simulation conducted by Takahashi and Ohnishi, the 2D axisymmetric Euler equation was numerically integrated to capture the shock wave propagation inside the rocket nozzle as follows:

$$\frac{\partial \tilde{\mathbf{Q}}}{\partial t} + \frac{\partial \tilde{\mathbf{E}}}{\partial x} + \frac{\partial \tilde{\mathbf{F}}}{\partial r} + \tilde{\mathbf{H}} = \tilde{\mathbf{S}}, \tag{6.48}$$

where $\tilde{\mathbf{Q}}$ is the conservation vector; $\tilde{\mathbf{E}}$ and $\tilde{\mathbf{F}}$ are the inviscid flux vectors in the axial and radial directions, $\tilde{\mathbf{H}}$ is the axisymmetric term; $\tilde{\mathbf{S}}$ is the energy source term evaluated by the PIC-MCC simulation. The CFD simulation indicated that a strong shock wave was induced inside the rocket nozzle when the ambient pressure was 1 atm because sufficient gas heating occurred by the microwave plasma. A high impulse was obtained at atmospheric pressure because the shock wave interacted strongly with the thruster wall. However, the shock wave weakened with a decrease in the ambient pressure because the energy absorption rate by the plasma decreased and the plasma-front propagation speed increased. This induced a decrease in the performance of the microwave rocket. As a countermeasure to the performance decrease, an external magnetic field was applied to the rocket nozzle. When ECR and plasma-front confinement were obtained by the magnetic field, the shock wave induced inside the nozzle became stronger, even at a lower pressure. The thrust performance at lower pressure was then improved by an external magnetic field.

Modeling for Subcritical Discharge

The microwave breakdown mechanism differed depending on the incident beam intensity. In particular, the electron production process drastically changed at the breakdown threshold in the air (approximately 2.6 MV/m). If only electron-impact ionization is considered at less than the breakdown threshold (subcritical condition), electron production cannot be maintained because the electron attachment process is dominant. Therefore, the plasma model proposed by Boeuf et al. could not reproduce the plasma-front propagation process under subcritical conditions.

Takahashi et al. conducted a 2D plasma-FDTD simulation to reproduce the breakdown process under subcritical conditions by introducing the Euler equation in the Cartesian coordinate system and considering the neutral gas expansion process (Takahashi et al. 2017). The TM electromagnetic wave propagation was numerically reproduced using the FDTD method in the 2D domain. The electron diffusion equation, which has the effective diffusion coefficient proposed by Boeuf et al., was numerically integrated to capture plasma diffusion dynamics. However, when only the electron-impact ionization process was considered, as in Boeuf's study, the electron production was not maintained at the subcritical condition; therefore, the neutral gas expansion was coupled with the plasma fluid model. When the gas expansion is considered, the transport coefficient of the electron can be a function of the reduced electric field E/N, where E is the local electric field intensity and N is the neutral gas density, because the electron energy and diffusivity are affected by the elastic collision process with the neutral particles in the atmospheric gas discharge. If sufficient gas expansion occurs in the ionization region, the electron-impact ionization frequency can be enhanced because of an increase in E/N, which contributes to electron production under subcritical conditions. In the Takahashi's model, a data map of the transport coefficient was constructed as a function of E_{eff}/N, where E_{eff} is the effective electric field intensity. The local reduced electric field was then calculated as E_{eff}/N by solving Euler and Maxwell's equations, which were used to evaluate the transport coefficients through interpolation of the data map. After updating the electron number density, the current density was obtained by solving the equations of motion for electrons. This current density was fed back into Maxwell's equations to reproduce the interaction between the wave and plasma. Moreover, the Joule heating term S_f was evaluated for one cycle of the incident microwave as follows:

$$S_f = \int \xi \mathbf{J} \cdot \mathbf{E} dt, \tag{6.49}$$

where ξ is the energy conversion rate from electrons to neutral gas. The Joule heating term S_f was substituted into the source term of the Euler equation to drive the shock and expansion waves.

When the coupling simulation had the plasma fluid, electromagnetic wave, and neutral gas dynamics modules, electron production was successfully maintained at the subcritical condition, similar to the previous experiment. Their simulation indicated that the mechanisms of electron production and plasma-front propagation differed depending on the incident beam intensity. When the incident electric field was lower than the breakdown threshold but close to the threshold intensity, a strong gas expansion occurred at the plasma front, and the conductivity of the plasma increased. Therefore, a high electric field region, whose intensity was higher than the breakdown threshold, was created in front of the bulk plasma because a strong wave reflection occurred owing to the high plasma conductivity. The seed electrons diffused from the bulk plasma to the high electric field region. The next plasma filament was then generated because electron-impact ionization was induced using the seed electrons diffused from the bulk region. However, the high electric field region, whose intensity was higher than the threshold, was not created when the incident electric field intensity was decreased, and another breakdown mechanism was required to maintain the electron production. At a much lower electric field intensity, strong gas expansion occurred because of Joule heating. In front of the bulk plasma, the reduced electric field intensity E_{eff}/N became higher than the breakdown threshold when the expansion wave propagated toward the incident beam source. The plasma front propagated toward the incident beam source because of the high E_{eff}/N-induced electron-impact ionization.

Numerical Analysis of "Comb-Shaped" Structures

As already mentioned, the plasma structure (Fig. 6.1) was observed in the ionization front propagating in an under-critical millimeter-wave beam (Yamaguchi et al. 2016; Fukunari et al. 2019). This structure was mainly observed on the plane containing the electric-field vector $\boldsymbol{E}$ and wave number vector $\boldsymbol{k}$ ($\boldsymbol{E}$–$\boldsymbol{k}$ plane). The plasmoids with a size of approximately 0.5λ were lined up with the same pitches of 0.9λ on the plane perpendicular to $\boldsymbol{k}$, where λ is the wavelength of the incident millimeter wave. The plasmoids propagated toward the beam source maintaining their distance from each other. Because the trajectory of a plasmoid has a shape like a comb, this structure is designated as a "comb-shaped" structure. This structure indicates the bias in the energy absorption and some energy concentration around the plasmoids. The propagation velocity of the ionization front in an under-critical condition is significantly fast that it cannot be explained with conventional propagation models. Such energy concentration may play an important role in fast propagation. In this section, a numerical investigation is conducted to reveal the condition in which the structure is formed. Furthermore, the effect of the structure on the ionization front is explained.

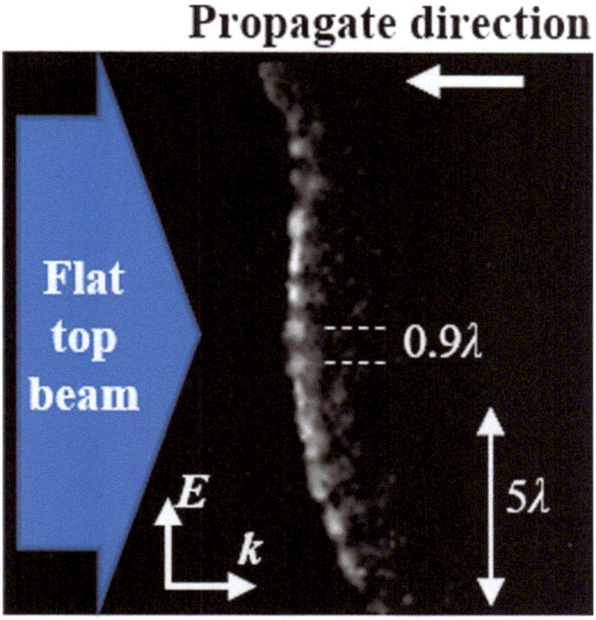

Fig. 6.1 High-speed camera image of ionization front with "comb-shaped" structure (Yamaguchi et al. 2016; Nakamura et al. 2018)

Numerical Reproduction and Structure Forming Mechanism

This structure was numerically reproduced by two-dimensional computations, and the forming mechanism has been already discussed in (Nakamura et al. 2018). Although the basic equations used for these computations are almost the same as those used in the computations for supercritical discharge (Kourtzanidis et al. 2014; Boeuf et al. 2010), two tuning parameters—α and β—were employed to the diffusion equation for electrons. α is a parameter that enhances the electric-field amplitude and β is a parameter that modifies the diffusion coefficient standardized by the ambipolar diffusion coefficient D_a. The diffusion equation used in this computation can be described as follows: where the ionization frequency ν_i and recombination coefficient r are functions of α times the root-mean-square of the electric field E_{rms}.

$$\frac{\partial n_e}{\partial t} - \beta D_a \nabla^2 n_e = \nu_i(\alpha E_{rms}) n_e - r(\alpha E_{rms}) {n_e}^2 \tag{6.50}$$

By controlling parameters α and β, the ionization-front propagation velocity and electron number density can be independently changed. Two-dimensional computations on the ***E–k*** plane were conducted for an incident beam amplitude of 1 MV/m. The parameters for reproducing the experimental value of 1 km/s propagation velocity (Oda et al. 2006) are listed in Table 6.1. The propagating ionization fronts computed using these sets of parameters are shown in Fig. 6.2. In the computations with $\alpha = 3.0$ and 3.3, the comb-shaped structure gradually appeared in the ionization front during a few microseconds of propagation from the initial

Table 6.1 Parameters reproducing experimental propagation velocity of 1 km/s, maximum electron number densities and plasma pitches in ionization front computed with the parameters (Nakamura et al. 2018)

α	β	Max. $n_e/(10^{21}\ \text{m}^{-3})$	Pitch/λ
2.7	1.23×10^2	0.17	No structure
3.0	2.94×10^1	0.98	0.85
3.3	1.06×10^1	1.85	0.89

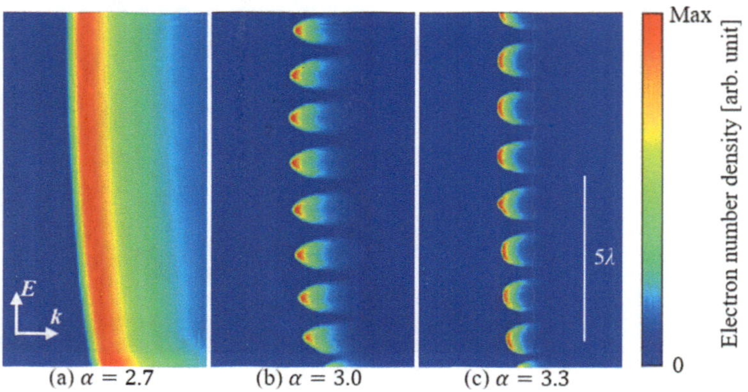

Fig. 6.2 Electron number-density distributions around propagating ionization fronts corresponded to the sets of parameters in Table 6.1 (Nakamura et al. 2018)

plasmoid. The pitch of the plasmoids was 0.85λ and 0.89λ, respectively, which is in agreement with that of experimentally observed plasmoids. However, this structure was not observed in the propagating plasmoid with $\alpha = 2.7$.

The maximum electron number density of each ionization front is listed in Table 6.1. The structure appeared when the maximum electron number density was higher than the cutoff density of collisional plasma, described in Eq. (6.51), using the frequency of the incident millimeter-wave f_{MMW} and momentum transfer frequency ν_{m}.

$$n_{\text{c}} = \frac{m_{\text{e}}\varepsilon_0 f_{\text{MMW}}\nu_{\text{m}}}{e^2} = 0.25 \times 10^{21}/\text{m}^3 \tag{6.51}$$

When the electron number density is higher than the cutoff density n_{c}, a part of the incident millimeter wave is reflected in the plasma. Therefore, the structure was formed owing to the interference between the incident and reflected millimeter waves. Fig. 6.3(a) shows the computed electric-field amplitude around a plasmoid with a uniform electron number density of $2.0 \times 10^{21}/\text{m}^3$ and a diameter of 0.4λ irradiated with a uniform milliner-wave beam coming from the left side. The fringes in front of the plasmoid were standing waves. Figure 6.3(b) shows the amplitude of the broken line through the center of the plasmoid in Fig. 6.3(a). The electric field reaches its bottom at approximately 0.35λ from the center and reaches its peak at

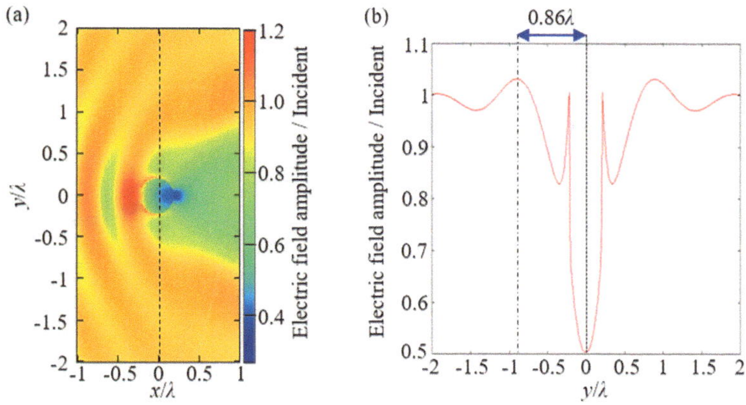

Fig. 6.3 (**a**) Electric-field amplitude around plasmoid with uniform milliner-wave beam coming from the left side computed with two-dimensional computation on the ***E–k*** plane. (**b**) Distribution on broken line on (**a**). The plasmoid has a uniform electron number density of $2.0 \times 10^{21}/m^3$ and a diameter of 0.4λ (Nakamura et al. 2018)

0.86λ from the center, which is illustrated in Fig. 6.3(b). The extension of the plasma along the electric field in the millimeter wave stops when the length along the electric field reaches approximately 0.5 because the amplitude decrement corresponds to the bottom at 0.35λ (Oda et al. 2006). This may be the reason of dividing the plasma into pieces. Because the neighborhood plasmoids reach the amplitude-enhanced position at 0.86λ, the plasmoids line up in the same pitches.

Effect of the Structure on Ionization-Front Propagation

The distributions of the electric-field amplitude and electron number density along the line parallel to the beam and through the center of one of the plasmoids in Fig. 6.2 (c) are shown in Fig. 6.4. Compared with the distributions obtained from the one-dimensional computation shown in the same figure, the peak value of the electric-field amplitude, which is the nearest peak of the standing wave to the ionization front, was only a few percentages larger. This caused only a small percentage difference in the propagation velocity. However, the peak value of the electron number density in the two-dimensional computation was almost twice as significant as that in the one-dimensional computation. This is because the electric-field amplitude in the two-dimensional computation maintained a higher value in the ionization front than that in the one-dimensional computation. Figure 6.5 shows the distribution of the power density absorbed by the plasma and the trajectory of the energy flow around one plasmoid, as shown in Fig. 6.2(c). The energy concentration shown in this figure causes the difference between one-dimensional and two-dimensional results shown in Fig. 6.4.

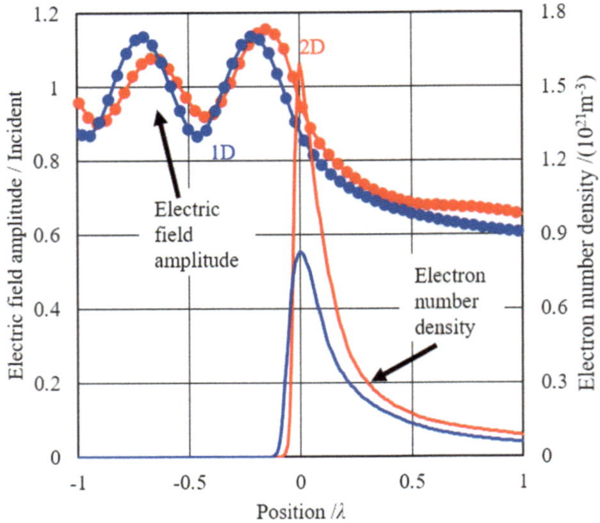

Fig. 6.4 Electric-field amplitude and electron number density along the line parallel to the incident beam through the center of one plasmoid in two-dimensional computation (red line) and one-dimensional computation (blue line). The parameter set of $\alpha = 3.3$ is used

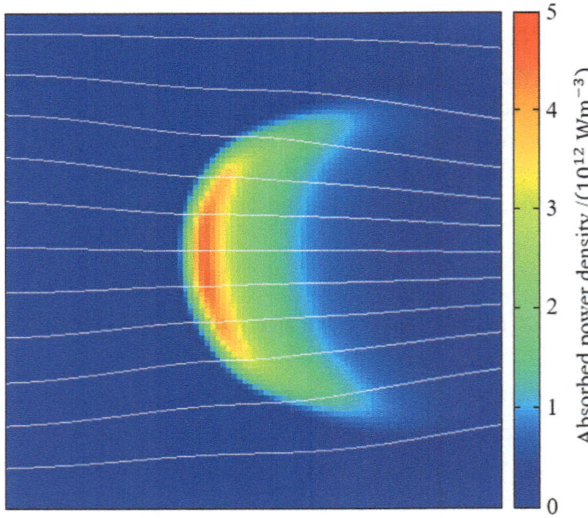

Fig. 6.5 Distribution of power density absorbed by the plasma around the single plasmoid in Fig. 6.2(c) and a trajectory of the energy flow (white line)

One-Dimensional Consideration of Ionization-Front Propagation in Under-Critical Condition

The discussion in the previous section revealed that the effect of the concentration of the beam energy caused by the structure is not sufficiently significant to explain the propagation velocity in the under-critical condition. This suggests that a one-dimensional discussion is required to reveal the propagation process. According to the steady propagation solution of the diffusion equation, the propagation velocity of the ionization front is determined by the peak value of $2\sqrt{Df_{\mathrm{i}}}$, where D and f_{i} are

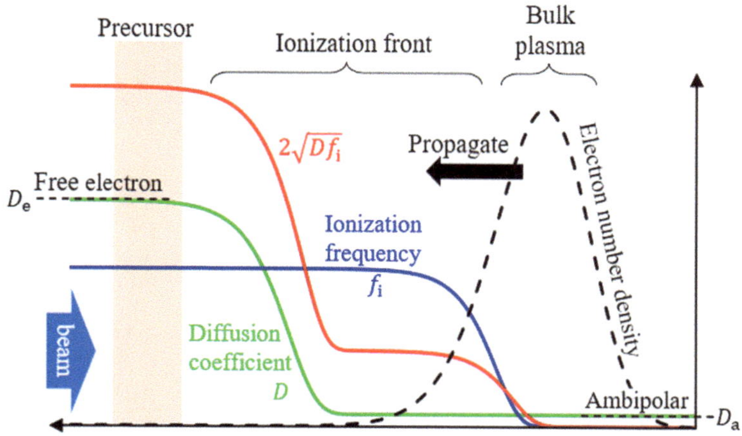

Fig. 6.6 Distributions of parameters around ionization front in supercritical condition

the diffusion coefficient of electrons and net ionization frequency, respectively (Ebert et al. 1997b).

First, the propagation under supercritical conditions is discussed. Figure 6.6 shows schematics of the distributions of the diffusion coefficient D, ionization frequency f_i, and $2\sqrt{Df_i}$ around the propagating ionization front under supercritical conditions. In this figure, the oscillations of these parameters caused by the standing wave with the interference between the incident and reflected waves are ignored. The diffusion coefficient D at a position sufficiently far from the ionization front, where the electron number density is significantly small, takes the value of the free-electron diffusion D_e. With an increase in electron number density, the diffusion coefficient decreases until it reaches the ambipolar diffusion coefficient D_a (Boeuf et al. 2010). Using a conventional ionization model for supercritical conditions (Kourtzanidis et al. 2014; Boeuf et al. 2010), the ionization frequency is determined only by the beam intensity. Therefore, the ionization frequency remained the same along the beam until the beam intensity attenuated with absorption via the plasma. Because of these parameter distributions, $2\sqrt{Df_i}$ has its peak value in the region sufficiently far from the ionization front, at which the diffusion coefficient is D_e. That is, this region determines the propagation velocity of the ionization front. This region will be referred to as "precursor" in the rest of this section.

Next, the propagation in the under-critical condition is discussed. Under this condition, the net ionization frequency at the region sufficiently far from the ionization front should be zero because of the definition of "under-critical." Therefore, some effect-enhancing ionization should spread out from the plasma or heated gas. Depending on the spreading of this enhancing effect, two cases can be considered. In the first case, the ionization-enhanced region is narrow compared with the scale of the electron number-density slope, as shown in Fig. 6.7(a). In this case, the propagation velocity was determined at the ionization-front slope, where the diffusion coefficient was ambipolar D_a. We refer to this case as "front driven." In the

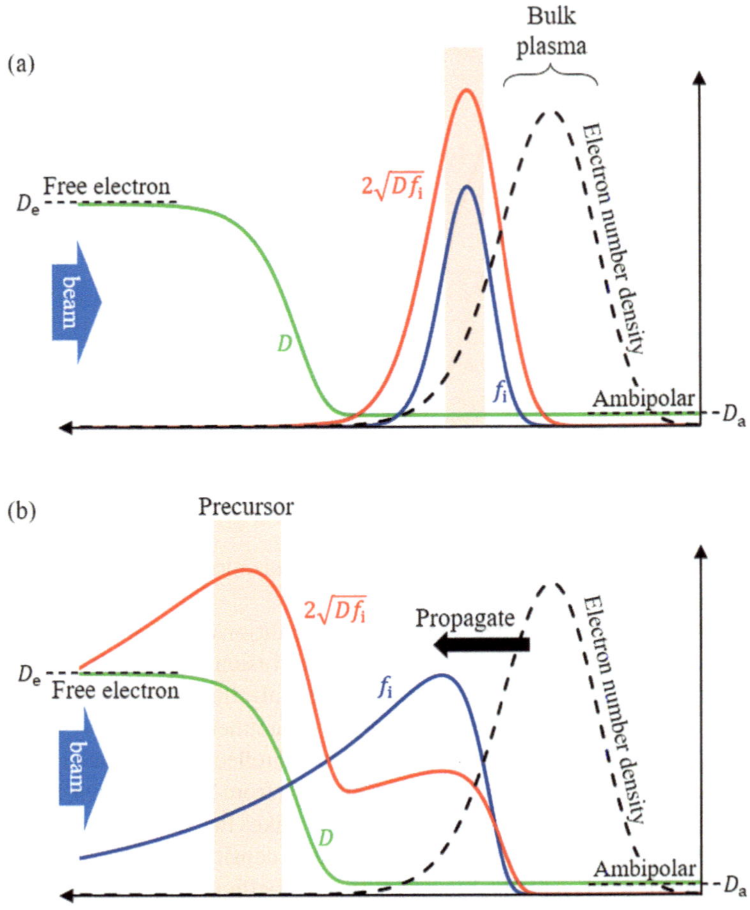

Fig. 6.7 Distributions of parameters around ionization front in under-critical conditions. (**a**) "Front-driven" case and (**b**) "precursor-driven" case

second case, the enhanced region is sufficiently wide and maintains a sufficient ionization frequency at the edge of the ionization front, as shown in Fig. 6.7(b). In this case, the propagation velocity is determined at the precursor, at which the diffusion coefficient is close to that of the free electrons D_e. We refer to this case as "precursor driven."

Which is the appropriate case for ionization-front propagation in the under-critical conditions? In the results of the atmospheric-air discharge experiment using a millimeter-wave beam with a frequency of 170 GHz, the ionization-front propagation velocity U_{ion} was 1 km/s for an incident power density of 2.5 GW/m^2 (Oda et al. 2006). Therefore, under this condition, the peak value of $2\sqrt{Df_i}$ should also be approximately 1 km/s. First, "front-driven" propagation was considered. In this case, the peak value is described as $2\sqrt{D_a f_{i,peak}}$, whereas $f_{i,\ peak}$ is the net ionization frequency at which $2\sqrt{Df_i}$ reaches its peak. The ambipolar diffusion coefficient D_a in the air plasma with an electron temperature of 1 eV was approximately 1×10^{-3} m^2/s. When this value is assumed as D_a, the ionization frequency $\nu_{i,\text{peak}}$ to realize the propagation velocity of 1 km/s with "front-driven" propagation is around 3×10^8 Hz. With this ionization frequency, a scale length of the ionization-front slope estimated from U_{ion}/f_i is around a few micrometers, which is significantly thin compared with the wavelength of the incident beam (1.76 mm). The energy from one electron to collisional ionization per unit time is $\varepsilon_i \nu_i \cong 5 \times 10^9$ eV/s when $\varepsilon_i = 15.6$ eV, the ionization energy of a nitrogen molecule. The energy absorbed by one electron with inverse bremsstrahlung is obtained as follows for a beam intensity of $S_{MMW} = 2.5$ GW/m^2, at which the electric-field amplitude E_0 is 1.4 MV/m:

$$\frac{e^2 E_0{}^2 \nu_m}{2m_e(\omega^2 + \nu_m{}^2)} = 6 \times 10^{10} \frac{\text{eV}}{\text{s}}$$

The momentum transfer frequency ν_m was 2×10^{12}/s. Therefore, the energy used for net ionization was approximately 8% of the absorbed energy. Assuming that this energy ratio can be applied to the total energy ratio, the maximum electron number density can be derived using the following equation:

$$\frac{0.08 \times S_{MMW}}{U_{ion}\varepsilon_i} = 8 \times 10^{22}/\text{m}^3$$

This value is approximately 350 times higher than the cutoff density of collisional plasma $n_c = 2.5 \times 10^{20}$/m^3. If the electron number density increases more to a high density in a very thin region than the wavelength, most of the beam energy should be reflected at the surface of the plasma. However, the experimental results suggest that such energy reflection does not occur. On the other hand, in precursor-driven propagation, the ionization frequency $f_{i,\text{ peak}}$ to realize the experimental propagation velocity is approximately 10^6/s because the diffusion coefficient at the precursor is close to the free-electron diffusion coefficient $D_e \cong 1 \times 10^{-1}$ m^2/s. This ionization frequency is more realistic than that desired for front-driven propagation. This result suggests that the precursor driven is more appropriate for the ionization-front propagation process in under-critical millimeter waves.

A decrease in the neutral particle density and an increase in excited-neutral particles can be considered to enhance the ionization frequency. In precursor-driven propagation, the effect of such factors should reach to the precursor region from bulk plasma. In the experiment, it was observed that the ionization front propagated at a speed exceeding the Chapman-Jouguet velocity, suggesting that the propagation velocity exceeded the acoustic velocity in the plasma region. Under this condition, the expansion wave generated in the heating region behind the bulk plasma cannot overtake the ionization front, and therefore, it cannot reach the precursor. For the same reason, it is difficult for the excited-neutral particles generated in the bulk plasma to reach the precursor via thermal diffusion. Therefore, it is considered that excited particles (or something else) are created in the precursor owing to certain influence.

Ionization-Front Propagation with Excited Neutral Particles Carried via Photons

As discussed in the previous section, mechanisms that increase the ionization frequency at the precursor are required to explain the ionization-front propagation in an under-critical millimeter wave. Therefore, a model considering excited-neutral particles carried via photons was proposed, and one-dimensional numerical computation was used for validation (Nakamura and Komurasaki 2020). With this computation, the ionization front propagating with a velocity of 50% of the experimental velocity was reproduced when the quenching frequency was lower than 10^6/s. However, this mechanism has not been experimentally proven, and further research is required. The propagation model is described in this section.

Transportation of Exited Neutral Particles via Photons

A high-energy electronically excited-neutral particle can be ionized by collision with an electron more easily in the ground state than in other states. Therefore, the ionization frequency of the precursor can be enhanced if a sufficiently high number of excited particles are present. However, excited-neutral particles cannot be carried with thermal diffusion to the precursor, as explained in the previous section; hence, alternative mechanism is required for this purpose. Figure 6.8 shows the schematics of the excited-neutral particle transfer process proposed in (Nakamura and Komurasaki 2020). The high-energy electronically excited-neutral species created in the bulk plasma quickly decay to the ground state, creating a photon with the same energy as the excitation energy. The timescale of the decay was of the order of 10^{-10} s (Wu et al. 2007). The created photon was reabsorbed by a neutral particle in the ground state after propagating a short distance with a mean free pass of approximately 10^{-5} m (Huffman et al. 1963), recreating an excited-neutral particle. Repeating this radiative decay and reabsorption process, the excited particles gradually spread to the precursor region. The ionization caused by the collisions between the excited particles reaching the precursor and electrons plays a key role in ionization-front propagation.

Diffusion Approximation of Transportation of Electronically Excited Particle via Photons

When the mean free pass of the photon is shorter than the scale length of the ionization front, the transportation of excited particles in a single state with the mechanisms explained in the previous subsection can be approximated to a diffusion equation. This diffusion equation is derived in this subsection. The number of

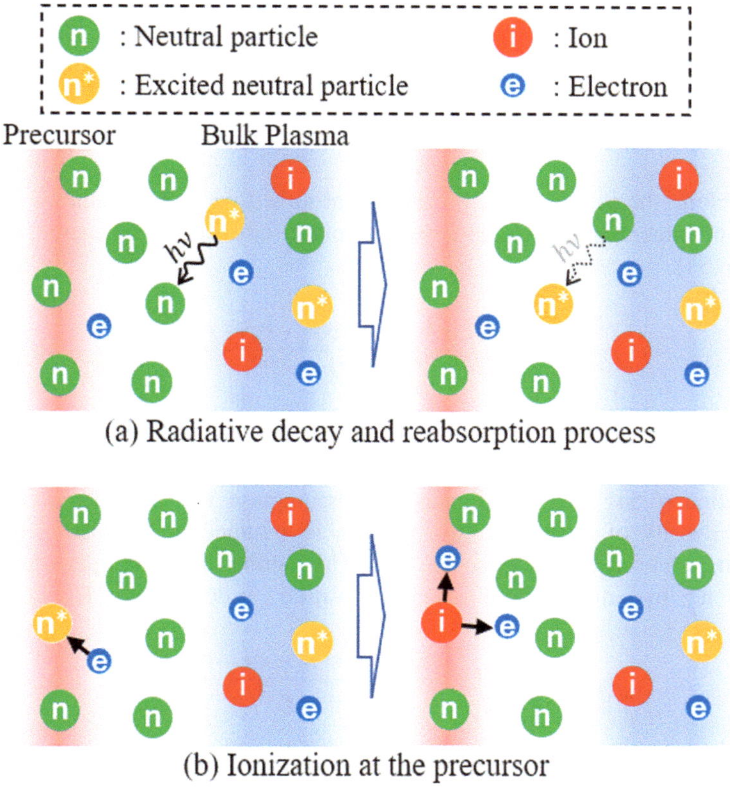

Fig. 6.8 Schematics of (**a**) transportation of excited-neutral particle via photon and (**b**) ionization caused by collisions between excited-neutral particles and electrons at the precursor (Nakamura and Komurasaki 2020)

photons radiating from the unit volume at an arbitrary position is described as $A_{\mathrm{Ei}} n_{\mathrm{n}}^{*}(\boldsymbol{r})$ using the Einstein coefficient A_{Ei} and number density of the excited particles $n_{\mathrm{n}}^{*}(\boldsymbol{r})$. If the radiation is isotropic, the number of photons radiated from the unit volume at $\boldsymbol{r}$ in an arbitrary direction (θ, φ) per unit solid angle is described by Eq. (6.52).

$$R_{\theta,\varphi}(\boldsymbol{r}) = \frac{A_{\mathrm{Ei}} n_{\mathrm{n}}^{*}(\boldsymbol{r})}{4\pi} \tag{6.52}$$

The number of photons decay from $A_{\mathrm{Ei}} n_{\mathrm{n}}^{*}(\boldsymbol{r})$ to $A_{\mathrm{Ei}} n_{\mathrm{n}}^{*}(\boldsymbol{r}) \exp(-\kappa l)$ after propagating a distance l because of absorption where κ is the absorption coefficent. The number of photons radiating from the unit volume at $\boldsymbol{r}$ and absorbed at $\boldsymbol{r}'$ is given by Eq. (6.53).

$$s(\boldsymbol{r},\boldsymbol{r}') = -\frac{1}{4\pi(\boldsymbol{r}-\boldsymbol{r}')^2}\frac{d}{dl_{\theta,\varphi}}\left\{A_{\mathrm{Ei}}n_{\mathrm{n}}^{*}(\boldsymbol{r})\exp(-\kappa|\boldsymbol{r}-\boldsymbol{r}'|)\right\}$$
$$= \frac{\kappa A_{\mathrm{Ei}}n_{\mathrm{n}}^{*}(\boldsymbol{r})\exp(-\kappa|\boldsymbol{r}-\boldsymbol{r}'|)}{4\pi(\boldsymbol{r}-\boldsymbol{r}')^2} \tag{6.53}$$

Therefore, the total number of photons absorbed per unit volume at $\boldsymbol{r}'$ can be derived by integrating Eq. (6.53) for every volume of the radiation source, as shown in Eq. (6.54).

$$S(\boldsymbol{r}') = \iiint \frac{\kappa A_{\mathrm{Ei}}n_{\mathrm{n}}^{*}(\boldsymbol{r})\exp(-\kappa|\boldsymbol{r}-\boldsymbol{r}'|)}{4\pi(\boldsymbol{r}-\boldsymbol{r}')^2}dV \tag{6.54}$$

This is same as the number of excited particles created by the reabsorption of photons, assuming that all the photons absorbed are used for it.

Equation (6.54) in a one-dimensional assumption was considered. Here, the number density of excited-neutral particles is assumed to be distributed in the direction z and can be described as $n_{\mathrm{n}}^{*}(z)$. The number of excited particles created at position z' by radiation from a plane at position z is described in Eq. (6.55).

$$s_{\mathrm{1D}}(z,z') = \iint \frac{\kappa A_{\mathrm{Ei}}n_{\mathrm{n}}^{*}(\mathrm{z})\exp(-\kappa|\boldsymbol{r}-\boldsymbol{r}'|)}{4\pi(\boldsymbol{r}-\boldsymbol{r}')^2}dxdy \tag{6.55}$$

This equation can be rewritten as Eq. (6.56), using $l_{z,\,z}$ and r defined as $l_{z,\,z'} \equiv |z - z'|$ and $r^2 \equiv (\boldsymbol{r} - \boldsymbol{r}')^2 - l_{z,\,z'}{}^2$, respectively.

$$s_{\mathrm{1D}}(z,z') = \int_0^{\infty} \frac{\kappa A_{\mathrm{Ei}}n_{\mathrm{n}}^{*}(\mathrm{z})\exp\left(-\kappa\sqrt{r^2+l_{z,z'}{}^2}\right)}{4\pi\left(r^2+l_{z,z'}{}^2\right)}2\pi r dr \tag{6.56}$$

Figure 6.9 shows the schematics of the photon propagating from a plane at z toward point z'.

Then, the total number of excited particles created by photon absorption per unit volume at position z' is expressed as Eq. (6.57).

$$S_{\mathrm{1D}}(z') = \kappa A_{\mathrm{Ei}}\int_{z=-\infty}^{z=\infty} n_{\mathrm{n}}^{*}(\mathrm{z})\int_{r=0}^{r=\infty}\frac{\exp\left(-\kappa\sqrt{r^2+l_{z,z'}{}^2}\right)}{2\left(r^2+l_{z,z'}{}^2\right)}rdrdz \tag{6.57}$$

Although the integration is conducted over the entire range of z in this equation, the effective range is only a few times larger than $1/\kappa$ from z' because the number of photons decays with a scale length of $1/\kappa$ during propagation. When the scale length

Fig. 6.9 Schematics of photon propagating from a plane at z toward point z'

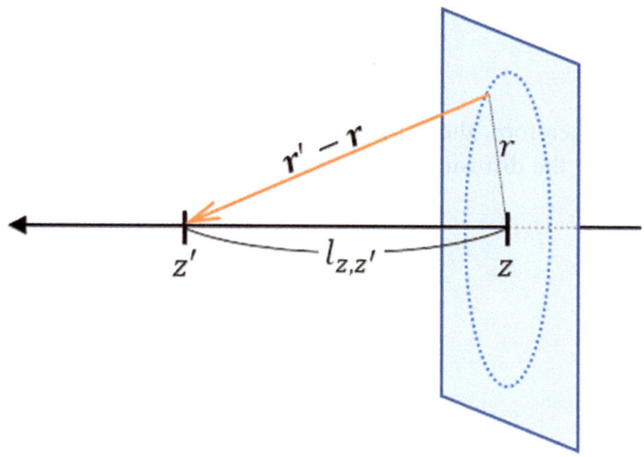

l_p of a distribution of n_n^* is sufficiently small compared with $1/\kappa$, n_n^* can be assumed as Eq. (6.58) using the Taylor expansion.

$$n_n^*(z) = n_n^*(z') + \left.\frac{dn_n^*}{dz}\right|_{z'}(z - z') + \frac{1}{2}\left.\frac{d^2 n_n^*}{dz^2}\right|_{z'}(z - z')^2 + O\left(\left(\frac{z - z'}{l_p}\right)^3\right) \tag{6.58}$$

Equation (6.59) is obtained by substituting this equation into Eq. (6.57), by neglecting terms more than the third order.

$$\begin{aligned}
S_{1D}(z') &= \kappa A_{Ei} n_n^*(z') \int \int_{r=0}^{r=\infty} \frac{\exp\left(-\kappa\sqrt{r^2 + l_{z,z'}{}^2}\right)}{2\left(r^2 + l_{z,z'}{}^2\right)} r dr dz \\
&+ \kappa A_{Ei} \frac{dn_n^*}{dz} \int (z - z') \int_{r=0}^{r=\infty} \frac{\exp\left(-\kappa\sqrt{r^2 + l_{z,z'}{}^2}\right)}{2\left(r^2 + l_{z,z'}{}^2\right)} r dr dz \\
&+ \frac{\kappa A_{Ei}}{2} \frac{d^2 n_n^*}{dz^2} \int (z - z')^2 \int_{r=0}^{r=\infty} \frac{\exp\left(-\kappa\sqrt{r^2 + l_{z,z'}{}^2}\right)}{2\left(r^2 + l_{z,z'}{}^2\right)} r dr dz = A_{Ei} n_n^*(z') \\
&+ \frac{A_{Ei}}{3\kappa^2} \frac{d^2 n_n^*}{dz^2}
\end{aligned} \tag{6.59}$$

The time derivative of n_n^*, treated as a function of time t and position z, can be derived as Eq. (6.60) considering the loss by radiative decay $A_{Ei} n_n^*$ and increment by collisional excitation S_{ex}.

$$\frac{\partial n_{\mathrm{n}}^{*}}{\partial t} = S_{\mathrm{1D}} - A_{n} n_{\mathrm{n}}^{*} + S_{\mathrm{ex}} = \frac{A_{\mathrm{Ei}}}{3\kappa^{2}} \frac{\partial^{2} n_{\mathrm{n}}^{*}}{\partial z^{2}} + S_{\mathrm{ex}} \tag{6.60}$$

Therefore, the following diffusion forms (Eqs. 6.61 and 6.62) can be derived using the diffusion coefficient D_{ex}.

$$\frac{\partial n_{\mathrm{n}}^{*}}{\partial t} - D_{\mathrm{ex}} \frac{\partial^{2} n_{\mathrm{n}}^{*}}{\partial z^{2}} = S_{\mathrm{ex}} \tag{6.61}$$

$$D_{\mathrm{ex}} \equiv \frac{A_{\mathrm{Ei}}}{3\kappa^{2}} \tag{6.62}$$

A three-dimensional equation can also be derived by substituting the Taylor-expanded n_{n}^{*} into Eq. (6.54).

$$\frac{\partial n_{\mathrm{n}}^{*}}{\partial t} - D_{\mathrm{ex}} \nabla^{2} n_{\mathrm{n}}^{*} = S_{\mathrm{ex}} \tag{6.63}$$

A_{Ei} and κ differed according to the excitation species. Typical values of these parameters for excited-neutral nitrogen molecules with excitation energies higher than 13 eV are $A_{\mathrm{Ei}} \sim 10^{10}$/s and $\kappa \sim 10^{5}$/m (Wu et al. 2007; Huffman et al. 1963). Such high-energy excited particles are essential because they can ionize efficiently by collision. Therefore, the typical value of D_{ex} for these excited species is of the order of 1 m^2/s.

One-Dimensional Model with Simple Assumption

The model used for the one-dimensional computation in (Nakamura and Komurasaki 2020) is explained in this subsection. The following five assumptions are made in this model:

1. Nitrogen molecule is the only component of air.
2. Motions of neutral particles are negligibly small compared to the ionization front.
3. Energy distribution of electrons obeys Boltzmann distribution with a temperature T_{e}.
4. The energy distribution of the vibrational excitation obeys a Boltzmann distribution with temperature T_{v} and can be described in Eq. (6.64) using a number density of all neutral particles n_{n}, number density of neutral particles with vibrational excitation level i, denoted as $n_{\mathrm{n},i}^{\mathrm{v}}$, and the excitation energy $\epsilon_{i}^{\mathrm{v}}$.

$$n_{\mathrm{n},i}^{\mathrm{v}} = \frac{n_{\mathrm{n}}}{Z_{\mathrm{v}}} \exp\left(-\frac{\epsilon_i^{\mathrm{v}}}{T_{\mathrm{v}}}\right) Z_{\mathrm{v}} \equiv \sum_i \exp\left(-\frac{\epsilon_i^{\mathrm{v}}}{T_{\mathrm{v}}}\right) \tag{6.64}$$

5. The energy distribution of electronic excitation, except the ground state, obeys the Boltzmann equation with the electron temperature at the bulk plasma $(T_{\mathrm{e}})_{\mathrm{B}}$ because most of the excitation particles are created in bulk. It is described in Eq. (6.65) using a number density of the electronically excited particle with an excitation level i, denoted as $n_{\mathrm{n},i}^{*}$, and excitation energy ϵ_i.

$$n_{\mathrm{n},i}^{*} = \frac{n_{\mathrm{n}}^{*}}{Z} \exp\left(-\frac{\epsilon_i}{(T_{\mathrm{e}})_{\mathrm{B}}}\right) Z_{\mathrm{ex}} \equiv \sum_{i \neq 0} \exp\left(-\frac{\epsilon_i}{(T_{\mathrm{e}})_{\mathrm{B}}}\right) \tag{6.65}$$

The total number of excited particles n_{n}^{*} is defined in Eq. (6.66).

$$n_{\mathrm{n}}^{*} \equiv \sum_{i \neq 0} n_{\mathrm{n},i}^{*} \tag{6.66}$$

The diffusion in Eq. (6.67), the same as that in Eq. (6.3), was used to obtain the electron number density:

$$\frac{\partial n_{\mathrm{e}}}{\partial t} - D_{\mathrm{eff}} \nabla^2 n_{\mathrm{e}} = f_{\mathrm{i}} n_{\mathrm{e}} \tag{6.67}$$

Net ionization frequency was determined by Eq. (6.68).

$$f_{\mathrm{i}} = n_{\mathrm{n},0}^{*} K_{0,\mathrm{ion}} + \sum_{i \neq 0} n_{\mathrm{n},i}^{*} K_{i,\mathrm{ion}} - n_{\mathrm{e}} r_{2\mathrm{B}} - {n_{\mathrm{e}}}^2 \sum_i r_i \tag{6.68}$$

The terms on the right-hand side describe ionization from the ground state, ionization from the excited state, two-body recombination, and three-body recombination, respectively.

The number densities of electronically excited particles are determined by Eqs. (6.65) and (6.69), which is the same as Eq. (6.61):

$$\frac{\partial n_{\mathrm{n}}^{*}}{\partial t} - D_{\mathrm{ex}} \nabla^2 n_{\mathrm{n}}^{*} = S_{\mathrm{ex}} \tag{6.69}$$

D_{ex} is the effective diffusion coefficient of the electronically excited particles. The source term S_{ex} is determined by Eq. (6.70).

$$S_{\text{ex}} = n_{\text{e}} n_{\text{n},0}^{*} \sum_{i \neq 0} K_{0,i} - n_{\text{e}} \sum_{i \neq 0} n_{\text{n},i}^{*} K_{i,0} - n_{\text{e}} \sum_{i} n_{\text{n},i}^{*} K_{n,\text{ion}} + {n_{\text{e}}}^{3} \sum_{i} r_{i} - \frac{n_{\text{n}}^{*}}{\tau_{\text{q}}} \quad (6.70)$$

The terms on the right-hand side describe excitation from the ground state, collisional deexcitation to the ground state, ionization from the excited state, recombination, and deexcitation with quenching, respectively.

The time evolution of the mean vibrational excitation energy ε_{v} was computed using the energy-conservation in Eq. (6.71).

$$n_{\text{n}} \frac{\partial \varepsilon_{\text{v}}}{\partial t} = n_{\text{e}} N_{0}^{\text{v}} \sum_{n \neq 0} \epsilon_{n}^{\text{v}} K_{0,n}^{\text{v}} - n_{\text{e}} \sum_{n \neq 0} N_{n}^{\text{v}} \epsilon_{n}^{\text{v}} K_{n,0}^{\text{v}} \quad (6.71)$$

The ionization, excitation, and deexcitation coefficients $K_{i,\ j}$ and $K_{i,j}^{\text{v}}$ and the recombination coefficient r_i are functions of the electron temperature. The electron temperature was determined by considering the residue of the reaction and absorbed energies, while the millimeter-wave energy absorbed by one electron per unit time with the inverse bremsstrahlung process was determined by Eq. (6.72).

$$s_{\text{IB}} = \frac{e^2 {E_0}^2 \nu_{\text{m}}}{2 m_{\text{e}} (\omega^2 + {\nu_{\text{m}}}^2)} \quad (6.72)$$

The energy-conservation Eq. (6.73) was used to obtain the electron temperature for electrons.

$$\begin{aligned} \frac{3}{2} \frac{\partial n_{\text{e}} k T_{\text{e}}}{\partial t} &= \frac{3}{2} D_{\text{en}} \nabla^2 (n_e k T_{\text{e}}) + s_{\text{IB}} n_{\text{e}} - n_{\text{e}} n_{\text{n},0}^{*} \sum_{i \neq 0} \epsilon_i K_{0,i} \\ &+ n_{\text{e}} \sum_{i \neq 0} n_{\text{n},i}^{*} \epsilon_i K_{i,0} - n_{\text{e}} \sum_{i} n_{\text{n},i}^{*} (\epsilon_{\text{ion}} - \epsilon_i) K_{i,\text{ion}} - \frac{3}{2} k T_{\text{e}} {n_{\text{e}}}^{2} r_{2\text{B}} \\ &+ {n_{\text{e}}}^{3} \sum_{i} (\epsilon_{\text{ion}} - \epsilon_i) r_i - n_{\text{e}} n_{\text{n},0}^{\text{v}} \sum_{i \neq 0} \epsilon_i^{\text{v}} K_{0,i}^{\text{v}} + n_{\text{e}} \sum_{i \neq 0} n_{\text{n},i}^{\text{v}} \epsilon_i^{\text{v}} K_{i,0}^{\text{v}} \\ &- n_{\text{e}} \sum_{i \neq 0} \sum_{j \neq 0} (\epsilon_i - \epsilon_j) n_{\text{n},i}^{\text{v}} K_{i,j} - \frac{m_{\text{e}}}{m_{\text{h}}} \frac{3}{2} n_{\text{e}} k T_{\text{e}} \nu_{\text{m}} \end{aligned} \quad (6.73)$$

This equation was combined with energy-conservation equation for electronical excitation (Eq. 6.74).

$$\frac{\partial}{\partial t}\sum_{i\neq 0} n_{\mathrm{n},i}^{*}\epsilon_i = D_{\mathrm{ex}}\nabla^2\sum_{i\neq 0} n_{\mathrm{n},i}^{*}\epsilon_i + n_{\mathrm{e}}\sum_{i\neq 0}\sum_{j\neq 0}\left(\epsilon_i-\epsilon_j\right)n_{\mathrm{n},i}^{*}K_{i,j}$$
$$+\left\{n_{\mathrm{e}}n_{\mathrm{n},0}^{*}\sum_{i\neq 0}\epsilon_i K_{0,i} - n_{\mathrm{e}}\sum_{i\neq 0} n_{\mathrm{n},i}^{*}\epsilon_i K_{i,0} - n_{\mathrm{e}}\sum_{i\neq 0} n_{\mathrm{n},i}^{*}\epsilon_i K_{i,\mathrm{ion}} + {n_{\mathrm{e}}}^{3}\sum_{i\neq 0}\epsilon_i r_i\right\}$$
$$-\frac{1}{\tau_{\mathrm{q}}}\sum_{i\neq 0} n_{\mathrm{n},i}^{*}\epsilon_i \tag{6.74}$$

where τ_{q} is the typical timescale of energy loss by quenching. The left-side term of Eq. (6.73) can be modified using Eq. (6.67) as follows:

$$\begin{aligned}\frac{\partial n_{\mathrm{e}}kT_{\mathrm{e}}}{\partial t} &= n_{\mathrm{e}}\frac{\partial kT_{\mathrm{e}}}{\partial t} + kT_{\mathrm{e}}\frac{\partial n_{\mathrm{e}}}{\partial t}\\ &= n_{\mathrm{e}}\frac{\partial kT_{\mathrm{e}}}{\partial t} + kT_{\mathrm{e}}\left\{f_{\mathrm{i}}n_{\mathrm{e}} + D_{\mathrm{eff}}\nabla^2 n_e\right\}\\ &= n_{\mathrm{e}}\frac{\partial kT_{\mathrm{e}}}{\partial t} + kT_{\mathrm{e}}\left\{n_{\mathrm{e}}\sum_{i} n_{\mathrm{n},i}^{*}K_{i,\mathrm{ion}} - {n_{\mathrm{e}}}^{2}r_{2\mathrm{B}} - {n_{\mathrm{e}}}^{3}\sum_{n} r_n + D_{\mathrm{eff}}\nabla^2 n_e\right\}\end{aligned} \tag{6.75}$$

Equation (6.74) can be rewritten as Eq. (6.76).

$$\begin{aligned}n_e\sum_{i\neq 0}\sum_{j\neq 0}\left(\varepsilon_i-\varepsilon_j\right)n_{\mathrm{n},i}^{*}K_{i,j} &= \frac{\partial}{\partial t}\sum_{i\neq 0} n_{\mathrm{n},i}^{*}\varepsilon_i - D_{\mathrm{ex}}\nabla^2\sum_{i\neq 0} n_{\mathrm{n},i}^{*}\varepsilon_i - \frac{1}{\tau_{\mathrm{q}}}\sum_{i\neq 0} n_{\mathrm{n},i}^{*}\varepsilon_i\\ &-\left\{n_{\mathrm{e}}n_{\mathrm{n},0}^{*}\sum_{i\neq 0}\varepsilon_i K_{0,i} - n_{\mathrm{e}}\sum_{i\neq 0} n_{\mathrm{n},i}^{*}\varepsilon_i K_{0,i} - n_{\mathrm{e}}\sum_{i\neq 0} n_{\mathrm{n},i}^{*}\varepsilon_i K_{i,\mathrm{ion}} + n_{\mathrm{e}}^{3}\sum_{i\neq 0}\varepsilon_i\gamma_i\right\}\end{aligned} \tag{6.76}$$

The first and second terms on the right side of this equation can be rewritten as in Eq. (6.77), assuming that the time derivative of T_{ex} is negligible.

$$\begin{aligned}\frac{\partial}{\partial t}\sum_{i\neq 0} n_{\mathrm{n},i}^{*}\varepsilon_i - D_{\mathrm{ex}}\nabla^2\sum_{i\neq 0} n_{\mathrm{n},i}^{*}\varepsilon_i &= \frac{1}{Z_{\mathrm{ex}}}\sum_{i\neq 0}\varepsilon_i\exp\left(-\frac{\varepsilon_i}{kT_{\mathrm{ex}}}\right)\frac{\partial}{\partial t}n_{\mathrm{n}}^{*}\\ -\frac{D_{\mathrm{ex}}}{Z}\sum_{i\neq 0}\varepsilon_i\exp\left(-\frac{\varepsilon_n}{kT_{\mathrm{ex}}}\right)\nabla^2 n_{\mathrm{n}}^{*} &= \frac{1}{Z_{\mathrm{ex}}}\sum_{i\neq 0}\varepsilon_i\exp\left(-\frac{\varepsilon_i}{kT_{\mathrm{ex}}}\right)\left(S_{\mathrm{ex}} + D_{\mathrm{ex}}\nabla^2 n_{\mathrm{n}}^{*}\right)\\ \frac{D_{\mathrm{ex}}}{Z}\sum_{i\neq 0}\varepsilon_i\exp\left(-\frac{\varepsilon_i}{kT_{\mathrm{ex}}}\right)\nabla^2 n_{\mathrm{n}}^{*} &= \frac{1}{Z_{\mathrm{ex}}}\sum_{i\neq 0}\varepsilon_i\exp\left(-\frac{\varepsilon_i}{kT_{\mathrm{ex}}}\right)S_{\mathrm{ex}}\end{aligned} \tag{6.77}$$

Z_{ex} is defined as Eq. (6.78).

$$Z_{\text{ex}} \equiv \sum_{i \neq 0} \exp\left(-\frac{\epsilon_i}{kT_{\text{ex}}}\right) \tag{6.78}$$

Equation (6.79) can be obtained by substituting Eqs. (6.75), (6.76), and (6.77) into Eq. (6.73).

$$\begin{aligned}\frac{3}{2}n_{\text{e}}\frac{dkT_{\text{e}}}{dt} &= \frac{1}{Z_{\text{ex}}}\sum_{i\neq 0}\epsilon_i \exp\left(-\frac{\epsilon_i}{kT_{\text{e}}}\right)S_{\text{ex}} + \frac{1}{\tau_{\text{q}}}\sum_{i\neq 0} n^*_{\text{n},i}\epsilon_i + \frac{3}{2}D_{\text{en}}\nabla^2(n_e kT_{\text{e}}) \\ &- \frac{3}{2}D_{\text{eff}}kT_{\text{e}}\nabla^2 n_e + s_{\text{IB}}n_{\text{e}} - 2n_{\text{e}}n^*_{\text{n},0}\sum_{i\neq 0}\epsilon_i K_{0,i} + 2n_{\text{e}}\sum_{i\neq 0} n^*_{\text{n},i}\epsilon_i K_{i,0} \\ &- n_{\text{e}}\sum_i n^*_{\text{n},i}\left(\epsilon_{\text{ion}} - 2\epsilon_i + \frac{3}{2}kT_{\text{e}}\right)K_{i,\text{ion}} + n_{\text{e}}{}^3\sum_i\left(\epsilon_{\text{ion}} - 2\epsilon_i + \frac{3}{2}kT_{\text{e}}\right)r_i \\ &- n_{\text{e}}n^{\text{v}}_{\text{n},0}\sum_{i\neq 0}\epsilon^{\text{v}}_i K^{\text{v}}_{0,i} + n_{\text{e}}\sum_{i\neq 0} n^{\text{v}}_{\text{n},i}\epsilon^{\text{v}}_i K^{\text{v}}_{i,0} - \frac{m_{\text{e}}}{m_{\text{h}}}\frac{3}{2}n_{\text{e}}kT_{\text{e}}\nu_{\text{m}}\end{aligned} \tag{6.79}$$

It was assumed that the sum of the diffusion terms was negligibly small as in Eq. (6.80).

$$\frac{3}{2}D_{\text{en}}\nabla^2(n_e kT_{\text{e}}) - \frac{3}{2}D_{\text{eff}}kT_{\text{e}}\nabla^2 n_e \cong 0 \tag{6.80}$$

In this case, Eq. (6.79) can be rewritten as Eq. (6.81).

$$\begin{aligned}\frac{3}{2}\frac{dkT_{\text{e}}}{dt} &= \frac{1}{Z_{\text{ex}}}\sum_{i\neq 0}\epsilon_i \exp\left(-\frac{\epsilon_i}{kT_{\text{e}}}\right)\frac{S_{\text{ex}}}{n_{\text{e}}} + \frac{1}{\tau_{\text{q}}n_{\text{e}}}\sum_{i\neq 0} n^*_{\text{n},i}\epsilon_i + s_{\text{IB}} \\ &- 2n^*_{\text{n},0}\sum_{i\neq 0}\epsilon_i K_{0,i} + 2\sum_{i\neq 0} n^*_{\text{n},i}\epsilon_i K_{i,0} - \sum_i n^*_{\text{n},i}\left(\epsilon_{\text{ion}} - 2\epsilon_i + \frac{3}{2}kT_{\text{e}}\right)K_{i,\text{ion}} \\ &+ n_{\text{e}}{}^2\sum_i\left(\epsilon_{\text{ion}} - 2\epsilon_i + \frac{3}{2}kT_{\text{e}}\right)r_i - n^{\text{v}}_{\text{n},0}\sum_{i\neq 0}\epsilon^{\text{v}}_i K^{\text{v}}_{0,i} + \sum_{i\neq 0} n^{\text{v}}_{\text{n},i}\epsilon^{\text{v}}_i K^{\text{v}}_{i,0} \\ &- \frac{m_{\text{e}}}{m_{\text{h}}}\frac{3}{2}kT_{\text{e}}\nu_{\text{m}}\end{aligned} \tag{6.81}$$

When it is assumed that the electron temperature reaches equilibrium faster than the timescale of the increment in the number of electrons and other particles because of the small heat capacity, the term of the time derivative can be treated as zero in Eq. (6.81).

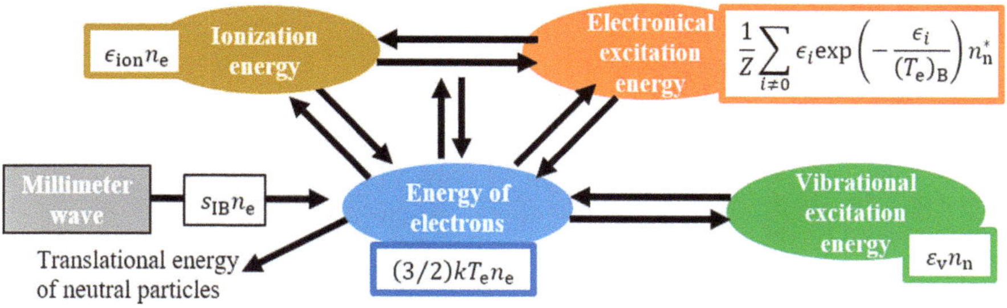

Fig. 6.10 Schematics of energy flow considered in the model

$$
\begin{aligned}
0 = & \frac{1}{Z_{\rm ex}}\sum_{i\neq 0}\epsilon_i \exp\left(-\frac{\epsilon_i}{kT_{\rm e}}\right)\frac{S_{\rm ex}}{n_{\rm e}} + \frac{1}{\tau_{\rm q} n_{\rm e}}\sum_{i\neq 0} n_{{\rm n},i}^* \epsilon_i + s_{\rm IB} - 2n_{\rm n,0}^*\sum_{i\neq 0}\epsilon_i K_{0,i} \\
& + 2\sum_{i\neq 0} n_{{\rm n},i}^*\epsilon_i K_{i,0} - \sum_i n_{{\rm n},i}^*\left(\epsilon_{\rm ion} - 2\epsilon_i + \frac{3}{2}kT_{\rm e}\right)K_{i,{\rm ion}} \\
& + {n_{\rm e}}^2\sum_i\left(\epsilon_{\rm ion} - 2\epsilon_i + \frac{3}{2}kT_{\rm e}\right)r_i - n_{\rm n,0}^{\rm v}\sum_{i\neq 0}\epsilon_i^{\rm v}K_{0,i}^{\rm v} + \sum_{i\neq 0} n_{{\rm n},i}^{\rm v}\epsilon_i^{\rm v}K_{i,0}^{\rm v} \\
& - \frac{m_{\rm e}}{m_{\rm h}}\frac{3}{2}kT_{\rm e}\nu_{\rm m}n_{\rm n}
\end{aligned}
\tag{6.82}
$$

The electron temperature can be obtained using Eq. (6.82). The schematics of the energy flow around the electrons are shown in Fig. 6.10. The electron temperature in the bulk is significantly higher than that in the ionization-front edge or precursor owing to the fact that the energy coming back from electronic and vibrational excitations is significant in bulk, while the electron temperature at the precursor is the highest in the previous supercritical models.

One-Dimensional Computation

A one-dimensional computation using the model explained above was conducted in (Nakamura and Komurasaki 2020). In this computation, the coefficients of ionization and excitation from the ground state were determined from cross sections in the database (Pancheshnyi et al. 2012). The ionization from each excited state was determined using a theoretical equation with classical assumptions (Lieberman and Lichtenberg 2005). The effects of the reverse reactions were determined to realize the Boltzmann distribution or Saha ionization equation when the plasma was in equilibrium. The computation was conducted by changing the diffusion coefficient of the excited-neutral particles $D_{\rm ex}$ and quenching frequency $\nu_{\rm q} \equiv 1/\tau_{\rm q}$ to determine the dependency on these parameters.

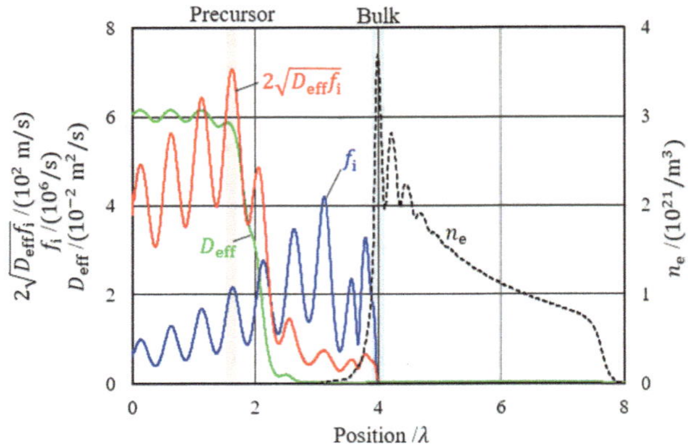

Fig. 6.11 Distributions of $2\sqrt{D_{\mathrm{eff}}f_{\mathrm{i}}}$, ionization frequency f_{i}, electron diffusion coefficient D_{eff}, and electron number density n_{e} around the computational ionization front. The parameters used for the computation are $D_{\mathrm{ex}} = 2$ and $\nu_{\mathrm{q}} = 0$. Incident beam intensity is 2.2 GW/m^2 (Nakamura and Komurasaki 2020)

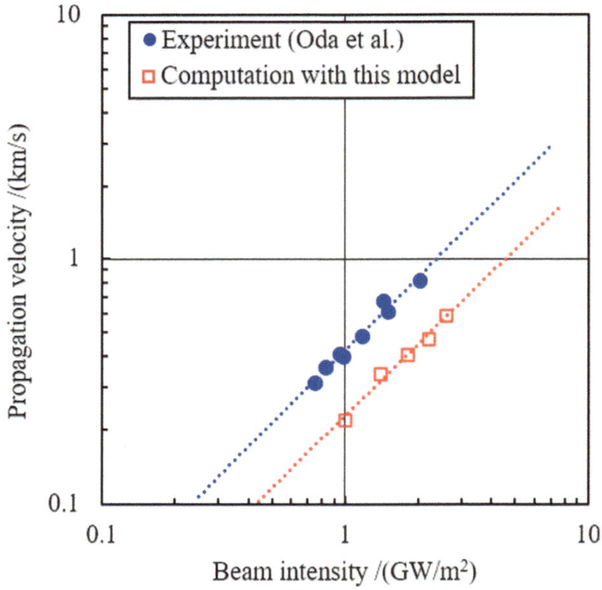

Fig. 6.12 Comparison between experimental (Oda et al. 2006) and computational (Nakamura and Komurasaki 2020) propagation velocities. The parameters used for the computations are $D_{\mathrm{ex}} = 2$ and $\nu_{\mathrm{q}} = 0$

Figure 6.11 shows the parameters that determine the propagation velocity around a propagating ionization front, computed for the incident beam intensity of 2.2 GW/m^2 with $D_{\mathrm{ex}} = 2$ and $\nu_{\mathrm{q}} = 0$. $2\sqrt{D_{\mathrm{eff}}f_{\mathrm{i}}}$ and reached its peak at the position where the diffusion coefficient is close to that of free electrons. It means that the precursor-driven propagation was successfully reproduced, as shown in Fig. 6.7(b). The propagation velocities obtained from the computation with the parameters $D_{\mathrm{ex}} = 2$ and $\nu_{\mathrm{q}} = 0$ are shown in Fig. 6.12. The propagation velocity of the computed ionization front was approximately 50% of the experimental velocity, and that is better than the results obtained at this intensity level for other computational models.

The computational results obtained for various diffusion coefficients without quenching are presented in Fig. 6.13(a). The differences in the propagation velocities were only approximately 20% when the diffusion coefficient changed from 0.1 to

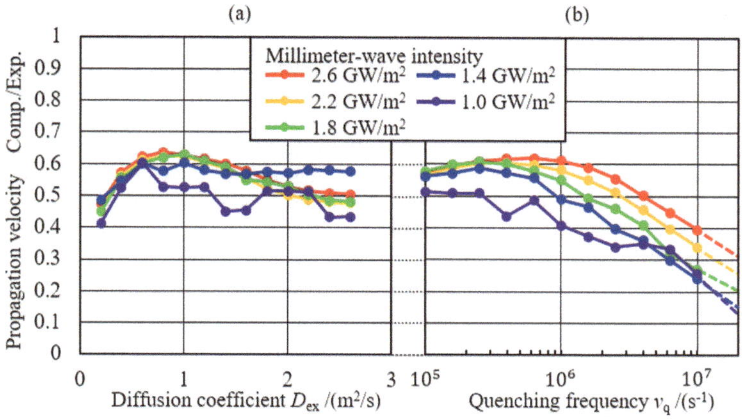

Fig. 6.13 Ionization-front propagation velocities normalized by experimental velocities (**a**) for various diffusion coefficients of excited particles D_{ex} ignoring quenching and (**b**) for various quenching frequency ν_q with constant diffusion coefficient $D_{ex} = 2$ (Nakamura and Komurasaki 2020)

2.6 m²/s. Figure 6.13(b) shows the computations conducted for various quenching frequencies ν_q with a constant diffusion coefficient $D_{ex} = 2$. When the quenching frequency was lower than 10^6/s, the propagation velocity was maintained at approximately 50% of the experimental value. However, the propagation velocity gradually decreased when the frequency increased above 10^6/s.

If high-energy excited species with sufficiently large diffusion coefficients and quenching frequencies lower than 10^6/s exist, the ionization front can propagate through this process. However, excited species with such features have not been identified, and further research is required to prove the propagation of this mechanism.

References

Beouf JP, Chaudhury B, Zhu GQ (2010) Theory and modeling of self-organization and propagation of filamentary plasma arrays in microwave breakdown at atmospheric pressure. Phys Rev Lett 104:015002

Boeuf JP, Chaudhury B, Zhu GQ (2010) Theory and modeling of self-organization and propagation of filamentary plasma arrays in microwave breakdown at atmospheric pressure. Phys Reavies Lett 104:015002

Boris, J. P. (1970). Relativistic plasma simulation-optimization of a hybrid code. Proc. 4th Conf. Numerical Sim. Plasmas, pp. 3–67

Chaudhury B, Boeuf JP, Zhu GQ, Pascal O (2011) Physics and modelling of microwave streamers at atmospheric pressure. J Appl Phys 110:11306

Ebert U, van Saarloos W, Caroli C (1996) Streamer propagation as a pattern formation problem. Phys Rev Lett 77:4178

Ebert U, van Saarloos W, Caroli C (1997a) Propagation and structure of planer streamer fronts. Phys Rev E 55:1530

Ebert U, Saarloos W, Caroli C (1997b) Propagation and structure of planar streamer front. Physical Reaview E 55(2):1530
Fukunari M, Tanaka S, Shinbayashi R, Yamaguchi Y, Tatematsu Y, Saito T (2019) Observation of a comb-shaped filamentary plasma array under subcritical condition in 303-GHz millimetre-wave air discharge. Sci Rep 9:17972
Hagelaar GJM, Pitchford LC (2005) Solving the Boltzmann equation to obtain electron transport coefficients and rate coefficients for fluid models. Plasma Sources Sci Technol 14(4):722
Hidaka Y, Choi EM, Mastovsky I, Shapiro MA, Sirigiri JR, Temkin RJ (2008) Observation of large arrays of plasma filaments in Air breakdown by 1.5-MW 110-GHz gyrotron pulses. Phys Rev Lett 100:035003
Hidaka Y, Choi EM, Mastovsky I, Shapiro MA, Sirigiri JR, Temkin RJ, Edmiston CF, Neuber AA, Oda Y (2009) Plasma structures observed in gas breakdown using a 1.5 MW, 110 GHz pulsed gyrotron. Phys Plasmas 16:055702
Huffman RE, Tanaka Y, Larrabee JC (1963) Absorption coefficients of nitrogen in the 1000-580 Å wavelength region. J Chem Phys 39:910
Kourtzanidis K, Boeuf JP, Rogier F (2014) Three-dimensional simulations of pattern formation during high-pressure, freely localized microwave breakdown in air. Phys Plasmas 21:123513
Lieberman MA, Lichtenberg AJ (2005) Principles of plasma discharges and materials processing. Wiley
MacDonald AD (1966) Microwave breakdown in gases. Wiley
Nakamura Y, Komurasaki K (2020) Theory and modeling of under-critical millimeter-wave discharge in atmospheric air induced by high-energy excited neutral-particles carried via photons. Plasma Sources Sci Technol 29(10):105017
Nakamura Y, Komurasaki K, Fukunari M, Koizumi H (2018) Numerical analysis of plasma structure observed in atmospheric millimeter-wave discharge at under-critical intensity. J Appl Phys 124:033303
Nam SK, Verboncoeur JP (2009) Theory of filamentary plasma array formation in microwave breakdown at near-atmospheric pressure. Phys Rev Lett 103:055004
Nanbu K (1980) Direct simulation scheme derived from the Boltzmann equation. I. Monocomponent gases. J Phys Soc Jpn 49:2042–2049
Oda Y, Komurasaki K, Takahashi K, Kasugai A, Sahamoto K (2006) Plasma generation using high-power millimeter-wave beam and its application for thrust generation. J Appl Phys 100:113307
Pancheshnyi S, Biagi S, Bordage MC, Hagelaar GJ, Morgan WL, Phelps AV, Pitchford LC (2012) The LXCat project: electron scattering cross sections and swarm parameters for low temperature plasma modeling. Chem Phys 398:148
Takahashi M, Ohnishi N (2014) Computational studies for plasma filamentation by magnetic field in an atmospheric microwave discharge. Appl Phys Lett 105:223504
Takahashi M, Ohnishi N (2016) Plasma filamentation and shock wave enhancement in microwave rockets by combining low-frequency microwaves with an external magnetic field. J Appl Phys 120:063303
Takahashi M, Ohnishi N (2018) Gas-species-dependence of microwave plasma propagation under external magnetic field. J Appl Phys 124:173301
Takahashi M, Kageyama Y, Ohnishi N (2017) Joule-heating-supported plasma filamentation and branching during subcritical microwave irradiation. AIP Adv 7:055206
Wu CYR, Fung H, Chang K, Singh TS, Mu X, Nee JB, Judge DL (2007) Fluorescence excitation spectra of the $b^1\Pi_u$, $b'^1\Sigma_u^+$, $c_n{}^1\Pi_u$, and $c'_n{}^1\Sigma_u^+$ states of in N_2 the 80-100 nm region. J Chem Phys 127:084314
Yamaguchi T, Fukunari M, Nakamura Y, Oda Y, Sakamoto K, Komurasaki K (2016) Fast-framing observation of filamentary plasma in atmospheric millimeter-wave breakdown. Front Appl Plasma Technol 9(2):79
Zh T (2001) Esirkepov.: exact charge conservation scheme for particle-in-cell simulation with an arbitrary form-factor. Comput Phys Commun 135:144–153

Chapter 7
Laser-Supported Detonation

Hiroshi Katsurayama and Kohei Matsui

Abstract This chapter takes a comprehensive look at laser-supported detonation (LSD), examining both experimental studies and numerical analyses. It investigates the propagation mechanism, the atmospheric pressure dependence, and the role of an electron density threshold in driving the LSD. It clarifies the influence of laser intensity, wavelength, focusing optics, and ambient gas conditions on LSD propagation, highlighting the importance of further investigations to elucidate its propagation structure. It also discusses the energy conversion efficiency, termination conditions, and the propagation velocity of the LSD wave.

Keywords Laser-supported detonation · Laser-supported combustion · High-power laser · Computational fluid dynamics · Inverse bremsshtrung · Plasma physics · Atmospheric propellant · Propagation mechanism · Shadowgraph measurement

Numerical Analyses of Laser-Supported Detonation Wave in Air

Introduction

An air-breathing pulse-laser launching system (Katsurayama et al. 2009) is a promising alternative to conventional chemical launch systems. The laser system can achieve a very large payload ratio because energy is provided from a ground-based laser and atmospheric air is used as a propellant. The fabrication cost of the launching vehicle is low because its body structure is very simple. The other cost is only electricity after the construction of a laser base. Although the cost of the laser

H. Katsurayama (✉)
Department of Mechanical and Physical Engineering, Tottori University, Tottori, Japan
e-mail: katsurayama@tottori-u.ac.jp

K. Matsui
Department of Space Systems Engineering, Kyushu Institute of Technology, Kitakyushu, Japan
e-mail: matsui.kohei437@mail.kyutech.jp

K. Mori et al. (eds.), *Beamed-mobility Engineering*, Springer Tracts in Electrical and Electronics Engineering, https://doi.org/10.1007/978-981-99-4618-1_7

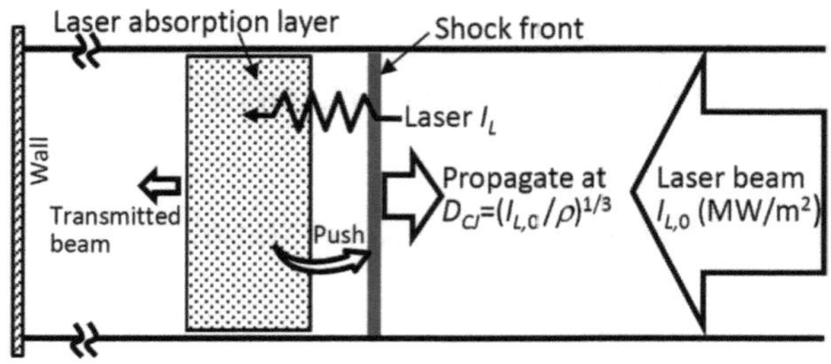

Fig. 7.1 Schematic of the propagation of a LSD wave

base is very expensive, it can be redeemed by increasing the number of the launchings.

The launching vehicle converts the laser energy into thrust through the laser-supported detonation (LSD) wave (Raizer 1977), which is driven in the breathed air. Figure 7.1 shows a schematic of the LSD wave, which is assumed to propagate quasi-one-dimensionally. A breakdown air plasma is formed by the laser-focusing through a parabolic nozzle or the direct laser irradiation to a nozzle wall. The plasma absorbs the following laser beam through the inverse bremsstrahlung radiation, and it grows to be a high-temperature laser absorption layer. The absorption layer expands drastically because the laser absorption is an isochoric process which can highly pressurize the layer. This drastic expansion forms a strong shock wave ahead of the absorption layer. Because the pressurized absorption layer pushes the shock wave, they supersonically propagate together, in the form of a structure similar to that of a popular chemical detonation.

Several experiments (Mori et al. 2002, 2004, 2006) and two-dimensional computations (Wang et al. 2002; Ogino et al. 2013) have been conducted to investigate the LSD propagation mechanism in air. For example, Ogino et al.'s (2013) computation reproduced the LSD propagation speed of Mori's et al. (2002) experiment in the atmospheric condition. However, these numerical studies did not explain the details of the laser absorption near the LSD front. The lack of this important topics may be because the grid resolution in their two-dimensional computations is not enough to clarify the laser absorption mechanism. Moreover, these computations did not discuss the LSD termination mechanism and the influence of the LSD propagation on the reduced pressure.

The final object of this study is to develop a computational fluid dynamic (CFD) code which can clarify the unknown mechanisms in the LSD propagation. In the present paper, we develop a one-dimensional (and fine-grid resolution) CFD code, which can capture the LSD propagation and laser absorption in the conditions of several atmospheric pressures, and compare the computed LSD propagation speed with theoretical Chapman-Jouguet (C-J) detonation speeds (Raizer 1977).

Computational Model and Method

We solve the one-dimensional Navier-Stokes equations with thermochemical nonequilibrium of 11 species air plasma. More details of the computational model and computational scheme are shown in Katsurayama and Abe (2013a, b). The laser heating is computed by integrating Beer's law:

$$\frac{dI_{\mathrm{L}}}{dx} = -k_{\mathrm{L}} I_{\mathrm{L}} \tag{7.1}$$

where I_{L} is the local laser intensity and k_{l} is the laser absorption coefficient, which is taken from Kemp and Lewis (1980).

Computational Condition

The present computation assumes the laser to be irradiated to the wall, which is located at the left side of the computational region, as shown in Fig. 7.1. Electrons are prepared in a narrow region (ten computational cells) near the wall to imitate the initial laser breakdown. Their number densities n_{e} and temperature T_{e} are set to 10^{24} m^{-3} and 15,000 K. Several of our numerical tests confirmed that the solution after the laser absorption occurs is insensitive to these values.

The laser wave length λ is set to 10.6 μm of the CO_2 laser, which is used in Mori et al.'s experiments (Mori et al. 2002, 2004, 2006). Their experiments reported that the initially irradiated laser intensity $I_{\mathrm{L},0}$ must be larger than 4 MW/cm^2 to form the LSD wave in the C-J state. The present computation uses $I_{\mathrm{L},0} = 10$ MW/cm^2, which will be large enough to drive the C-J LSD.

The width of the computational cells is set to λ for resolving the absorption process because the laser absorption occurs in the scale of one wave length.

Results and Discussion

We computed the LSD propagations in the standard atmosphere and the reduced-pressure atmospheres of which pressure p_{a} is from 1 atm to 0.1 atm.

Results of the Standard Atmosphere of $p_{\mathrm{a}} = 1$ atm

Figure 7.2 shows the LSD propagation in the standard atmosphere, in which p, T_{h}, and T_{e} are static pressure, heavy particle translational temperature and electron-vibrational temperature, respectively. The influence of the initial condition is

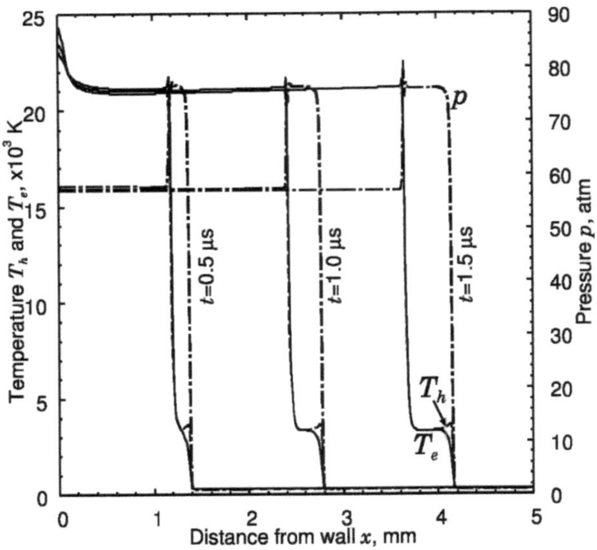

Fig. 7.2 LSD propagation in the atmosphere of $p_a = 1$ atm

confined in the narrow region ($x < 0.3$ mm) near the wall. A wave like the chemical detonation, which consists from the strong shock wave and the following sharp temperature jump, appears at time $t = 0.5$ μs. After this time, the detonation-like wave propagates at a steady speed.

Figure 7.3 shows the structure of this wave, the distributions of electron number density n_e, laser absorption coefficient k_l, and the ratio of I_L to $I_{L,0}$. The laser drives a strong shock wave whose pressure is over 80 atm. The value of T_h rises to 4000 K by the strong shock heating. The T_e gradually increases by collisional energy relaxation with the shock-heated heavy particles, and it becomes equilibrium with T_h at the location of $x = -0.2$ mm from the shock front (see Fig. 7.3a). The ionization reactions slowly proceed by this moderate increase of T_e, and it increases n_e during the width of 0.5 mm (see Fig. 7.3b). The gradually ionizing region will be called the "ionization induction zone" which is similar to the chemical reaction induction zone of the chemical detonation.

The value of n_e reaches 10^{22} m^{-3}at the end of the ionization induction zone. Behind the location, the value of k_l increases rapidly because it is proportional to n_e^2 (see Fig. 7.3b). This rapid increase of k_l triggers the laser absorption, and I_L rapidly decreases from 100% to 0% (see Fig. 7.3c). This results in the formation of a thin laser absorption layer, which absorbs100% of the laser energy. As a result of this rapid and large laser absorption, the of T_e and n_e drastically increases in the absorption layer, and their values, respectively, reach 20,000 K and 10^{24} m^{-3} behind the absorption layer. The T_h in the absorption layer is equilibrium with T_e because the energy transfer from the laser heated electrons is very rapid in this highly pressurized region.

Fig. 7.3 LSD structure in the atmosphere of $p_a = 1$ atm: the distributions of (**a**) p, T_h, and T_e, (**b**) n_e and k_l, and (**c**) $I_L/I_{L,0}$

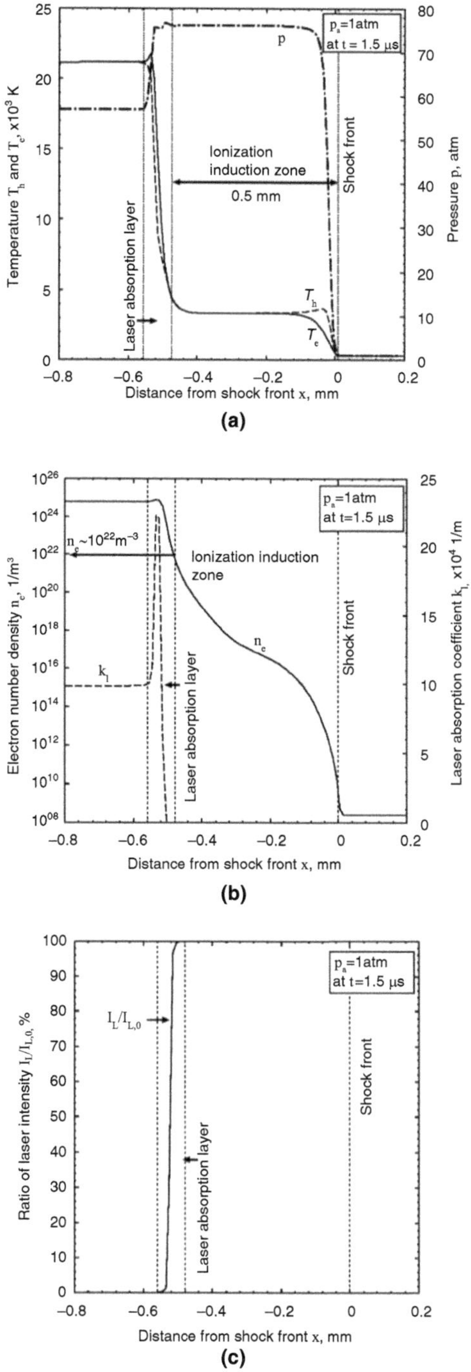

As a result, the laser-supported wave is found to have a chemical detonation-like structure, which consists of a strong shock wave, an induction zone for ionization, and a following thin laser absorption layer.

Moreover, it is found that the electron density of $n_e \sim 10^{22}\ \mathrm{m}^{-3}$ is necessary to trigger the laser absorption.

Results of the Reduced-Pressure Atmosphere

Figure 7.4 shows the wave structure, the distributions of n_e and the ratio of I_L to $I_{L,0}$ in the reduced-pressure atmosphere of $p_a = 0.2$ atm. The width of the induction zone decreases with the decrease of p_a, because the shock speed increasing with the decrease of p_a enhances the shock heating which prompts the ionization (see Figs. 7.3a and 7.4a). The width in $p_a = 0.2$ atm becomes 2/5 of that in $p_a = 1$ atm because the strong shock heating moves the location of the laser absorption threshold $n_e \sim 10^{22}\ \mathrm{m}^{-3}$ forward (see Fig. 7.4b).

Figure 7.5 shows the wave structure, the distributions of n_e, and the ratio of I_L to $I_{L,0}$ in the reduced-pressure atmosphere of $p_a = 0.1$ atm. When p_a is reduced to 0.1 atm, the induction zone vanishes, and the value of T_e (and also T_h) immediately increases to 20,000 K just behind the shock front (Fig. 7.5a). The drastic increase of k_l occurs at the shock front because the value of n_e exceeds the laser absorption threshold $n_e \sim 10^{22}\ \mathrm{m}^{-3}$ just behind the shock front (Fig. 7.5b). As a result, the laser absorption occurs at the same instant of the shock heating, and 100% of the laser energy is absorbed in the thin absorption layer, which is located just behind the shock front (Fig. 7.5b).

Comparison of the Wave Speeds

To verify whether the computed laser-supported wave is a stable detonation (i.e., the detonation in the C-J state, which are measured in experiments (Mori et al. 2002, 2004, 2006), we compare the computed shock front speed D_s and absorption layer speed D_a with the theoretical C-J detonation speed (Raizer 1977) D_{CJ}

$$D_{CJ} = 1.07(I_{L,0}/\rho)^{1/3} \tag{7.2}$$

where ρ is the atmospheric density.

Table 7.1 shows the wave speeds in $p_a = 1$ and 0.1 atm. In $p_a = 1$ atm, D_a is slower than D_s, and the difference is 300 m/s. Therefore, the absorption layer gradually separates from the shock front, and the LSD structure will be eventually broken. Moreover, both of D_s and D_a are much smaller than D_{CJ}. As a result, the present computation will not be able to reproduce the stable C-J detonations measured in the experiments (Mori et al. 2002, 2004, 2006).

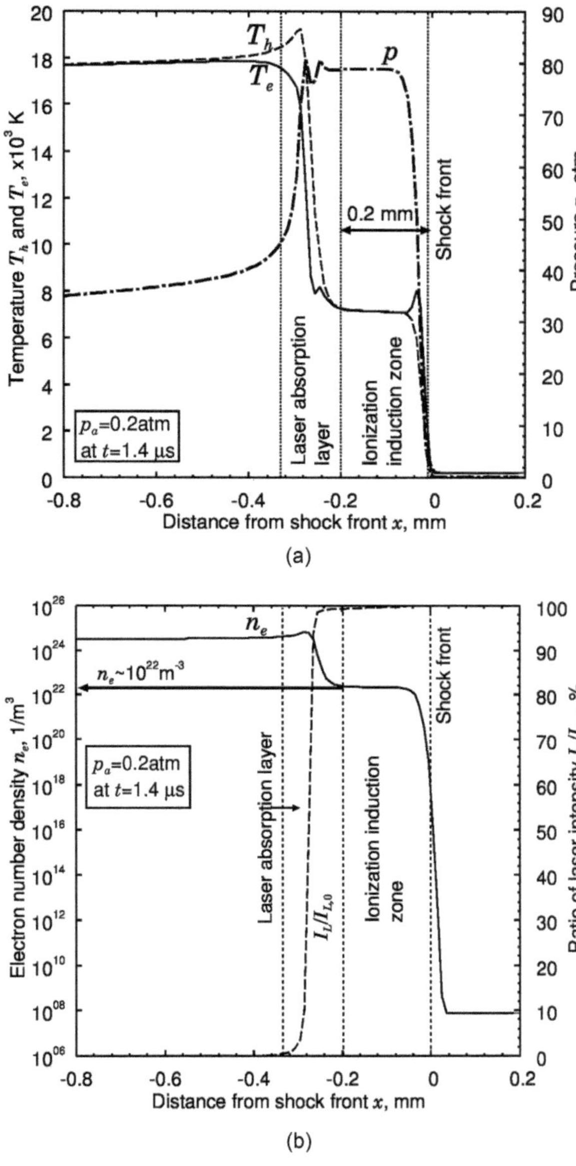

Fig. 7.4 LSD structure in $p_\mathrm{a} = 0.2$ atm: the distributions of (**a**) p, T_h, and T_e, and (**b**) n_e and $I_\mathrm{L}/I_{\mathrm{L},0}$

In $p_\mathrm{a} = 0.1$ atm, the LSD structure is stably kept because D_a is equal to D_s. However, the speeds are again much smaller than D_CJ.

As a result, the computed LSD waves are not on the C-J state on any atmospheric pressure conditions. However, the computation showed that the pressure dependency of the LSD wave qualitatively agrees with that of the C-J detonation, because $dD_\mathrm{s}/dp_\mathrm{A}$ and $dD_\mathrm{a}/dp_\mathrm{A}$ (=$5.2 \sim 5.5 \times 10^{-2}$ (m/s)/Pa) are identical to the value of dD/dp_A(=4.7×10^{-2} (m/s)/Pa).

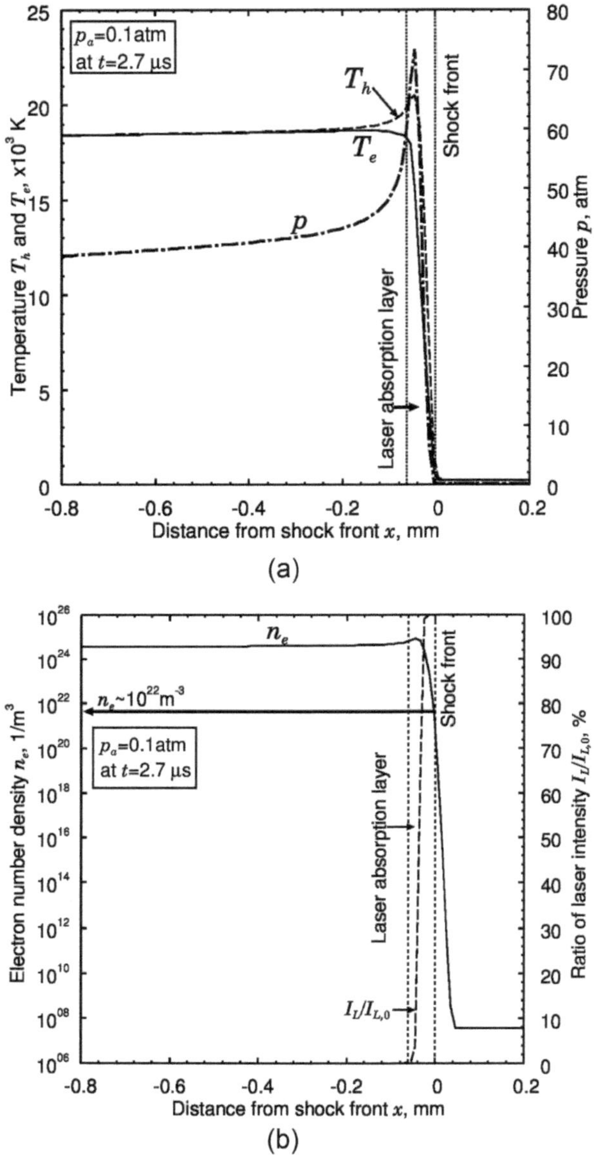

Fig. 7.5 LSD structure in $p_a = 0.1$ atm: the distributions of (**a**) p, T_h, and T_e, and (**b**) n_e and $I_L/I_{L,0}$

Table 7.1 Comparison of wave speeds

Atmospheric pressure P_a, atm	1	0.1
Shock front speed D_s, m/s	2766	7537
Absorption layer speed D_a, m/s	2466	7525
Theoretical C-J speed D_{CJ}, m/s	4711	9028

Summary

We developed a 1-D CFD code to clarify the LSD propagation mechanism and investigated the atmospheric pressure dependency on the LSD propagation. The computation successfully produced a laser-supported wave, which has a structure similar to that of the chemical detonation with a strong shock wave, an induction zone for ionization, and a following thin laser absorption layer. Moreover, it is found that electron density of $n_e \sim 10^{22}\ \mathrm{m}^{-3}$ is necessary to trigger the laser absorption. The computations in the reduced-pressure atmosphere showed that the width of the ionization induction zone decreases with the decrease of the atmospheric pressure because of the enhanced shock heating. However, the computed speed of the laser-supported wave was found to be much smaller than the stable C-J detonation speed. Nevertheless, the computation showed that the pressure dependency of the laser-supported wave speed qualitatively agrees with that of the C-J detonation.

Experimental Studies of Laser-Supported Detonation

Propagation Velocity of Laser-Supported Detonation Wave

This chapter reviews experimental studies on laser-supported detonation (LSD) conducted from the 1960s to the present. Various experiments have been conducted to elucidate the physical processes of LSD. Experiments are performed using a pulsed laser with focusing optics to obtain sufficient intensity to initiate breakdown. The generated plasma absorbs the subsequent laser and propagates toward the laser source. The timescale of the LSD depends on the laser's pulse duration, which is typically nanoseconds to microseconds. LSD is generally maintained at laser intensities of tens of $\mathrm{GW/m^2}$ or higher. On the other hand, the laser intensity is not sufficient to maintain LSD, and the laser power is absorbed in laser-supported combustion (LSC) regime. The propagation velocity of the LSD wave was preliminarily discussed by Ramsden and Davies (1964) and Ramsden and Savic (1964). The plasma was produced by a Q-spoiled ruby laser whose power, wavelength, and pulse duration were 5–10 MW, 694.3 nm, and 20 ns, respectively. The laser was focused to a diameter of 0.2 mm by a lens with a focal length of 8 mm. The propagation velocity was obtained from a streak photograph of the spark. Ramsden calculated the Chapman-Jouguet detonation velocity by replacing the reaction energy with the absorbed laser energy and agreed with the experimental result. As described by Raizer, the propagation velocity V in the Chapman-Jouguet condition is proportional to the cubic root of laser intensity S (Raizer 1965),

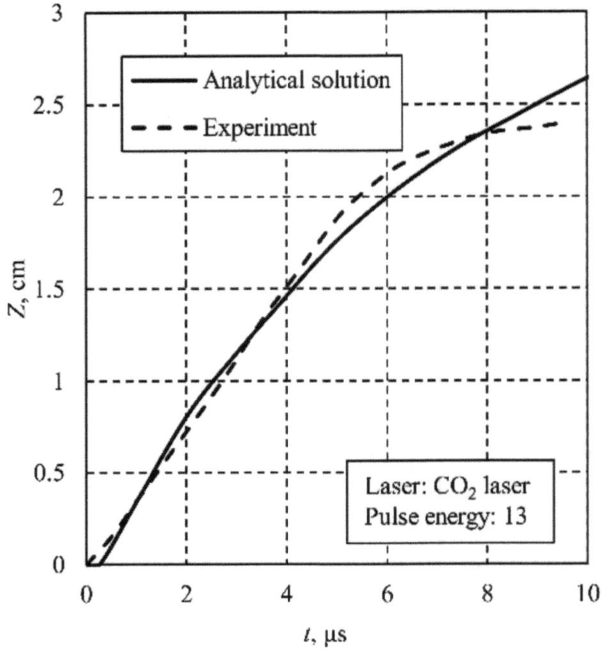

Fig. 7.6 Analytical LSD wave motion for the experimental condition compared with results from the experiment using aluminum specimens

$$V = \left\{ \frac{2(\gamma - 1)^2}{\rho} S \right\}^{\frac{1}{3}}.$$

Here, γ and ρ are specific heat ratio behind the LSD wave and gas density in front of the wave. Much research focused on the relationship between laser intensity and the propagation velocity to examine whether the Chapman-Jouguet theory is appropriate. Maher investigated the propagation velocity of LSD waves in the air using a CO_2 laser with a maximum pulse energy of 15 J and a wavelength of 10.6 µm (Maher et al. 1974). The laser beam was focused by a lens with a focal length of 24 cm onto aluminum and lead specimens to initiate gas breakdown. The displacements of the wave were obtained from a streak camera photograph and compared with the analytical solution calculated by integrating the C-J velocity:

$$Z = \int \left\{ \frac{2(\gamma - 1)^2}{\rho} S(t, Z) \right\}^{\frac{1}{3}} \mathrm{d}t.$$

Since the laser power attenuates with time and the irradiated area changes as the wavefront move toward the laser beam source, the laser intensity incident on the LSD wave is a function of time and the position of the wavefront. Figure 7.6 shows the comparison between the experimental result and the analytical solution in the case of pulse energy of 13 J. The difference between the analytical solution and the experimental data was speculated to be due to the nonuniformity of the laser

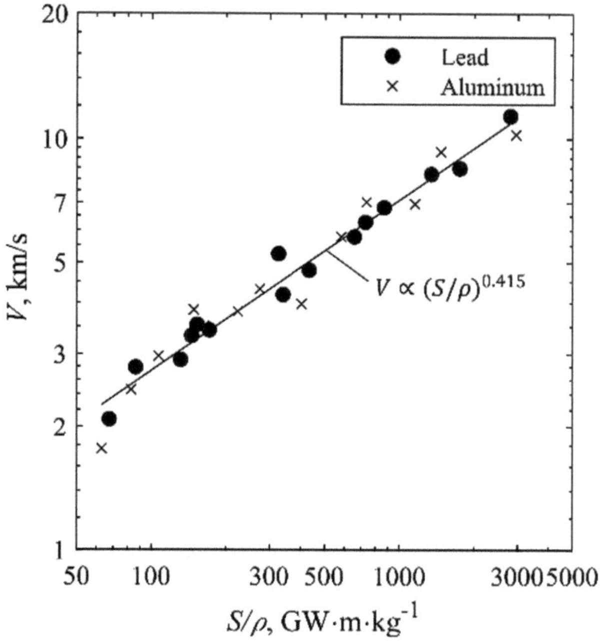

Fig. 7.7 Relationship between S/ρ and propagation velocity for lead and aluminum specimens (Maher et al. 1974)

intensity as a function of time, the finite absorption length, and unknown ignition processes. Figure 7.7 shows a relationship between S/ρ and measured propagation velocity when the laser beam focused on lead and aluminum specimens. The velocity with lead and aluminum specimens were consistent with each other, which suggests that the LSD wave propagates independently of the surface of the ignition. The exponent of the fitting curve of $(I/\rho)^A$ was found to be 0.415, which is definitely different from 1/3 of the C-J theory. On the other hand, Bournot showed good agreement with C-J theory, except for the velocity with low laser intensity (Bournot et al. 1979). The propagation velocity of the LSD was observed using a CO_2 laser in the laser intensity range of $(10^2 - 10^3)$ GW/m^2. The results showed that the propagation velocity at an intensity of 10^2 GW/m^2 is not consistent with the C-J theory due to the smaller absorption length.

Some experimental results have also shown that the exponents of a power function differ significantly from 1/3. Boiko performed the experiment using a CO_2 laser with a pulse energy of 100 J and measured the propagation velocity of the laser absorption wave in the air by taking a streak photograph (Boiko et al. 1976). The results showed that a much higher velocity than that of C-J detonation in the region of laser intensity higher than 2×10^{12} W/m^2. The exponent of the fitting with power law was 2/3. This discrepancy with the C-J theory is explained by the fact that the laser absorption layer is produced by radiation from the bulk plasma, not by the shock wave. Boiko also observed the laser-absorption wave in Xe and Ar atmosphere at a pressure of 740 torr using the CO_2 laser (Boiko et al. 1978). The measured velocity was faster than 100 km/s at laser intensity $>(3 - 4) \times 10^{11}$ W/m^2 and

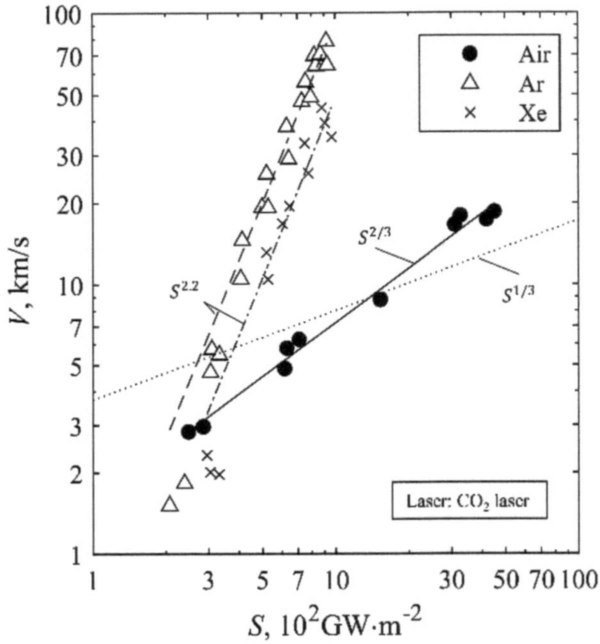

Fig. 7.8 Propagation velocity of superdetonation in air, argon, and xenon atmosphere (Boiko et al. 1976, 1978)

proportional to $S^{2.2 \pm 0.2}$, which considerably differs from 1/3. They concluded that this was the first study to observe such extremely fast "superdetonation" using a CO_2 laser. The propagation velocity of superdetonation observed in (Boiko et al. 1976, 1978) is shown in Fig. 7.8. Zakharchenko used a Q-switched neodymium glass laser with a pulse energy of 30 J and a wavelength of 1.06 μm to observe the superdetonation wave (Zakharchenko 1984). The laser was focused using a lens with a focal length of 5 m and irradiated to thin tungsten filaments with diameters of 5, 15, and 45 μm. The measured propagation velocity was faster than that of C-J and depended on $S^{1.37}$ with 5 and 15 μm diameter of the filament and $S^{0.82}$ with a diameter of 45 μm. The experimental results were compared with the calculation conducted by Iliyn (Ilyin et al. 2008). Shimano compared the propagation velocities in air, Ar, and He atmosphere at a pressure of 1 atm with various focusing diameters (Shimano et al. 2017). The exponents obtained by power function fitting showed 0.46 for air, 0.23 for argon, and 1.18 for helium. The C-J theory could not explain this significant difference in the exponent depending on gaseous species.

Termination Conditions of LSD Wave

When a pulsed laser beam generates LSD, the expansion of the plasma transitions from LSD to LSC as the laser power decays. The investigation of the termination

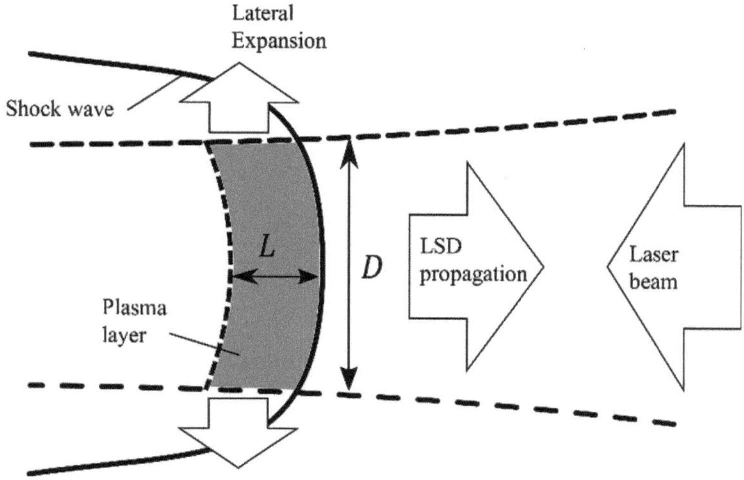

Fig. 7.9 Schematic of lateral expansion of LSD wave

condition of LSD is essential because the strength of the resulting shock wave depends on a regime-transition threshold. Typically, the threshold is laser intensity of 10–100 GW/m^2. Mori investigated the influence of focusing *f*-number on the LSD termination threshold (Mori et al. 2002). Transversely excited atmospheric (TEA) CO_2 lasers with pulse energies of 4 J and 10 J were utilized in the experiment. The cross section of the laser was 30 mm × 30 mm, whose equivalent beam diameter was 34 mm. Parabolic mirrors with focal lengths of 38.1 mm and 76.2 mm were used for focusing the laser. The corresponding f were 1.1 and 2.2. Time histories of the displacements of the shock and ionization wavefronts were obtained by laser shadowgraphs using a He-Ne laser. The laser intensity at the separation of the shock wavefront and the ionization wavefront was the threshold value. The results showed that the threshold of the laser intensity for the termination S_{th} increased with the focusing f number and was described as $S_{th} \propto f^{0.2-0.5}$.

Raizer predicts that the shock wave propagates in a direction perpendicular to LSD propagation, as shown in Fig. 7.9, resulting in energy loss, which affects the termination condition and the propagation velocity of LSD (Shimamura et al. 2015). The energy loss due to the lateral expansion is formulated by the ratio of the diameter of the plasma channel D to the thickness of the plasma layer L. Several experiments have been conducted using confinement to restrict the radial motion of the shock wave and evaluate the effect of the lateral expansion. Ushio confined the plasma with a wedged nozzle to observe quasi-1D propagation (Ushio et al. 2008). A CO_2 laser with a power of 10 J was utilized and focused on an aluminum plate by a line-focusing parabolic mirror with a focal length of 48 mm. The LSD propagation was observed with the wedge nozzle and without it to compare the effects of the lateral expansion. The experimental results showed that the confinement reduced the laser intensity threshold for termination. The laser intensities at termination were 34 GW/m^2 for the unconfined and 17 GW/m^2 for the quasi 1-D confined. Shimamura

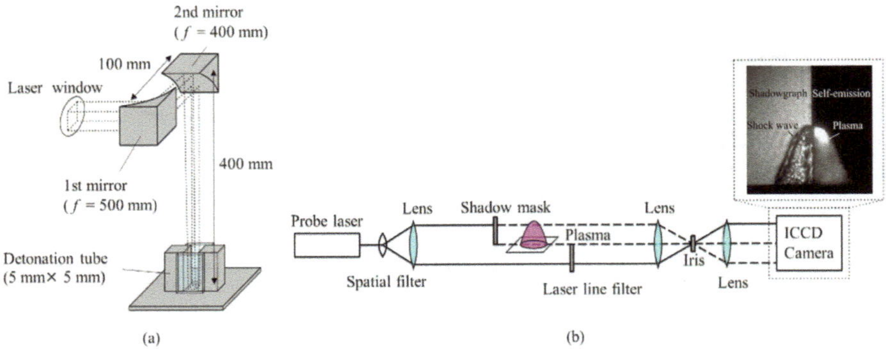

Fig. 7.10 Experimental apparatus used in (Matsui et al. 2017). (**a**) Focusing optics of CO2 laser and detonation tube. (**b**) Visualization optics method and a typical image of HSHS method

also observed LSD termination with a quasi 1-D tube (Shimamura et al. 2015). The tube had a cross section of 2 × 1 mm at the closed end and 2 × 3 mm at the open end. A CO_2 laser was focused on the closed end by a line-focusing parabolic mirror with a focal length of 150 mm. In the study, Hugoniot analysis was conducted to investigate the termination condition using the propagation velocity measured by laser shadowgraph. A pressure gauge was equipped at the closed end of the tube to measure the stagnation pressure and verify the result of the Hugoniot analysis. The results showed that the Rayleigh line and the Hugoniot curve had two intersections, which indicate that LSD is an overdrive detonation. Under the termination condition, the Rayleigh line and Hugoniot curve had a single intersection, which means the Chapman-Jouguet detonation. The computed stagnation pressure behind the LSD wave showed good agreement with the measured pressure. Matsui performed experiments with a 5 mm × 5 mm diameter tube and observed the termination using the half self-emission and half shadowgraph (HSHS) method (Matsui et al. 2017). The ionization front emission and the shock wave front were simultaneously obtained by the HSHS method. The shock wave front was projected by a diode-pumped solid state (DPSS) laser with a wavelength of 532 nm. The experimental setup is shown in Fig. 7.10. The CO2 laser was focused by two cylindrical mirrors with focal lengths of 400 and 500 mm. The laser beam was irradiated into the detonation tube with a cross section of 5 mm × 5 mm. Figure 7.11 shows the HSHS image of the LSD propagation in the detonation tube. In the experiments, the effect of the confinement on the termination condition and propagation velocity was discussed by comparing the propagation of the LSD with and without the detonation tube. The results suggest that the effect of lateral enthalpy dissipation on the LSD termination condition and the propagation velocity could be less influential by utilizing a relatively large beam diameter and straight laser beam channel. Bufetov utilized a quartz tube to restrict the lateral expansion of the LSD wave in the air atmosphere (Bufetov 1984). In that experiment, gas breakdown was initiated by a Q-switched laser. Then, a millisecond neodymium laser was irradiated into the tube to maintain the wave propagation. The diameters of the tubes were 7.3, 11.3, and

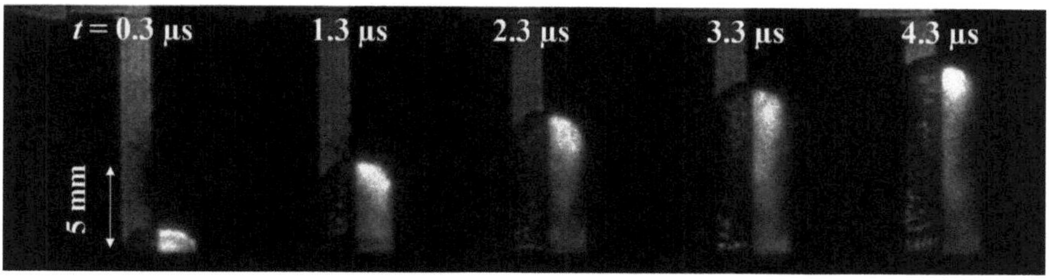

Fig. 7.11 Propagation of the LSD wave in the detonation tube. Elapsed time from the start of laser irradiation is shown at the top of the figure

Fig. 7.12 Energy conversion flow of LSD

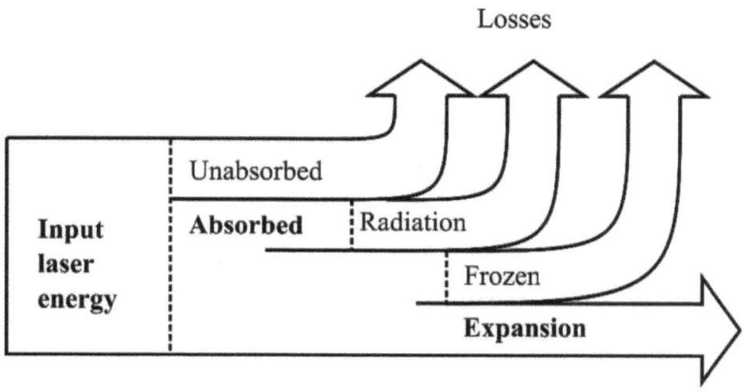

16.6 mm. As a result, the restriction of the lateral expansion made propagation velocity faster than without the confinement and reduced the threshold for maintaining the LSD.

Energy Conversion Efficiency in LSD

During the propagation process of LSD, the optical energy of the laser is converted to fluid energy. Figure 7.12 shows the schematic of the energy conversion flow of the LSD. After the formation of the plasma, the laser beam is absorbed through an inverse bremsstrahlung process. Some portion of the laser is transmitted without being absorbed. The rapid expansion of the plasma generated a blast wave. On the other hand, part of the absorbed energy becomes internal energy frozen in plasma or radiation loss, which does not contribute to the expansion of the gas. The energy conversion efficiency from the laser to the blast wave energy is vital in engineering applications such as laser propulsion. Mori studied the energy conversion efficiency using Sedov's self-similar solution and investigated the effect of the atmospheric condition and the input laser energy on the efficiency (Mori et al. 2004). The blast wave energy E_{bw} is the energy defined as the bulk energy of gas motion in the volume of the blast wave V_{bw}:

Table 7.2 Energy efficiencies of glass laser and CO_2 laser

	Nd:glass laser				TEA-CO_2 laser (Mori et al. 2004)	
	$E_i = 1.0$ J		$E_i = 2.0$ J		$E_i = 10$ J	
f	3.75	6.27	3.75	6.27	2.2	3.3
$\eta_{bw} = E_{bw}/E_i$ (%)	31	59	40	34	47	44
E_{bw}/E_{ab}	90	78	81	62	48	45

$$E_{bw} = \int_{V_{bw}} \left((p - p_a)/(\gamma - 1) + \frac{1}{2}\rho u^2 \right) dV.$$

p, ρ, u, and γ represent pressure, density, velocity, and the specific heat ratio of the fluid element, respectively. p_a is the ambient pressure. The blast wave conversion efficiency is a ratio of the blast wave energy E_{bw} to input laser energy E_i,

$$\eta_{bw} = \frac{E_{bw}}{E_i}.$$

In the study, plasmas were produced using a CO_2 laser at an atmospheric pressure in the range of 10–100 kPa. The input energy was changed from 4.0 to 12.8 J. Temporal variations of the displacement and the Mach number of elliptically expanding blast waves were obtained from a shadowgraph image. The blast wave energy was calculated using the Sedov-Taylor solution (Sedov 1959). The result showed that the efficiency was almost insensitive to the input laser energy and approximately 40–50% at a pressure of 100 kPa. On the other hand, as the ambient pressure decreased, the conversion efficiency decreased and was about 20% at a pressure of 10 kPa.

Wang investigated the effect of a laser wavelength on the blast wave conversion efficiency (Wang et al. 2010). The LSD was produced using a Q-switched neodymium glass (Nd:glass) laser with a wavelength and a pulse width of 1.053 μm and 33 ns, respectively. The laser was focused by focusing optics with *f*-numbers of 3.75 and 6.27. The input energy is 1.0 and 2.0 J. In the experiment, the transmitted laser energy without being absorbed by the plasma was measured. Same as Mori's method, the blast wave efficiency was obtained using the Sedov-Taylor solution. The results are presented in Table 7.2, with the efficiency measured in (Mori et al. 2004). The blast wave energy conversion efficiency with the glass laser is not significantly different from that of the CO_2 laser, though the absorption coefficient of the inverse bremsstrahlung is proportional to the cube of the wavelength of the input laser. This result suggests that the plasma density is of the glass laser higher than that of the CO_2 laser-induced plasma. The efficiency of the glass laser depends on the *f*-number compared to the CO_2 laser, which implies that the optimum *f*-number exists for the energy conversion.

Summary

This chapter reviews the experimental results of LSD propagation velocity, termination condition, and energy conversion efficiency. Experimental results vary greatly depending on laser intensity, wavelength, focusing optics, and ambient gas conditions. Much more research is still needed to elucidate the propagation structure of LSD. With the development of measurement equipment, more experimental results will be obtained in the future, which will help in modeling LSD.

References

Boiko VA, Danilychev VA, Zvorykin VD et al (1976) Investigation of gasdynamic processes and recoil impulse produced by optical breakdown of air near a target surface by radiation of an electorn-beam-controlled CO2 laser. Sov J Quantum Electron 6:1065

Boiko VA, Danilychev VA, Duvanov BN et al (1978) Observation of supersonic radiation waves in gases generated by CO2 laser. Sov J Quantum Electron 8:134

Bournot P, Pincosy PA, Inglesakis G et al (1979) Propagation of a laser-supported detonation wave. Acta Astronaut 6:257–267

Bufetov IA (1984) Optical discharge accompanying a restriction imposed on lateral expansion of gas and a reduction in the threshold of light-induced detonation. Sov J Exp Theor Phys Lett 39: 258–261

Ilyin AA, Bukin OA, Bulanov AV (2008) Regimes of laser plasma expansion at optical breakdown in the normal atmosphere. Tech Phys 53:693–696

Katsurayama H, Abe T (2013a) Thermochemical nonequilibrium modeling of a low-power argon arcjet wind tunnel. J Appl Phys 113:053304

Katsurayama H, Abe T (2013b) Numerical prediction of a critical altitude on electrodynamic aerobraking in a hypersonic continuum-rarefied transitional regime. AIAA Paper 2013-3003

Katsurayama H, Komurasaki K, Arakawa Y (2009) A preliminary study of pulse-laser powered orbital launcher. Acta Astronaut 65:1032–1041

Kemp H, Lewis P (1980) Laser-heated thruster interim report. NASA CR-161665

Maher WE, Hall RB, Johnson RR (1974) Experimental study of igition and propagation of laser-supported detonation waves. J Appl Phys 45:2138–2145

Matsui K, Shimano T, Ofosu JA et al (2017) Accurate propagation velocity measurement of laser supported detonation waves. Vacuum 136:171–176

Mori K, Komurasaki K, Arakawa Y (2002) Influence of the focusing F number on the heating regime transition in laser absorption waves. J Appl Phys 92:5663–5667

Mori K, Komurasaki K, Arakawa Y (2004) Energy transfer from a laser pulse to a blast wave in reduced-pressure air atmospheres. J Appl Phys 95:5979–5983

Mori K, Komurasaki K, Arakawa Y (2006) Threshold laser power density for regime transition of a laser absorption wave in a reduced-density air atmosphere. Appl Phys Lett 88:121502

Ogino Y, Nagano A, Ishihara T, Ohnishi N (2013) A fitting formula for radiative cooling based on non-local thermodynamic equilibrium population from weakly-ionized air plasma. J Phys Conf Ser 454:012080

Raizer Y (1965) HEATING OF A GAS BY A POWERFUL LIGHT PULSE. Sov Phys JETP 21: 1508–1519

Raizer YP (1977) Laser-induced discharge phenomena, Chap. 6. Consultants Bureau, New York and London

Ramsden SA, Davies WER (1964) Radiation scattered from the plasma produced by a focused ruby laser Beam. Phys Rev Lett 13:227–229

Ramsden SA, Savic P (1964) A radiative detonation model for the development of a laser-induced spark in air. Nature 203:1217–1219

Sedov LI (1959) Similarity and dimension methods in mechanics. Academic Press, New York

Shimamura K, Ofosu JA, Komurasaki K, Koizumi H (2015) Predicting propagation limits of laser-supported detonation by Hugoniot analysis. Jpn J Appl Phys 54:016201

Shimano T, Ofosu JA, Matsui K et al (2017) Laser-induced discharge propagation velocity in helium and argon gases. Trans Jpn Soc Aeronaut Space Sci 60:378–381

Ushio M, Komurasaki K, Kawamura K, Arakawa Y (2008) Effect of laser supported detonation wave confinement on termination conditions. Shock Waves 18:35–39

Wang T-S et al (2002) Advanced performance modeling of experimental laser lightcraft. J Propuls Power 18:1129–1138

Wang B, Komurasaki K, Yamaguchi T et al (2010) Energy conversion in a glass-laser-induced blast wave in air. J Appl Phys 108:124911

Zakharchenko SV (1984) Superdetonation optical discharge waves in air ($\lambda = 1.06\ \mu$). Sov J Quantum Electron 14:1429–1429

Chapter 8
High-Power Beam Source and Beam Transmission

Yasuhisa Oda

At Ama-no-iwato, Omoikane-no-kami descripted four equations
for Amaterasu-o-okami. Then there appeared the great source radiating radio wave
and light to provide beamed energy for everything on the earth.
A modernized description of ancient Japanese tail

Abstract This research explores high-power beam sources and their transmission, crucial components for applications involving beamed energy propulsion and power transmission. This paper introduces key aspects of electromagnetism, including the properties of Gaussian beams and the correlation between beam diameter and frequency. Various oscillation sources are discussed, with emphasis on devices like the magnetron, klystron, gyrotron, and free-electron laser. The exploration of high-power beam source technology examines gyrotron tubes, high-voltage power supply, superconducting magnets, matching optical units, and auxiliary systems. The study further looks into transmission technology, reviewing waveguides, miter bends, dummy loads, and power selection and control equipment. An electronic cyclotron heating system for fusion plasma serves as an illustrative example of high-power beam sources.

Keywords High-power beam source · Beam transmission · Electromagnetism · Gaussian beam · Oscillation source · Gyrotron · High-voltage power supply · Superconducting magnet · Waveguide · Electronic cyclotron heating system

---

Y. Oda (✉)
Department of Mechanical Engineering, Setsunan University, Neyagawa, Japan
e-mail: yasuhisa.oda@mec.setsunan.ac.jp

K. Mori et al. (eds.), *Beamed-mobility Engineering*, Springer Tracts in Electrical and Electronics Engineering, https://doi.org/10.1007/978-981-99-4618-1_8

Introduction of Electromagnetism

The beamed energy is the energy transportation which utilizes electromagnetic waves and its characteristics are described by electromagnetism.

Derivation of electromagnetic waves from Maxwell's equations.

The Maxwell equation is a combination of four equations for electric and magnetic fields, using electric field $\boldsymbol{E}$, magnetic flux density $\boldsymbol{B}$, current density $\boldsymbol{j}$, charge density ρ, permittivity in vacuum ε_0, and light velocity c as follows:

$$\nabla \cdot \boldsymbol{E} = \frac{\rho}{\varepsilon_0}$$

$$\nabla \times \boldsymbol{E} = -\frac{\partial \boldsymbol{B}}{\partial t}$$

$$\nabla \cdot \boldsymbol{B} = 0$$

$$c^2 \nabla \times \boldsymbol{B} = \frac{\boldsymbol{j}}{\varepsilon_0} + \frac{\partial \boldsymbol{E}}{\partial t}$$

In a vacuum, the charge ρ and the current $\boldsymbol{j}$ become 0, so the Maxwell equation in the vacuum can be rewritten as follows:

$$\nabla \cdot \boldsymbol{E} = 0$$

$$\nabla \times \boldsymbol{E} = -\frac{\partial \boldsymbol{B}}{\partial t}$$

$$\nabla \cdot \boldsymbol{B} = 0$$

$$c^2 \nabla \times \boldsymbol{B} = \frac{\partial \mathbf{E}}{\partial t}$$

By combining the Maxwell equation, the following differential equation can be obtained. Its solution is the famous formula which presents the traveling wave caused by the interaction of the electric field and the magnetic field in speed of light, namely, electromagnetic wave.

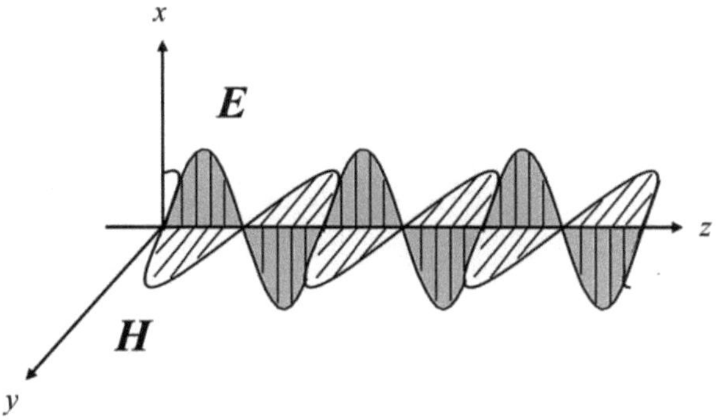

Fig. 8.1 Schematics of electromagnetic wave, transverse wave with orthogonal relation of the electric field, and the magnetic field

$$\nabla^2 \boldsymbol{E} = \frac{1}{c^2}\frac{\partial^2 \boldsymbol{E}}{\partial t^2}$$

The solution of a plane wave propagating in one dimension in the z-axis direction is described as the following equation:

$$\boldsymbol{E} = E_{\mathrm{x}} \cdot \boldsymbol{e}_{\mathbf{x}} = A e^{j(\omega t - kz)} \cdot \boldsymbol{e}_{\mathbf{x}}$$

In this equation, $\boldsymbol{e}_{\mathbf{x}}$ and $\boldsymbol{e}_{\mathbf{y}}$ represent unit vectors. ω and k represent the angular frequency and the propagation constant of a plane wave, respectively. They have the following relationship with the permittivity ε_0 and magnetic permeability μ_0 in the vacuum.

$$k = \omega\sqrt{\varepsilon_0 \mu_0} = \frac{2\pi}{\lambda}$$

As shown in Fig. 8.1, this phenomenon is a transverse wave, and the electric field and the magnetic field are orthogonal to each other.

As is well known, the propagation velocity of the electromagnetic wave in the vacuum is presented as follows and corresponds to the speed of light $c = 2.979925 \times 10^8$ [m/s]:

$$c = \frac{1}{\sqrt{\varepsilon_0 \mu_0}}$$

From the electromagnetic wave equation to the beam distribution equation.

Propagation of electromagnetic waves is a three-dimensional phenomenon. Now the description of electromagnetic wave propagating in one-dimensional direction is represented. When the wavefront of an electromagnetic wave spreads infinitely, the propagation plane becomes a plane. However, the energy of the actual

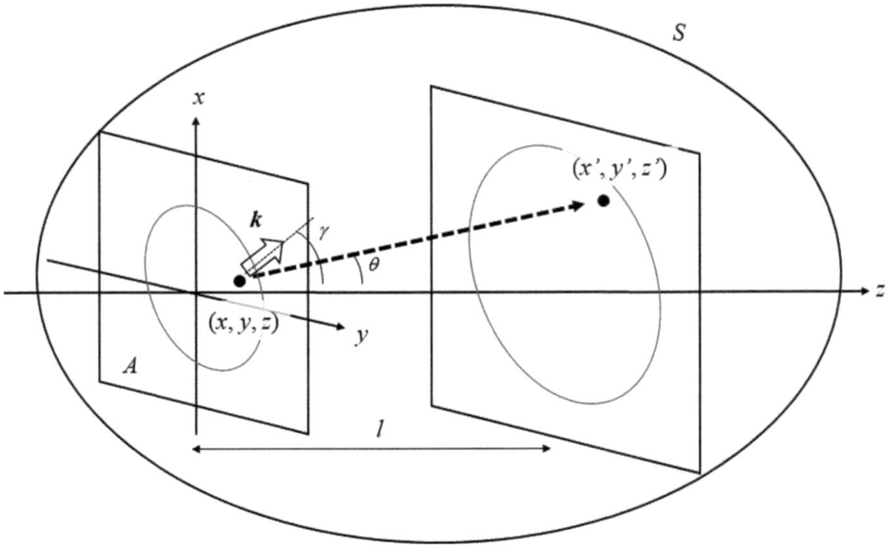

Fig. 8.2 Coordinate and plane relation for the analysis of electromagnetic wave propagation

electromagnetic wave has a limited region existing on the radiation plane, and the description in this case is given [1].

First, for electromagnetic waves distributed in space, the electric field $\boldsymbol{E}$ is described using the intensity U at each point, the vector $\boldsymbol{r}$ indicating the point, the wave vector k containing information on the propagation direction, and the unit vector $\boldsymbol{e}_\mathbf{p}$ in the propagation direction. It becomes

$$\boldsymbol{E} = U(\boldsymbol{r})\boldsymbol{e}_\mathbf{p}\ \exp[j(\omega t - \boldsymbol{k}\cdot\boldsymbol{r})]$$

By substituting this equation into the wave equation obtained in the previous section, a differential equation that does not include the following time items can be obtained. That is, it leads to finding the spatial intensity distribution without time change.

$$\nabla^2 U(\boldsymbol{r}) + k^2 U(\boldsymbol{r}) = 0$$

When describing beam propagation, we will consider it based on the Huygens principle. Huygens' principle shows that the action of electromagnetic waves at each location on the beam appears in superposition. Therefore, as shown in Fig. 8.2, the state of the electromagnetic field at a distant place can be obtained by integrating the action of the elements at each place on an arbitrary beam distribution. The

[1] ファインマン, レイトン, サンズ, ファインマン物理学 電磁気学.

distribution on the plane away from the radial plane can be obtained, and this formula is known as Kirchhoff's Huygens diffraction integral.

$$U(x',y',z') = \frac{1}{4\pi} \iint_S \left[\frac{\partial U}{\partial n} \frac{e^{-jkr}}{r} - U \frac{\partial}{\partial n} \left(\frac{e^{-jkr}}{r} \right) \right] \mathrm{d}S$$

More practically, the following shapes are convenient:

$$U_2(x_2,y_2,l) = \frac{jk}{4\pi} \iint_A (\cos\gamma + \cos\theta) U_1(x_1,y_1) \frac{e^{-jkr}}{r} \mathrm{d}S$$

For the above equation, if the wave vector of the beam is close to the z axis and the spread d of the beam distribution on the x and y planes is small with respect to the propagation distance l, cos γ and cos θ are 1. It can be described as follows:

$$U(x',y',l) = \frac{j}{\lambda} \iint_{S1} U(x,y) \frac{e^{-jkr}}{r} \mathrm{d}x\mathrm{d}y$$

$$r = \sqrt{(x-x')^2 + (y-y')^2 + d^2}$$

Such a situation is called paraxial approximation and is a condition commonly used in beam propagation evaluation.

The most common beam propagation form has a Gaussian distribution, and the beam optics propagating in normal space is assumed to be a Gaussian beam. This means that the following equation, which is obtained by incorporating the spatial distribution of the intensity U into the wave equation, becomes the Helmholtz equation, and as its solution, the form of beam propagation with the Gaussian distribution shown in Fig. 8.3 can be obtained.

$$\nabla^2 U - \frac{n^2}{c^2} \frac{\partial^2 U}{\partial t^2} = 0$$

$$U(x,y) = A \exp\left(-\frac{x^2+y^2}{\omega^2} \right) = A \exp\left[-\left(\frac{r}{\omega} \right)^2 \right]$$

In reality, the ideal Gaussian must have a beam distribution up to infinity, and the beam emitted from an artificial wave source such as an antenna does not have a perfect Gaussian distribution, and sidelobes appear. Even in such a case, Kirchhoff's

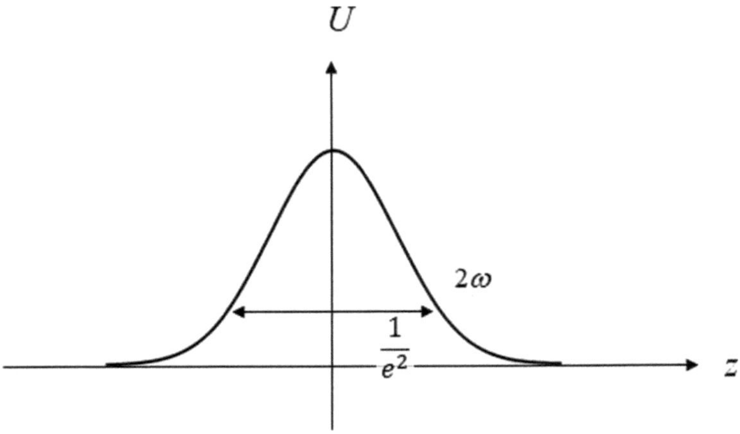

Fig. 8.3 Gaussian distribution and its radius and intensity

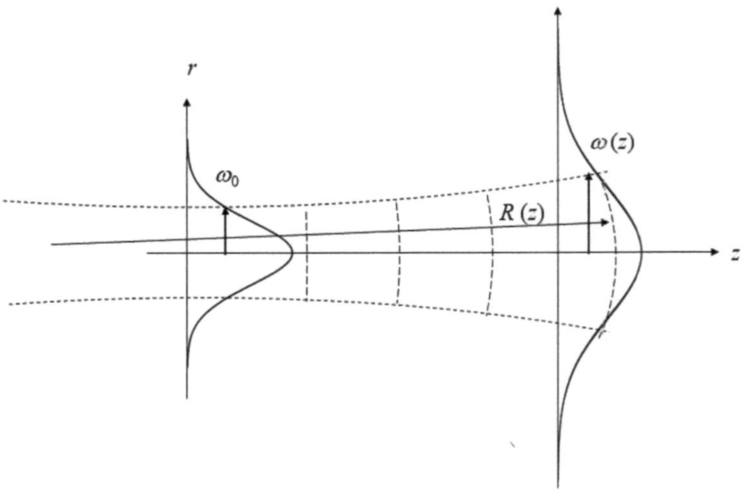

Fig. 8.4 Characteristics change of Gaussian beam propagation

Huygens diffraction integration is effective, and it is possible to reproduce the experimental data with sufficient accuracy [2].

[2] 小山次郎, 西原浩, 光波電子工学, コロナ社.

Features of the Gaussian Beam

As shown in Fig. 8.4, the Gaussian beam has the smallest beam diameter position called beam waist, and its profile plane is perpendicular to the beam axis. When the beam profile plain is perpendicular to the beam axis, phase distribution is constant. With propagation of beam from the beam waist, phase profile of the beam has a curved distribution, and the beam diameter increases. Eventually, when the phase distribution surface becomes spherical, the beam diameter spreads at a certain angle, and the size of the beam after long-distance propagation can be easily estimated.

$$\omega(z) = \omega_0 \sqrt{1 + \left(\frac{2z}{k\omega_0^2}\right)^2} = \omega_0 \sqrt{1 + \left(\frac{z\lambda}{\pi\omega_0^2}\right)^2}$$

$$R(z) = z\left[1 + \left(\frac{k\omega_0^2}{2z}\right)^2\right] = z\left[1 + \left(\frac{\pi\omega_0^2}{z\lambda}\right)^2\right]$$

$$\psi(z) = \cos^{-1}\left[\frac{\omega_0}{\omega(z)}\right]$$

Relationship Between Beam Diameter and Frequency

Beam diameter after long-distance propagation depends on the beam waist diameter and wavelength. In general, the longer the wavelength, which corresponds to the lower frequency, the beam has the larger diameter after long-distance propagation. To focus the beam power to a small diameter area after long-distance transmission of the beam, the initial diameter of the beam needs to be large and a large antenna system is required for such application.

In the use of beam energy transmission to move objects, it is necessary to keep a high-energy transmission efficiency toward moving objects of which size is small, so it is required to focus the beam on the target object, and the wavelength and beam station size are considered to adapt the requirement [3].

[3] 國分康雄, 光波工学, 共立出版.

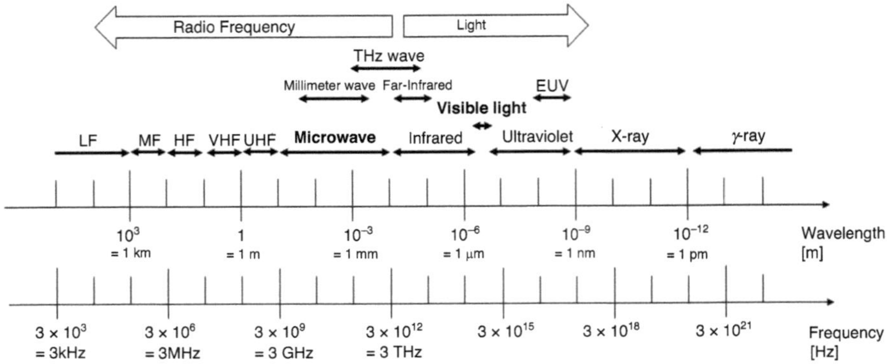

Fig. 8.5 Category of electromagnetic waves

Frequency of Electromagnetic Waves and Power Source

Electromagnetic waves are phenomena that appeared as a solution to the Maxwell's equations, but their characteristics and generation methods differ greatly depending on their wavelength or frequency. Generally, they are categorized into radio wave, light, and radiation in descending order of wavelength, and the details are shown in Fig. 8.5.

Radio wave is an electromagnetic wave with a relatively long wavelength obtained by radiating resonance of an electric circuit from an antenna. It is classified into long frequency (LF), medium frequency (MF), high frequency (HF), very high frequency (VHF), ultrahigh frequency (UHF), microwave, and millimeter wave according to the length of the wavelength, but the radio wave itself is defined as an electromagnetic wave with a frequency of 3000 GHz or less by the international technical standard. Radio waves in the high frequency range of microwaves and above have properties similar to those of light, which will be described later.

Oscillation Caused by Electric Circuit

The most basic method of generating electromagnetic waves is antenna radiation of an AC waveform, which can be obtained by resonance of an electric circuit composed of an inductor, such as coil, and a capacitor. Radios that appeared in the early twentieth century had a frequency of about several hundred kHz, but in line with the development of broadcasting, communication, and radar technology, microwaves and millimeter waves from the MHz region to the GHz region have also been oscillated. There is. In broadcasting and communication applications where stability is required, in most cases, a stable frequency source such as a crystal oscillator with modulator and amplifier is utilized. Semiconductors are used for these components, and the reachable frequency and power level are improving with the evolution of semiconductors. Further, in order to obtain a particularly high frequency, a method such as multiplying the fundamental frequency is also used.

In the radio wave region, frequencies of at least several GHz or higher are applied from the viewpoint of focusing beam property. In this frequency region, microwave electron tubes are suitable power source. In a microwave electron tube, electrons traveling in a vacuum are accelerated or decelerated by an electromagnetic field, and this interaction exchanges kinetic energy of electron with an electromagnetic wave power. Microwave electron tubes can obtain good performance as oscillator and amplifier in microwaves and millimeter waves. There are several types of microwave electron tubes, and the applicable frequencies and power levels of each type are different.

Microwave Electron Tube

In the oscillation of radio waves in the GHz band, microwave electron tubes are important oscillation method that utilizes the interaction between electrons and electromagnetic fields in space. Magnetrons are good example of microwave electron tube which is widely spread as a key component of household microwave ovens. Magnetrons can generate strong electromagnetic waves with a simple high-voltage circuit, and it has a simple structure and a possibility of mass production. Microwave electron tubes can cover high-frequency bands such as tens to hundreds of GHz and high output power of hundreds of kW to MW levels. The electron tube is a kind of vacuum tube, and it was often used even at low frequencies in the past, but when realizing high-power oscillation in the high-frequency band, the electron tube is an only-one choice, and various systems are studied and developed for many applications.

Magnetron

A magnetron is a microwave oscillation device, which converts the kinetic energy of electrons into electromagnetic waves in several cavities settled around a circular electron beam formed by a magnetic field of permanent magnet. Magnetrons are widely used in household microwave ovens as cheap microwave oscillator because they can be mass produced and can be driven by a simple high-voltage power supply system. In addition, the energy conversion efficiency from the electric power to microwaves reaches as high as 80%. In the pulse operation used for radar applications, the peak output power reaches 10 MW in the microwave to millimeter wave

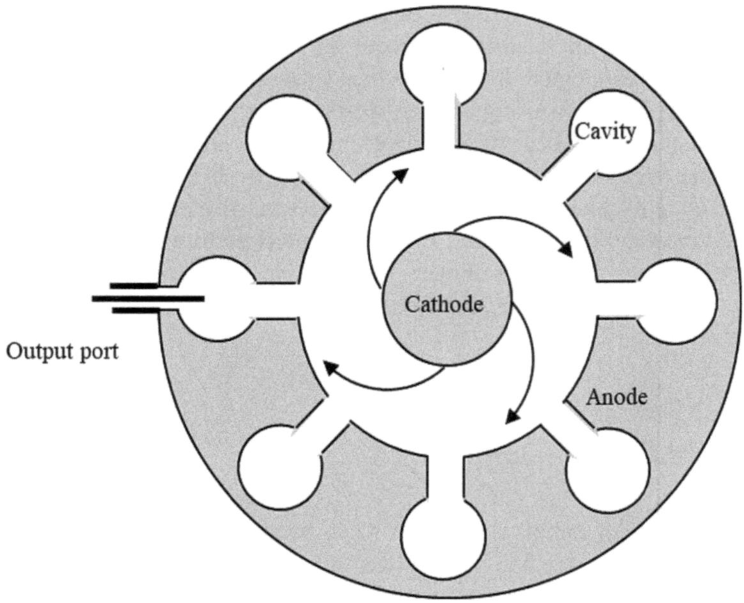

Fig. 8.6 Schematics of magnetron confinement

band region. In the CW operation used for heating applications, the average output power reaches up to about 20 kW [4] (Fig. 8.6).

Klystron

Klystron settles two or more cavities around the straight electron beam. Klystron applies velocity modulation to the electron beam with the cavity resonator in the front stage and electromagnetic waves from the cavity resonator in the rear stage by periodic density distribution of electron beam. This density distribution is formed by velocity modulation while electron beam traveling between the cavities. In principle, klystron works as an amplifier, but klystron can be operated as an oscillator by injecting some of the output power from the rear cavity resonator to the front cavity resonator. The electric power to electromagnetic wave conversion efficiency is improved by arranging multiple cavities, and the efficiency of more than 70% can be theoretically obtained. There are products with an efficiency of 65%. Regarding the output power of klystrons, in pulse operation, peak power reaches 100 MW, and even in CW operation average power exceeds 1 MW. It is widely used as a final stage amplifier for broadcasting and communication stations and as a high-frequency

[4] 桜庭一郎, 電子管工学 第2版,森北出版.

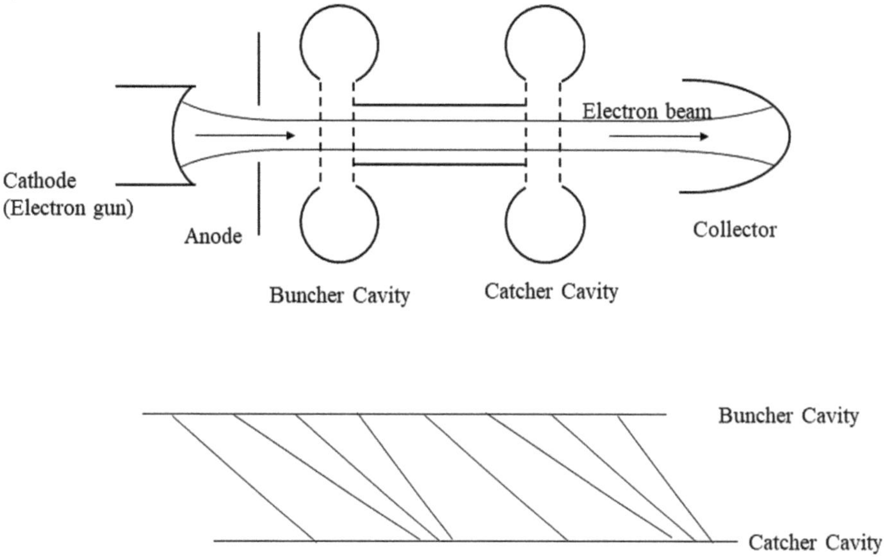

Fig. 8.7 Schematics of klystron confinement. Velocity modulation obtained in buncher cavity forms density distribution in catcher cavity

source for accelerators. The typical operating frequencies of klystron are hundreds of MHz to exceeding 10 GHz (Fig. 8.7)[4].

Gyrotron

The gyrotron is a millimeter wave oscillation system that uses the cyclotron resonance maser effect to convert the kinetic energy of the cyclotron rotational motion of the electron beam on the magnetic field line into electromagnetic waves. Unlike magnetrons and klystrons, the oscillation frequency is strongly regulated not only by the size of the cavity but also by the orbital speed of the cyclotron rotational motion due to the magnetic field lines, so the ultrahigh-order mode can be applied as the resonance mode in the cavity. Then, the gyrotron can produce electromagnetic wave with wavelengths in millimeter range. The MW level output is possible even in the millimeter wave region. For the purpose of heating plasma in a magnetic field confinement fusion experimental device, continuous oscillation of more than 1 MW is realized even in the frequency range exceeding 100 GHz, and it is also applied to research on beaming propulsion (Oda et al. 2019) (Fig. 8.8).

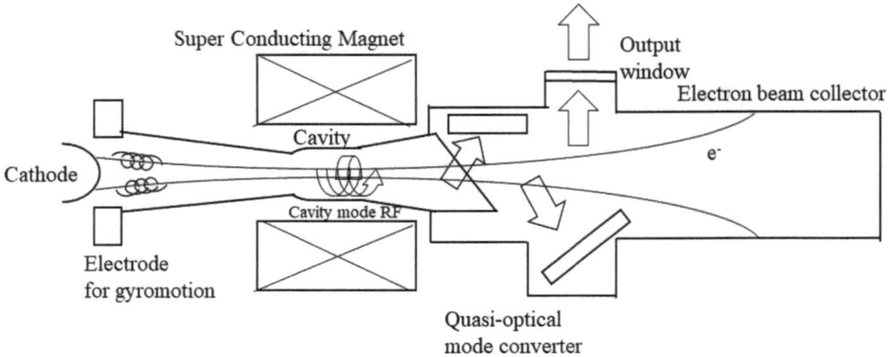

Fig. 8.8 Schematics of gyrotron confinement. Electron beam with gyromotion produced from cathode travels through cavity converting its kinetic energy to microwave, and rest energy is absorbed in the collector. Microwave in cavity forms cavity mode, and it is converted to Gaussian-like beam through quasi-optical mode converter

Synchrotron Radiation

When flying electrons lose energy, lost energy is converted into electromagnetic waves. By changing the amount of lost energy, radiation can obtain any wavelength range. This system is realized in an accelerator system. In general, the light sources have a limitation in the wavelength range. In particular, a powerful light source is very limited in the region where the frequency is located between radio waves and light, the so-called terahertz waves, and the region with a short wavelength from vacuum ultraviolet to X-ray. Indeed the light source using accelerators can generate light in these range and can play a main role. Now, it has become a popular light source and has been used in many applications.

A deflection magnet installed in the beamline of the accelerator allows the electron beam in the accelerator to bend its orbit. This change in beam trajectory gives white light with a broad spectrum of directivity, which is called synchrotron radiation. On the other hand, stronger light can be obtained by inserting a circuit that periodically changes the magnetic field for the purpose of generating strong light, which is called undulator synchrotron radiation. Undulator synchrotron radiation has a sharp peak spectrum, and a stronger peak intensity than the Wiggler synchrotron radiation can be obtained. Further, since the peak wavelength obtained by the arrangement shape of the undulator circuit can be changed, it can be widely used as a light source having a wavelength according to the purpose and application. In accelerators intended for general synchrotron radiation utilization, the equipment may be large, and many insertion light sources are arranged to secure a large number of beamlines to meet a wide variety of usage requirements. For example, Spring-8 has 62 beamlines. The standard Spring-8 undulator has a total power of 13.7 kW.

Free-Electron Laser

A free-electron laser is designed to enhance the coherence of light in the orbital change caused by the undulation magnet for the accelerated electron beam. As with normal synchrotron radiation, the wavelength can be freely selected by arranging the magnets, but the power level can be extremely large. In order to use it as a powerful light source, the accelerator is not a ring-type accelerator often used in synchrotron radiation equipment, but a linear accelerator that can obtain larger acceleration is used as a beam source. It has been developed as a high-power beam source in the hundreds of GHz to THz region, which is difficult to oscillate in an electron tube and as a laser source in the X-ray region without a practical resonator mirror. In the THz region, Jefferson-lab's FEL in the United States has achieved a peak output of 10 MW and an average output of 100 W and has achieved great results in bridging the gap between radio waves and light.

Technology for High-Power Beam Source

Beamed energy propulsion and any other application using beamed energy power transmission require a beam power source. The beam power source shall produce large output power and focused beam diameter on the target, which corresponds to proper selection of beam wavelength. The required output level varies depending on the application, but here, we will introduce sources with a level of several tens of kW to several MW.

Beam Source: Gyrotron Tube

The gyrotron generates a hollow-type electron beam from an electron gun called magnetron injection gun (MIG), and the electrons starts the cyclotron motion by the magnetic field produced by superconducting magnet. The kinetic energy of the rotation motion of electrons is maintained by the relation of magnetic field and electric field formed by gyrotron electrodes. Electrode confinement has two variations, namely, diode type and triode type. Diode-type MIG has cathode and body electrodes as shown in Fig. 8.9. The shape of electrodes forms electric field to accelerate electrons both gyromotion and translational motion. The pitch factor, ratio of gyromotion and translational motion, is fixed in diode-type MIG confinement. Triode-type MIG has cathode, anode, and body electrodes as shown in Fig. 8.10. Electric field by anode and cathode works for the acceleration of gyromotion and electric field by the body, and cathode works for acceleration of translational motion. Hence, the pitch factor is controllable by changing anode voltage. This is very useful to achieve a highly efficient operation using hard

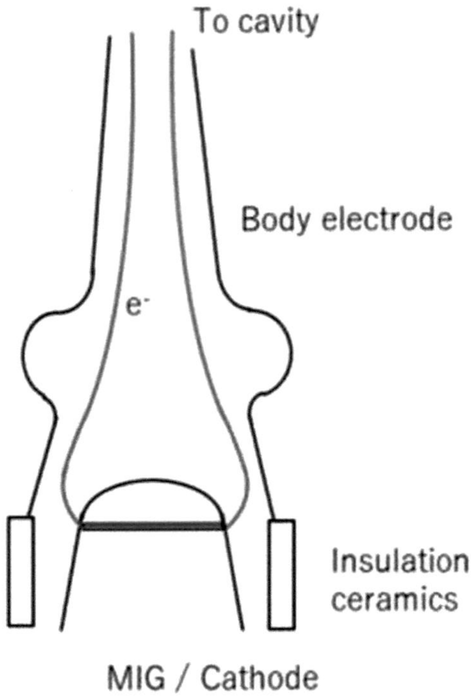

Fig. 8.9 Diode-type MIG confinement

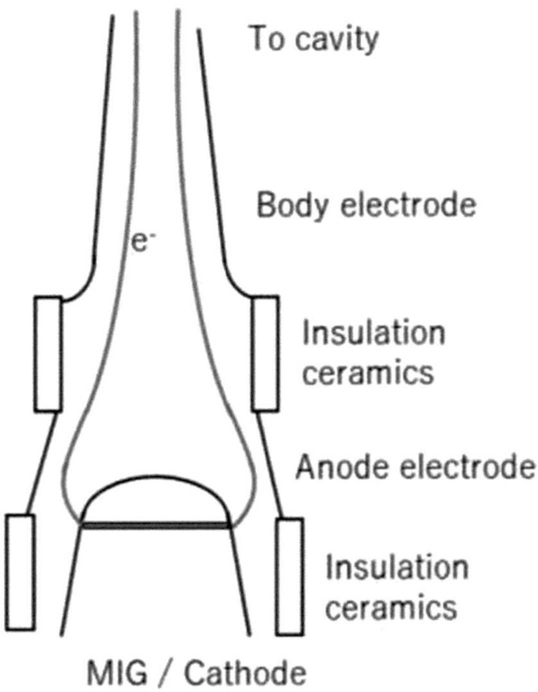

Fig. 8.10 Triode-type MIG confinement

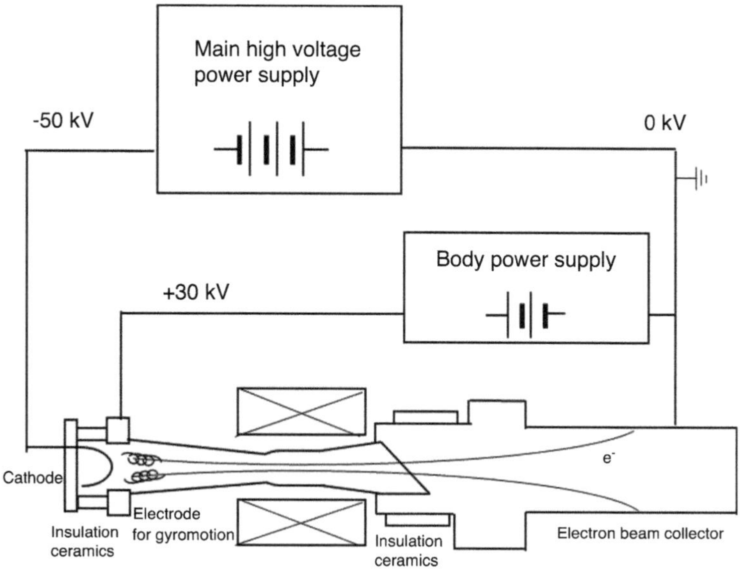

Fig. 8.11 Power supply configuration for diode-type MIG

excitation region and high-efficiency multifrequency operation; both are important usage in modern gyrotron development. The difference in MIG confinement affects power supply configuration. For diode type, the MIG needs the main high-voltage power supply for cathode feeding and body power supply for CPD operation as shown in Fig. 8.11. CPD operation is a method to improve the operation efficiency by decelerating electron's translational motion of which component cannot be not converted to electromagnetic wave energy in the cavity, and this can reduce the operating voltage of the main HVPS. For triode type, the MIG needs additional power supply for anode electrode feeding as shown in Fig. 8.12.

The speed of electron's rotation is accelerated, and the electron beam is introduced into the open-end resonant cavity. When the speed of electron approaches the speed of light, the interaction of electron motion and electric field exchanges energy by cyclotron resonance maser process as shown in Fig. 8.13. Since the operation frequency of cyclotron resonance maser depends on the magnetic field as represented in following equation, the high-order mode cavity which can have large size can be utilized. For example, the resonance mode of the cavity of the 170 GHz gyrotron is $TE_{31,11}$ which is applicable for 1 MW production.

Since it is difficult to transmit the high-order mode generated in the cavity over a long distance, a quasi-optical mode converter is utilized to convert the cavity mode into a Gaussian beam with a single peak as shown in Fig. 8.14. The quasi-optical mode converter has a simple wall inside, which maintains mode conversion of high-order cavity mode forming superposition of multiple modes, which produces a single-peak profile at the output of the converter. Several mirrors maintains X–Y radius and focusing of the single-peak beam, and the gyrotron can produce a

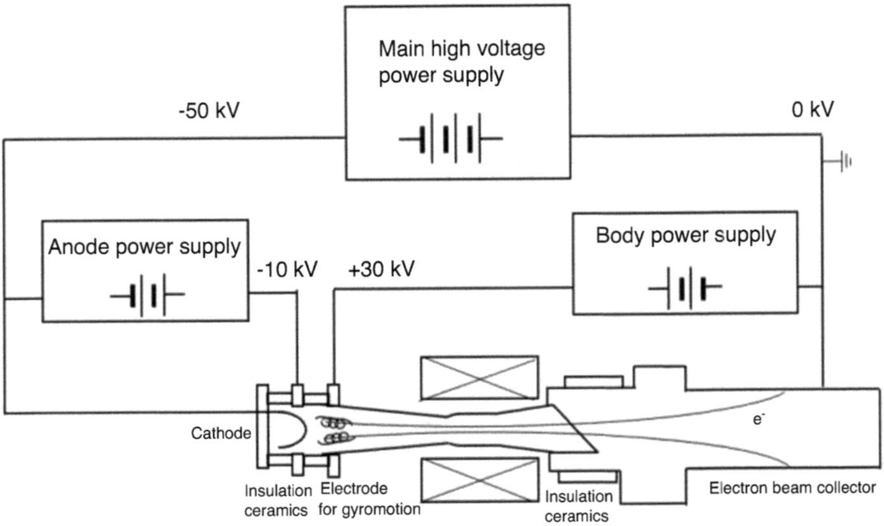

Fig. 8.12 Power supply configuration for triode-type MIG

Gaussian beam as an output beam profile. The gyrotron is an electron tube, the inside is kept in an ultra-high vacuum, and the output window is necessary. Since the window produces heat due to the dielectric loss when passing through the window, the window is a major constraint for continuous operation in high-power operation. In a high-power gyrotron, a steady output has been achieved at the 1 MW level by using diamond, which has a low dielectric loss and a high heat transfer performance for cooling, as a window (Oda et al. 2019; Denisov et al. 2008; Sakamoto et al. 2007; Minami et al. 2006).

High-Voltage Power Supply for Gyrotron

Driving a gyrotron requires a high voltage of several tens of kV and several tens of A, and the stability of the power supply determines the operating performance of the gyrotron. In addition, active electron beam control is performed at the anode electrode near the MIG for accelerating the electron beam (triode) and a body power supply for CPD operation. CPD operation is a method to improve the operation efficiency by decelerating electron's translational motion of which component cannot be not converted to electromagnetic wave energy in the cavity, and this can reduce the operating voltage of the main HVPS. It requires a complicated high-voltage power supply, but it achieves an energy conversion efficiency of over 50%.

For main HVPS, several types are utilized for various systems. The capacitor bank type is a simple system as shown in Fig. 8.15. An HV power supply which produces low-current-charge capacitor bank, and a large current is supplied to the

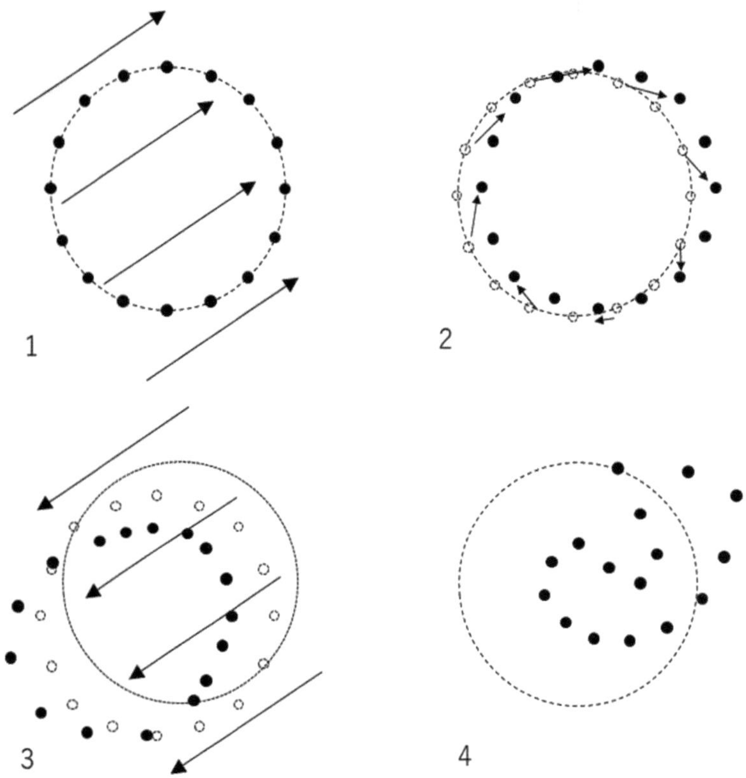

Fig. 8.13 Schematic of cyclotron resonance maser. At the initial state, electrons are rotating on the same radius cyclotron motion. (1) By acceleration and deceleration by electric field, electrons get larger or smaller radius orbit from the initial orbit. (2) The numbers of accelerated electrons and decelerated electrons become unbalanced. For gyrotron operation, the numbers of decelerated electrons are higher by electric field cycle. (3) Finally, most electrons go down to small radius orbit losing kinetic energy (4)

gyrotron from the capacitor bank when HV switch is turned on. This system has limitation in pulse duration. By selecting small HV power supply, normal electric power plug can operate a high-power gyrotron in small-scale laboratory.

For long pulse and CW operation, direct HV power supply from HV grid or motor generator is necessary. For such configuration, DC-generator-type system, as shown in Fig. 8.16, and multi-stuck-type system, as shown in Fig. 8.17, are used. Recent gyrotron system, like the ITER system, utilizes multi-stuck-type HVPS which has wide controllability by PSM control (Rathod et al. 2019).

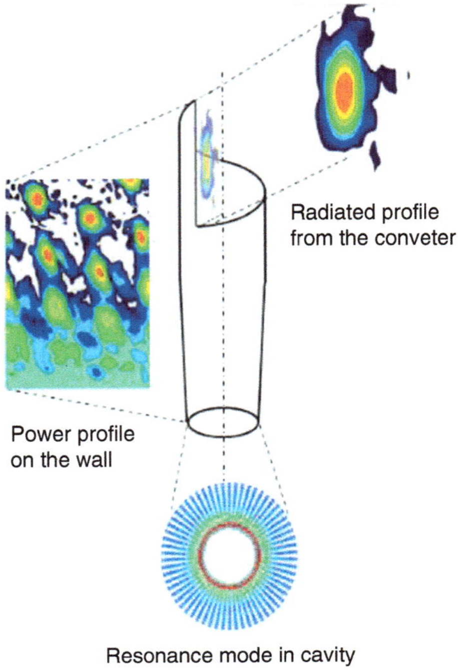

Fig. 8.14 Quasi-optical mode converter for cavity mode conversion to Gaussian profile

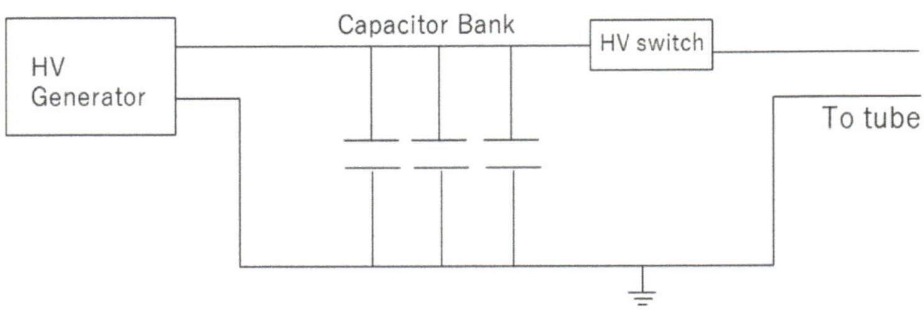

Fig. 8.15 Capacitor-bank-type power supply for gyrotron

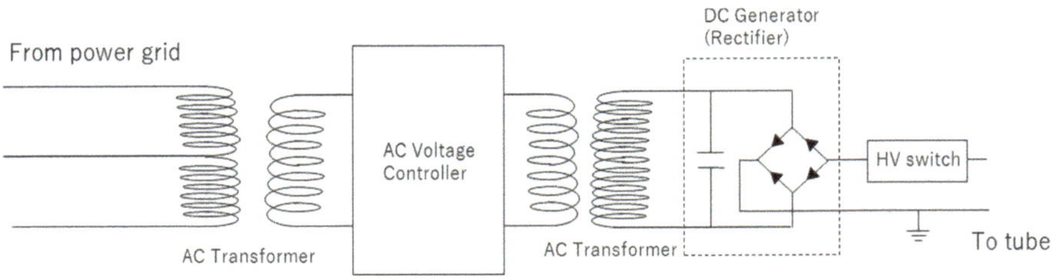

Fig. 8.16 DC-generator-type power supply for gyrotron

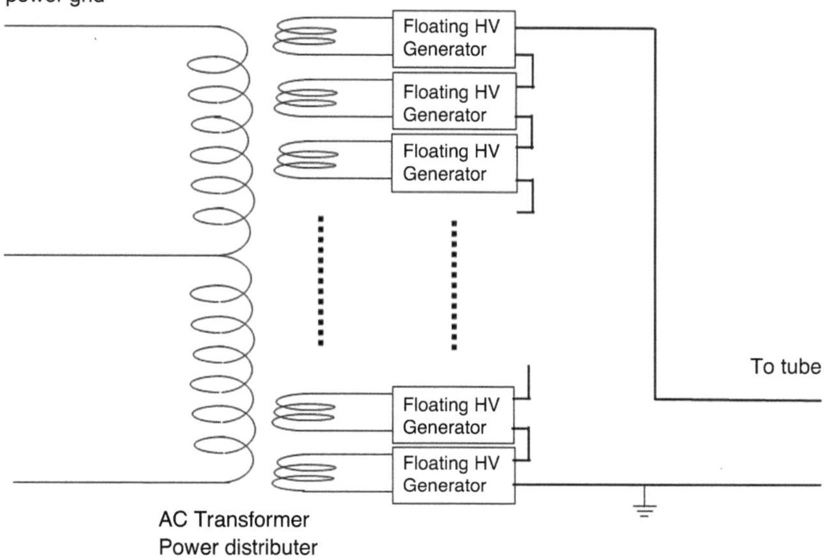

Fig. 8.17 Multi-stack HV unit-type power supply for gyrotron

Superconducting Magnet

A superconducting magnet is used to drive a millimeter-wave band gyrotron because it requires a magnetic field of several T levels at the cavity of gyrotron to start CRM. The electron beam orbit and its gyromotion in the gyrotron is controlled by the magnetic field lines, and the cusp magnetic field is suitable. To produce a cusp magnetic field, a combination of single superconducting solenoid coil and normal conducting coil or combined coils in the cryostat of the superconducting magnet is utilized. In recent years, instead of the conventional liquid helium for cooling superconducting materials, magnets using a refrigerator and conduction heat transfer have become widespread, and the complexity of operation has been improved (Kamikado et al. 1999).

Matching Optical Unit

A matching optical unit (Fig. 8.18) is a device which couple RF beam from gyrotron window to waveguide of transmission line. It is necessary to maintain RF beam profile to couple to waveguide propagation mode; a couple of phase correcting mirrors, as shown in Fig. 8.19, are utilized for beam coupling. For highly efficient coupling, mirrors in the matching optical unit shall be adjusted by position and angle control stage according to output RF beam profile optimization process. For

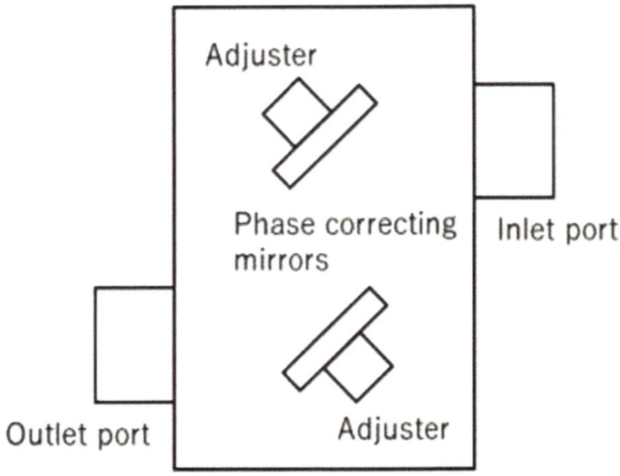

Fig. 8.18 Configuration of matching optical unit for gyrotron

Fig. 8.19 Phase correcting mirror for gyrotron beam coupling in matching optical unit

high-power beam profile measurement, thermal measurement of screen settled on the beam line using an IR camera is widely utilized. Since the evacuated waveguide system is utilized in high-power system, the matching optical unit is designed as vacuum component, and cooling water for mirrors is provided through vacuum feedthrough (Oda et al. 2012, 2013).

Auxiliaries

Since the efficiency of high-power gyrotrons are around 50%, it requires the same cooling power level as the RF output power. Table 8.1 shows the power balance of ITER gyrotron, although the collector loss is dominant and other component still needs a large amount of cooling which defines the design of auxiliary cooling system.

Table 8.1 Gyrotron loss power for 1 MW gyrotron system (Oda et al. 2019)

Measurement items	Value
MOU loss power	30 kW
In-tube diffraction loss	20 kW
In-tube Ohmic loss (except cavity)	22 kW
Cavity loss	33 kW
Collector loss	1000 kW

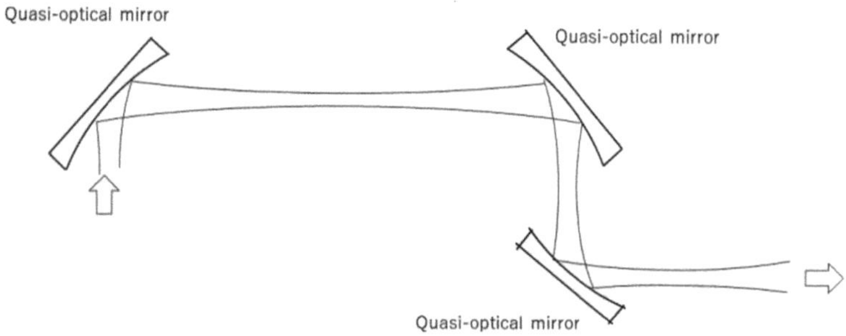

Fig. 8.20 Quasi-optical beam transmission system

There are various engineering equipment such as mechanical support structures, high-voltage power feed system, safety system, and control system. In the fusion heating system, the development of a beam source of several tens of MW class is actually progressing, and the study of the system of the level exceeding 100 MW is already underway.

Technology for High-Power Beam Transmission

In order to transmit beamed energy from the beam station to the target, it is necessary to transmit and maintain the high-power electromagnetic wave energy from the power source to the antenna system. In this section, technical components for handling strong electromagnetic energy ranging from several tens of kW to several MW is described.

To deliver microwave or millimeter-wave power from the beam source to some points, a quasi-optical beam transmission (Fig. 8.20) or waveguide transmission (Fig. 8.21) can be applicable for high-power transmission. Quasi-optical beam transmission is a method of transmitting in space using a Gaussian beam, but since the beam diameter increases with long-distance transmission due to the influence of diffraction, it is necessary to adjust the transmission path with an optical system such as a mirror. In waveguide transmission, by confining an electromagnetic field in a metal tube, it is possible to transmit in a narrower space than in beam transmission, but a loss occurs on the surface of the waveguide.

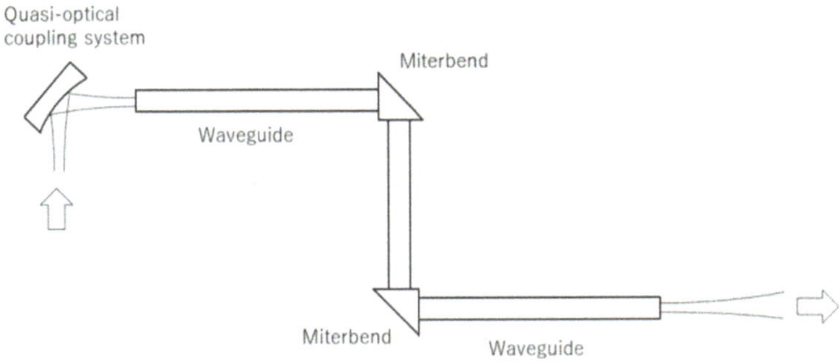

Fig. 8.21 Waveguide beam transmission system

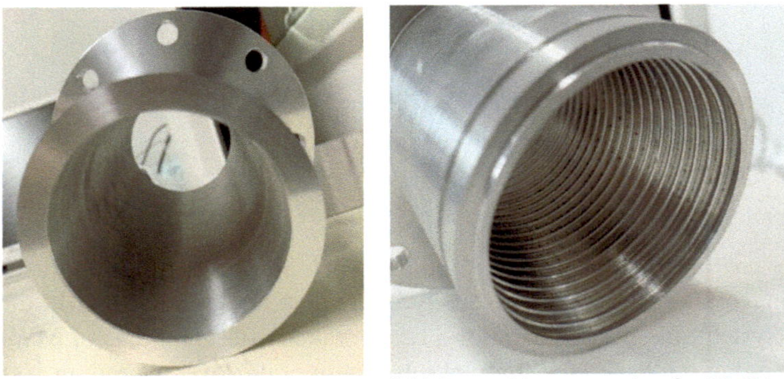

(a) Waveguide corrugation for 70 GHz

(b) Waveguide corrugation for 28 GHz

Fig. 8.22 Corrugated waveguide of 63.5 mm inner diameter for millimeter wave. (**a**) Waveguide corrugation for 70 GHz. (**b**) Waveguide corrugation for 28 GHz

Waveguide

In waveguides in microwave band, the scale of waveguide is selected as half or quarter of the length of a wavelength and such waveguide can transmit microwave power in the fundamental transmission mode and such waveguide is called fundamental wave waveguide. On the other hand, in the short-wavelength wave such as a millimeter-wave band, the scale of a typical fundamental wave waveguide becomes smaller, resulting in unacceptable losses during high-power transmission. Therefore, a waveguide having a diameter larger than the wavelength, which is called an oversized waveguide, is used. Since the wavelength is shorter than the diameter of the waveguide, the electromagnetic field in the waveguide becomes complicated, and an appropriate control beam profile is required.

For high-power millimeter-wave system, corrugated waveguides (Fig. 8.22) are widely used. The fundamental transmission mode of corrugated waveguide is HE_{11}

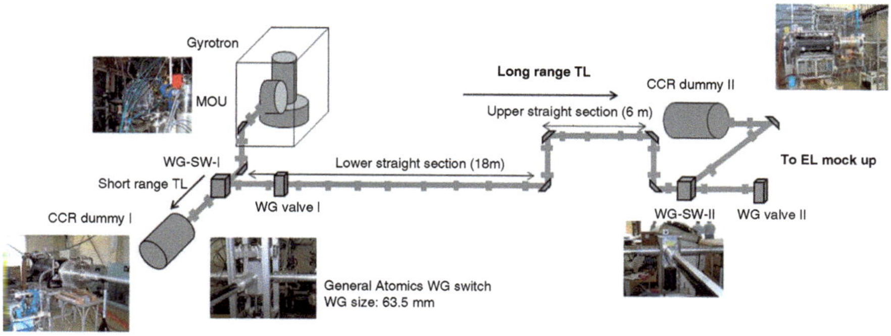

Fig. 8.23 Corrugated waveguide system and components for ITER RF transmission line test stand (Oda et al. 2018)

mode, which is a similar profile of the TEM mode (transverse electromagnetic), and it is well coupled to Gaussian profile.

Loss of HE_{11} mode transmission is very low in corrugated waveguide.

Miter bend is a direction change component of waveguide, which utilize a mirror. Due to reflection loss on the mirror, 0.5%–1% of power loss occurs in the miter bend.

The corrugated waveguide system achieved a power transmission efficiency of higher than 95% in 40 m length transmission experiment in 170 GHz/1 MW system as shown in Fig. 8.23 (Oda et al. 2018; Doane and Olstad 2008).

Miter Bend

Since the transmission mode in the waveguide can only propagate in straight line, a miter bend is utilized to change the direction of the transmission line. Figure 8.24 shows the schematic of the miter bend. Miter bend has a mirror inside to reflect the beam into desired direction. A 90° miter bend is utilized for fusion system. Since mirror reflection is utilized, miter bend has a transmission loss caused by ohmic loss of mirror reflection. In addition, the miter bend generates a higher-order transmission mode which also causes the transmission loss. Typically the transmission loss by a single miter bend is 1–2% in an experiment.

Dummy Load

To measure the RF power delivered through the transmission line, a dummy load is utilized for a high-power transmission line. Since power density in the waveguide is very high, a beam profile transformer is applied to reduce power density for power absorber. The simplest profile transformer is a conical mirror which can expand the

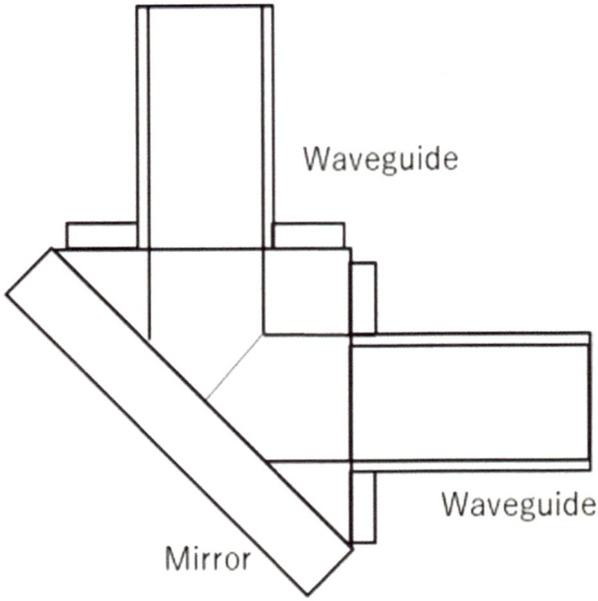

Fig. 8.24 Miter bend for corrugated waveguide

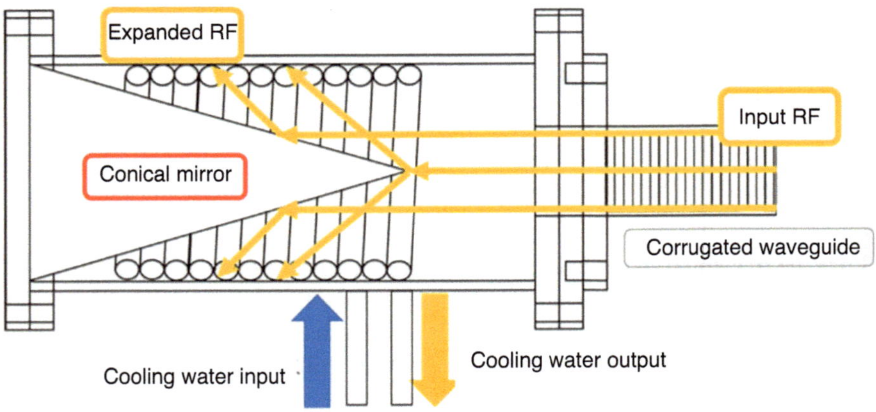

Fig. 8.25 Configuration of dummy load with conical mirror

beam to the surrounding wall as shown in Figs. 8.25 and 8.26. The expanded beam is absorbed by the cooling water, and the RF power can be estimated from the cooling water temperature change. For long pulse operation, various types of expanding mirrors with cooling circuit are used, for example, a rotating mirror.

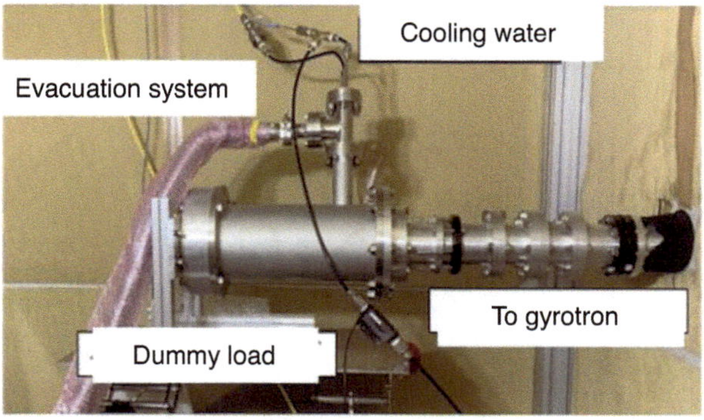

Fig. 8.26 Example of short pulse dummy load for RF power measurement. The device was manufactured in Setsunan University

Power Selection and Control Equipment

There are active waveguide components to change beam properties. In a corrugated waveguide system, the miter bend can be configured as a switch which changes the direction of the beam by direction of the mirror. As another configuration, a polarizer can control the polarity of the beam by changing the corrugation of the miter-bend mirror surface by mirror rotation. These are mechanical control components.

There are also non-mechanical control components. A system that switches the beam direction by controlling the wavelength of electromagnetic waves by a system using a resonance ring has also been proposed as shown in Fig. 8.27. Such device utilizes a resonance condition change which is determined by the relation of the wavelength and ring length to change the output port. Since non-mechanical control component is promising for a high-speed beam control, research on such components are underway (Saigusa et al. 2017).

Example of High-Power Beam Source

In the early 2020s, there is no practical system that supplies propulsion energy to flying objects such as beamed energy propulsion. However, a large beam source that is comparable to the power level required in beamed energy propulsion is already under development, and this section introduces them.

Electronic Cyclotron Heating System for Fusion Plasma

In magnetic confinement fusion, the electrons behave in cyclotron motion due to the magnetic field. The electron cyclotron heating (ECH) system utilizes radio frequency power of which frequency is resonant to the electron's cyclotron motion. In order to improve the plasma performance for fusion reaction, the confinement magnetic field

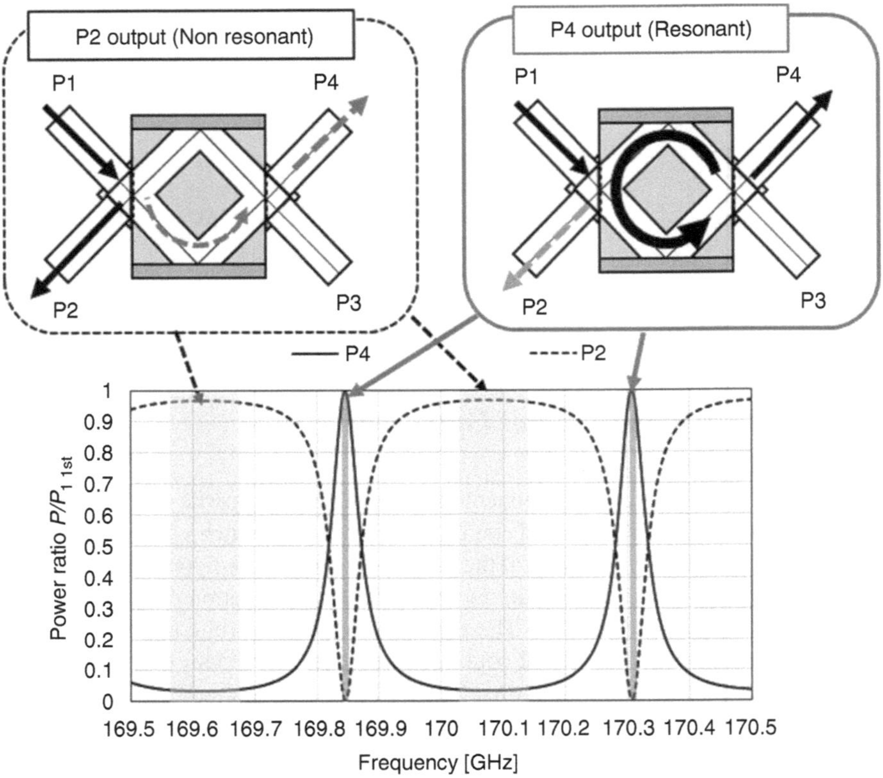

Fig. 8.27 Frequency dependence of P2 and P4 output power. Reflection coefficient $a = 0.7$ and resonator length $L = 650$ mm

is several T in the large fusion device, and the operation frequency of ECH system is in the millimeter wave band, which exceeds several tens of GHz to 100 GHz. In addition, the RF power required to heat a large fusion device ranges from several MW to several tens of MW.

As of the early 2020s, in the ITER project under construction in France, the specifications of the ECH system are 170 GHz/20 MW, and the system is designed to produce the required output power by a cluster of 24 gyrotrons. This system includes a high-voltage power supply system for driving gyrotrons, waveguide transmission lines of over 100 m for introducing the beam output power from each gyrotron into the ITER fusion reactor, and various auxiliaries like cooling of the gyrotron, support structure, and etc. Technology and design developed for huge ECH system can be used for other applications, like beamed energy propulsion (Darbos et al. 2016).

References

Darbos C et al (2016) Status of the ITER electron cyclotron heating and current drive system. J Infrared Millim Terahertz Waves 37(1):4–20

Denisov GG, Litvak AG, Myasnikov VE, Tai EM, Zapevalov VE (2008) Development in Russia of high-power gyrotrons for fusion. Nucl Fusion 48:054007

Doane J, Olstad R (2008) Transmission line technology for electron cyclotron heating. Fusion Sci Technol 53:39–53

Kamikado T, Itoh S, Hamada M, Hayashi S, Kawata Y, Shibutani K (1999) A high raping-rate superconducting magnet cooled by a 4K GM refrigerator. 低温工学 34:227–229

Minami R, Kasugai A, Takahashi K et al (2006) The Designs of high efficiency launcher of quasi-optical mode converter for high power gyrotrons. Int J Infrared Milli Waves 27:13–24

Oda Y, Kajiwara K, Takahashi K, Sakamoto K (2012) A method for gyrotron beam coupling into a corrugated waveguide. Fusion Sci Technol 61:203–208

Oda Y, Kajiwara K, Takahashi K, Mitsunaka Y, Sakamoto K (2013) High efficiency coupling of radio frequency beams from the dual frequency gyrotron with a corrugated waveguide transmission system. Rev Sci Instrum 84:013501

Oda Y, Anderson J, Doane J, Kajiwara K, Ikeda R, Takahashi K, Sakamoto K (2018) A study on multi frequency operation of RF transmission system for EC H&CD system. Nucl Fusion 58: 106041

Oda Y, Ikeda R, Kajiwara K, Kobayashi T, Hayashi K, Takahashi K, Moriyama S, Sakamoto K, Eguchi T, Kawakami Y, Mitsunaka Y, Darbos C, Henderson MA (2019) Development of the first ITER gyrotron in QST. Nucl Fusion 59:086014

Rathod V, Dilip ES, Shah R, Mandge D, Yadav A, Sharma A, Parmar R, Singh NP, Rao SL (2019) Developments towards 1MW Gyrotron Test Facility at ITER India. In: EC-20, EPJ web of conferences 203, p 04013

Saigusa M, Sekiguchi K, Honzu T, Mori J, Shozu H, Oda Y, Ikeda R, Takahashi K (2017) High power test of a temperature controlled diplexer for electron cyclotron current drive system. Fusion Eng Des 123:385–389

Sakamoto K, Kasugai A, Takahashi K, Minami R, Kobayashi N, Kajiwara K (2007) Achievement of robust high-efficiency 1 MW oscillation in the hard-self-excitation region by a 170 GHz continuous-wave gyrotron. Nat Phys 3:411–414

Chapter 9
Retrodirective Wireless Optical Energy Transmission Using Optical Phase Conjugation

Kotomi Kawakami and Hideki Okamura

Abstract The increasing demand for wireless optical energy transmission in various aerospace applications necessitates efficient laser beam propagation over long distances, potentially across turbulent atmospheres. This study delves into retrodirective wireless optical energy transmission, employing optical phase conjugation to mitigate beam spreading and wandering caused by atmospheric disturbance. The phase conjugate mirror (PCM), a device reflecting phase conjugate light (PCL), enables automatic targeting/tracking action, and near-diffraction-limited focusing of the laser beam, making it ideal for remote moving targets. This chapter reviews existing wireless optical power transfer techniques, such as gimbal tracking and adaptive optics, while introducing optical phase conjugation technology as a new method for retrodirective optical power transfer using phase correction. Principles of optical phase conjugation and PCL generation techniques are also explored.

Keywords Wireless optical energy transmission · Laser beam propagation · Atmospheric turbulence · Optical phase conjugation · Phase conjugate mirror · Phase conjugate light · Beam spreading · Beam wandering · Adaptive optics · Gimbal tracking

The demand for wireless optical energy transmission is increasing in various fields of aerospace, such as fuel-free aircraft including drones, space solar power systems, laser propulsion, and small space-debris removal using laser aberration. For these applications, it is necessary to irradiate a high-energy laser beam on a remote target, which can be moving. The direction of propagation of the laser is not always toward the sky (in the direction of the zenith). Because laser beams are severely attenuated in haze, fog, and rain, it is necessary to prepare several separate locations and

K. Kawakami (✉)
College of Liberal Arts and Sciences, Kitasato University, Sagamihara, Kanagawa, Japan
e-mail: k-kawa@kitasato-u.ac.jp

H. Okamura
Division of Arts and Sciences, International Christian University, Mitaka, Tokyo, Japan
e-mail: hokamura@icu.ac.jp

K. Mori et al. (eds.), *Beamed-mobility Engineering*, Springer Tracts in Electrical and Electronics Engineering, https://doi.org/10.1007/978-981-99-4618-1_9

irradiate the laser beams horizontally from the locations with good conditions. Therefore, a long-distance propagation in horizontal or slant paths is expected.

In general, although a laser has excellent directionality, when the beam propagates near the surface of the earth, atmospheric turbulence widens the laser beam beyond its natural spread and deflects it from its original path, resulting in severe energy loss. Therefore, it is necessary to overcome the difficulty in maintaining an efficient power feed. To maintain the pointing accuracy of the laser beam for a long distance, it is necessary to correct the distortion of the optical wavefront as the beam propagates in turbulent atmosphere.

The phase conjugate mirror (PCM) is a device that reflects a replica of the incident light in the direction opposite to the incident light. Because this replica called phase conjugate light (PCL) has the same wavefront as the incident light but travels backward, the phase lead is reflected as phase lag. Therefore, experiencing the same distortion effect twice, the phase lag/lead is compensated as if rewinding time, and the original optical wavefront is restored at the position of the light source. This time-reversal property enables near-diffraction-limited focusing of the laser beam in the presence of atmospheric disturbance. Moreover, the time reversibility can be used to track a moving object because PCL returns without control (only by optical phenomena) to the position of the light source, even if it changes. This automatic tracking/focusing action is suitable for the optical transmission of remote moving objects (Schäfer et al. 2007a, b; Giuliano 1981; Kruzhilin 1978).

This chapter explains the unique properties of PCL, and the application of optical phase conjugation for automatic target acquisition, tracking, and optical energy transfer in free space. Section "Atmospheric Disturbance" presents the effects of atmospheric turbulence. Section "Wireless Optical Power Transfer Technologies" briefly describes the gimbal system and adaptive optics, which are existing long-distance wireless optical power transfer and phase correction techniques, respectively, and introduces optical phase conjugation technology as the third possible method for retrodirective optical power transfer using the phase correction effect. The principles of optical phase conjugation are elucidated in section "Principles of Optical Phase Conjugation." The method for PCL generation is presented in section "Phase-Conjugate Light Generation Technique."

Atmospheric Disturbance

Turbulence-Induced Beam Spreading and Beam Wandering

The possible methods for long-distance wireless energy transmission are limited to microwaves or lasers (Ahmad et al. 2017). In long-distance transmission, the presence of air causes several problems, even with the laser having an excellent directivity. Similar to heat haze on a hot day, air at different temperatures sways in the atmosphere with the wind and convection. Therefore, a laser beam experiences refractive index fluctuations in air, resulting in wavefront distortion and loss of directivity. If the sizes of the air parcels at different temperatures are smaller than the beam diameter, the beam is divided into multiple cells, and additional diffraction broadens the beam diameter beyond its natural diffraction spread (Weichel 1990). The beam radius a_t at which the intensity decreases to 1/e or 0.368 as its peak value is described as a function of the wavelength λ and propagation distance L:

$$a_t = 2.01\left(\lambda^{-1/5} C_n^{6/5} L^{8/5}\right),$$

where the atmospheric structure constant C_n indicates the turbulence strength, which varies considerably depending on path conditions, such as the time and altitude.

As the beam passes through the boundaries of air parcels larger than the beam diameter, refraction changes the direction of the light path. The constant movement of wind causes the arrival point of the beam to shift, which is called beam wandering. The rate of fluctuation is relatively low and of the order of 0.5 Hz to a few kilohertz (Weichel 1990; Murty 1979). The radial variance due to beam wander σ_r can be expressed as

$$\sigma_r^2 = 1.83\ C_n^2 \lambda^{-1/6} L^{17/6}.$$

C_n^2 typically ranges from $10^{-17}\text{m}^{-2/3}$ for weak turbulence to $10^{-13}\text{m}^{-2/3}$ or more for strong turbulence. The variance in the x-direction is

$$\sigma_x^2 = \sqrt{\frac{\sigma_r^2}{2}}.$$

The effect of beam spreading and wandering is stronger when the propagation distance is considerable (Fig. 9.1). This tendency is stronger when the path is horizontal and near the surface of the earth, where atmospheric turbulence is strong. Therefore, a technique to compensate for distorted wavefronts is necessary to prevent energy loss in the receiver.

Wireless Optical Power Transfer Technologies

Gimbal Tracking

The gimbal method is commonly used for shining light on a moving object, where the mirror is moved by a motor (Lee et al. 2000; Tidwell et al. 2012; Hayes 1977). For example, in an experiment in which light energy was transmitted from a 200-W laser light source on the ground to a kite plane 60-m above the ground, the reflected light from the corner cube in the center of the solar panel of the kite plane was observed using a four-segment photodetector to capture the target object

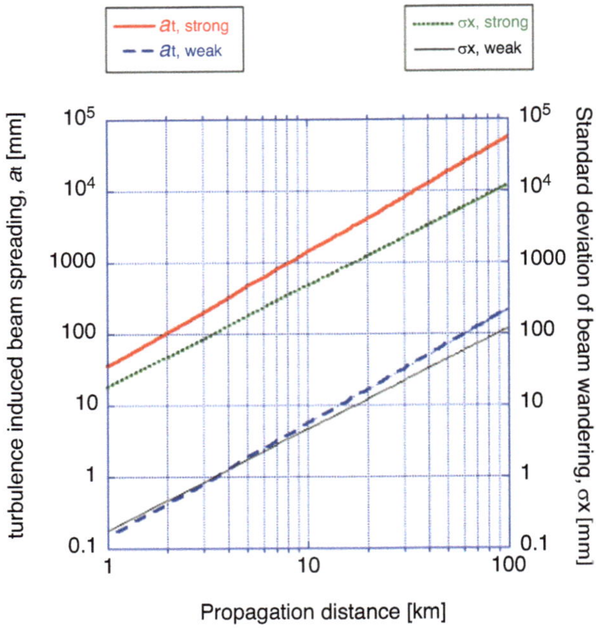

Fig. 9.1 Turbulent beam spreading a_t and standard deviation of the beam wandering in the x-direction σ_x as a function of the propagation distance for strong ($C_n^2 = 10^{-13}$) and weak ($C_n^2 = 10^{-17}$) turbulence

(Kawashima et al. 2007). The probe light that was initially spread was focused by reducing the divergence angle using two lenses after capturing the target object, and tracking was performed by controlling the two rotation stages. However, such a mechanical beam deflection method involves a tradeoff between the weight, direction accuracy, and speed. In addition, the generation of vibration and torque by the rotating stage becomes a problem when mounting the system on a spacecraft.

Adaptive Optics

Adaptive optics (AO) is another well-established wavefront correction technique mainly used in astronomical fields. AO, which compensates the images of stars distorted by atmospheric turbulence, is necessary for telescopes located on the ground where the influence of atmospheric turbulence is unavoidable. Correction is performed by measuring the phase errors and performing iterative control of a deformable mirror. Thus, the system requires a wavefront sensor, deformable mirror, and control computer. For example, in the Subaru Telescope, the incoming wavefront shape is measured by a wavefront sensor comprising 188 microlens arrays, and the surface shape of the deformable mirror is controlled through 188 actuators to cancel wavefront distortion. Of late, AO precompensation has been mainly researched for free-space optical communication (Bonnefois et al. 2019; Brady et al.

2017). The system measures the wavefront distortions of a detected beacon beam and mechanically controls a deformable mirror to create a conjugate light wave. The usage of a wavefront sensor that cannot cope with strong turbulence is one of the constraints. Moreover, the number of actuators controlling the deformable mirror is small, ranging from several tens to hundreds. In lunar laser ranging, which measures the distance between the earth and moon by shining a laser on a retroreflector on the surface of the moon and detecting the reflected light, low-order adaptive correction using a tip–tilt mirror that compensates only for beam wandering is being studied based on the idea that tip–tilt error accounts for 87% of wavefront distortions (Xiong and Feng 2003).

Optical Phase Conjugation

The advantage of the PCL is that beam shaping and steering can be performed only by optical phenomena, enabling automatic target acquisition, tracking, and pointing. In addition, compensation for birefringence and simultaneous multitarget focusing is possible (Fisher 1983). Because light returns with no loss of beam quality after passing through distorting optical elements such as amplifiers with a thermal lens, the PCM works synergistically with the master oscillator power amplifier (MOPA) system. Table 9.1 summarizes previous studies on light energy transmission using PCL. Most of the experiments involve field tests on wavefront-distortion compensation; however, Markov et al. generated PCL for a target moving at 800 mm/s.

Table 9.1 Pointing and tracking experiments using optical phase conjugation

First author	Nonlinear effect	Description	Ref.
N.F. Pilipetskii	SBS	Experimental demonstration of time reversibility	Lebow and Ackerman (1989)
R.C. Lind	Saturable absorption	Wavefront correction after 2 km of propagation	Lebow and Ackerman (1989)
P.S. Lebow	BEFWM	Wavefront correction after 560 m of propagation	Lebow and Ackerman (1989)
H. Bruesselbach	SBS	Wavefront correction after 6 km of propagation	Bruesselbach et al. (1995)
V. Kmetik	SBS	Wavefront correction after 100 km of propagation	Kmetik and Niino (2002)
V. Markov	Thermal effect	Tracking experiment (800 mm/s)	Markov et al. (2003)
		Wavefront correction after 6.4 km of propagation	Markov et al. (2006)
K. Kawakami	Saturable amplification	Tracking experiment (80 mm/s) Field of view >20° Simultaneous two-point focusing	Kawakami et al. (2015, 2017)

Principles of Optical Phase Conjugation

A device called a phase conjugate mirror produces PCL which has an identical phase front as the incident light traveling in the exact opposite direction (180°) regardless of the angle of incidence. Consider incident light of frequency ω propagating in the z-direction:

$$E(r,t) = \mathrm{Re}\left[\psi(x,y,z)e^{i\omega t}\right] = \mathrm{Re}\left[A(x,y)e^{i(-kz+\varphi(x,y))}e^{i\omega t}\right],$$

where the slowly varying amplitude $A(x,y)$ is real. The PCL wave can then be expressed as follows:

$$E_{\mathrm{PCL}}(r,t) = \mathrm{Re}\left[\psi^*(x,y,z)e^{i\omega t}\right] = \mathrm{Re}\left[A(x,y)e^{i(kz-\varphi(x,y))}e^{i\omega t}\right].$$

In the above equation, only the space-dependent term has a complex conjugate relationship. Therefore, PCL has a reversed phase front of the incident light wave. PCL is also called time-reversed light because it travels back in exactly the same path as that of the incident light, similar to rewinding time. This property can be observed from the fact that PCL satisfies the relationship, $E_{PCL}(r,t) = E(r.-t)$. Therefore, the diverging light emitted from the point light source is reflected by the PCM and becomes a convergent light focusing on the position of the point light source. Note that the focal spot size cannot be smaller than the diffraction limit because part of the incident light is lost owing to the finite aperture size of the PCM, resulting in incomplete phase conjugation. PCMs have been arrayed to increase the aperture size, in a study (Schäfer et al. 2007a).

PCL is generated by recording the incident light in a nonlinear medium in the form of a diffraction grating and reading it out. Recording is performed through various nonlinear processes such as saturable gain, the optical Kerr effect, stimulated Brillouin scattering (SBS), and photorefraction. Considering saturable gain as an example, the gain of the medium is modulated by the incident light waves to form a gain diffraction grating. Specifically, light from the target object (object light) and the reference light prepared separately are incident on a uniformly excited gain medium such as an Nd:YAG crystal. The incoming object light and reference light interfere in the medium. Stimulated emission then occurs actively at locations of constructive interference, and the excited atoms fall to the ground level. In contrast, at locations of destructive interference, the excited atoms remain at the upper level. Therefore, the number of excited atoms is modulated according to the brightness and darkness of the interference fringes. This corresponds to the spatial and periodical modulations of the amplification factor of the medium, which behaves as an amplitude modulation diffraction grating. The gain grating diffracts the light, and the diffracted light becomes the PCL. This process is called saturable gain four-wave mixing, in which light is introduced to the medium from three directions to generate a fourth PCL (Fig. 9.2).

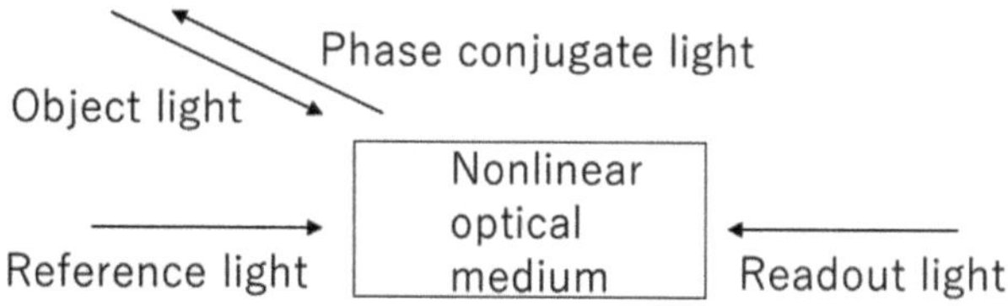

Fig. 9.2 Illustration of four-wave mixing (the reference and readout light should be counter-propagated)

The method for PCL generation is similar to the principle of hologram recording and reproduction. The difference is that in holograms, a virtual image is observed, whereas the PCM generates a real image. The hologram records the object information in the interference fringes by simultaneously irradiating the recording medium (film) with the object and reference light. When the film is developed and exposed to the reference light alone, the interference fringes function as an amplitude-modulation-type diffraction grating. The diffracted light has the same properties as the original object light. Therefore, an image can be seen behind the film even in the absence of a physical object. In the case of the PCM, the interference fringes are illuminated by light traveling in a direction opposite to the reference light. This can be expressed as follows:

Let the object light E_{obj} and reference light E_{ref} be

$$E_{\text{obj}}(\boldsymbol{r}, t) = E_{\text{obj}} \exp\left(\omega t - \boldsymbol{k}_{\text{obj}} \cdot \boldsymbol{r}\right) i,$$
$$E_{\text{ref}}(\boldsymbol{r}, t) = E_{\text{ref}} \exp(\omega t - \boldsymbol{k}_{\text{ref}} \cdot \boldsymbol{r}) i,$$

where $\boldsymbol{r}$ is the position, t is the time, ω is the frequency, and $\boldsymbol{k}$ is the wave number vector.

Then, the interference pattern $I(\boldsymbol{r})$ by the two waves is given by

$$\begin{aligned} & I(\boldsymbol{r}) \\ & = < E_{\text{obj}}(\boldsymbol{r}, t) + E_{\text{ref}}(\boldsymbol{r}, t) > \\ & = \left|E_{\text{obj}}(\boldsymbol{r}) + E_{\text{ref}}(\boldsymbol{r})\right|^2 \\ & = \left|E_{\text{obj}}(\boldsymbol{r})\right|^2 + \left|E_{\text{ref}}(\boldsymbol{r})\right|^2 \\ & + E_{\text{obj}}(\boldsymbol{r}) E_{\text{ref}}(\boldsymbol{r})^* \exp\left(-\boldsymbol{k}_{\text{obj}} \cdot \boldsymbol{r} + \boldsymbol{k}_{\text{ref}} \cdot \boldsymbol{r}\right) i \\ & + E_{\text{obj}}(\boldsymbol{r})^* E_{\text{ref}}(\boldsymbol{r}) \exp\left(\boldsymbol{k}_{\text{ref}} \cdot \boldsymbol{r} - \boldsymbol{k}_{\text{obj}} \cdot \boldsymbol{r}\right) i, \end{aligned}$$

where $< >$ indicates time averaging. By illuminating the readout light E_{read}, we get

$$E_{\text{read}}(\boldsymbol{r}, t) = E_{\text{read}} \exp(\omega t - \boldsymbol{k}_{\text{read}} \cdot \boldsymbol{r}) i,$$

On this interference pattern, the intensity distribution I of the interference fringes becomes a mask with different transmittances for the readout light, and the amplitude of the readout light is modulated. Therefore, the transmitted light E_t is given by

$$
\begin{aligned}
&E_t(\boldsymbol{r}, t) \\
&= I(\boldsymbol{r}) E_{\text{read}}(\boldsymbol{r}, t) \\
&= \left(\left| E_{\text{obj}}(\boldsymbol{r}) \right|^2 + \left| E_{\text{ref}}(\boldsymbol{r}) \right|^2 \right) E_{\text{read}}(\boldsymbol{r}, t) \\
&+ E_{\text{obj}}(\boldsymbol{r}) E_{\text{ref}}(\boldsymbol{r})^* E_{\text{read}}(\boldsymbol{r}, t) \exp\left(\omega\, t - \boldsymbol{k}_{\text{obj}} \cdot \boldsymbol{r} + 2\boldsymbol{k}_{\text{ref}} \cdot \boldsymbol{r} \right) i \\
&+ E_{\text{obj}}(\boldsymbol{r})^* E_{\text{ref}}(\boldsymbol{r}) E_{\text{read}}(\boldsymbol{r}, t) \exp\left(\omega\, t + \boldsymbol{k}_{\text{obj}} \cdot \boldsymbol{r} \right) i.
\end{aligned}
$$

Here, because the readout light is in the opposite direction of the reference light, the phase-matching condition $\boldsymbol{k}_{\text{read}} = -\boldsymbol{k}_{\text{ref}}$ is applied. The first term of the above equation for the transmitted light is the 0th-order diffracted light transmitted by the readout light without being diffracted, whereas the second term is the first-order diffracted light that diverges after transmission. It can be observed that the third term is the first-order diffracted light, which is the PCL of the original object light. Most PCMs are volume holograms that use thick media, such as nonlinear crystals, and their angular selectivity is excellent. In addition, unlike general holograms, the PCM is a dynamic hologram in which the interference fringes are recorded and readout simultaneously. Because the incident waves exchange energy in between, the reflectivity can exceed unity depending on the mechanism, which is another unique property of the PCM.

Phase-Conjugate Light Generation Technique

Nonlinear Optical Element

The first study on PCL dates back to 1972 (Fisher 1983). B. Ya. Zel'dovich et al. observed that the stimulated Brillouin backscattering process in CS_2 gas corrected the distorted wavefront after passing twice through a ground glass distorting object (Zel'dovich et al. 1995). Since then, PCL generation through various nonlinear optical effects has been reported. Table 9.2 summarizes the typical PCL generation mechanisms and their principles and characteristics. Among them, the photorefractive effect and stimulated Brillouin scattering (SBS) are commonly used. Although the photorefractive effect has no threshold and can generate PCL using weak light, the response speed depends on the intensity of the incident light waves and can range from microseconds to minutes. SBS can generate PCL with only one incident light; however, the PCL reflected by the diffraction grating propagating at the velocity of sound undergoes a frequency shift (several GHz). Therefore, the light generated by SBS is not PCL in a strict sense. In addition, the incident light with a high-peak power is required owing to the existence of a threshold. The Brillouin-enhanced four-wave mixing (BEFWM) is a four-wave mixing method that uses induced Brillouin scattering. Unlike SBS, it has no threshold and can realize high reflectance; however, as in SBS, frequency shift occurs.

Table 9.2 Optical phase conjugation mechanisms

Nonlinear effect	Response speed [s]	Principle	Feature
Optical Kerr effect (Sakai 1992)	10^{-12} to 10^{-11}	Refractive-index change due to molecular reorientation occurring in a medium with anisotropic polarizability	Fast response. High-peak intensity required (>10 GW/cm^2)
Stimulated Brillouin scattering (Brignon and Huignard 2003)	10^{-9}	Refractive-index change due to the elastic optical effect	Doppler shift. High-peak power required (10–100 kW)
Photorefraction (Brignon and Huignard 2003)	10^{-6} to few minutes	Refractive-index change due to Pockels effect	Slow response speed. Light absorption
Free carriers in semiconductors (Sakai 1992)	10^{-12} to 10^{-7}	Permittivity-change due to the generation of an electron-hole pair	Large fringe spacing required (>10 μm)
Saturable absorption	>10^{-6} (Hemmer et al. 1995)	Refractive-index change in a two-level gain medium	High saturation intensity required
Molecular reorientation in liquid crystals	10^{-4} to 10^{-2} (Johnson et al. 1990)	Change in transmittance due to Fredericks transition	Slow response speed. Temperature control required
Thermal effect (Sakai 1992)	10^{-1} to 1	Refractive-index change due to the change in molecular density	Slow response speed. Diffraction grating elimination by thermal diffusion
Brillouin enhanced four-wave mixing	10^{-9}	Four-wave mixing in an SBS cell	Frequency shift

To utilize PCL for wireless optical energy transfer, it is desirable to meet the following requirements: (1) a medium with a large aperture to suppress diffraction loss due to long-distance propagation, (2) a wide viewing angle, (3) a fast response that is crucial for tracking high-speed moving objects, and (4) the absence of frequency shifts and thresholds for ease of operation. It is also desirable that the operating wavelength be within the atmospheric window. The gain saturation method is a good candidate that meets the above requirements. The first advantage of using gain saturation is its high response speed (ns). A wide viewing angle can be expected because the pitch of the gain modulation can be narrowed. The highly regarded ceramic laser medium can be used in which any shape, size, and doping concentration can be realized at low cost. Because the thermal-lens effect generated within the gain medium cannot be canceled, it is suitable for application as an MOPA system.

Digital Phase Conjugation

An alternative method is available for generating PCL using digital devices. In this scheme, interference patterns are recorded by a camera and displayed on a spatial light modulator (SLM), such as a liquid crystal panel (Cui and Yang 2010) or digital mirror device (Wang et al. 2015). PCL is generated by illuminating the readout light on the SLM. The positions of the camera and SLM are symmetrical with respect to the beam splitter; therefore, the optical path of the PCL coincides with the object light after passing through the beam splitter (Fig. 9.3). As recording and reading of the interference patterns are performed in separate positions, unwanted interaction between the incident light waves can be prevented. This renders the system robust because independent laser oscillators can be selected for each process. For example, a highly coherent laser beam can be used for recording, whereas a high-power laser oscillator such as CO_2, LD, fiber, or even solar-pumped solid-state lasers can be used for reading. Because of the unrestricted operating wavelength, utilization of the atmospheric windows become easy.

The duration between image acquisition and display should be short for generating PCL before the atmospheric conditions change. The fluctuation speed of atmospheric turbulence is represented by time $\tau = r_0/v$ at which the velocity of the turbulence (wind speed) v passes through the Fried (coherence) length r_0, which is the length within which the wavefront can be regarded as a plane wave. For example, Sergeyev et al. measured Fried lengths of 2 mm to 15 cm in paths over 3 km at an elevation of 250 m (Sergeyev and Roggemann 2011). Assuming a wind speed of 10 m/s, the fluctuation time $\tau = 0.2$–15 ms. Therefore, a response speed of the order of milliseconds or less is required to freeze the fluctuations in the atmospheric turbulence. Liquid crystal devices are extensively used because of their high resolutions and ease of use. Liquid crystal devices have been installed on a small solar power sail demonstrator called the Interplanetary Kite-craft Accelerated by Radiation of the Sun (IKAROS) to show their usage in harsh environments. Most of the

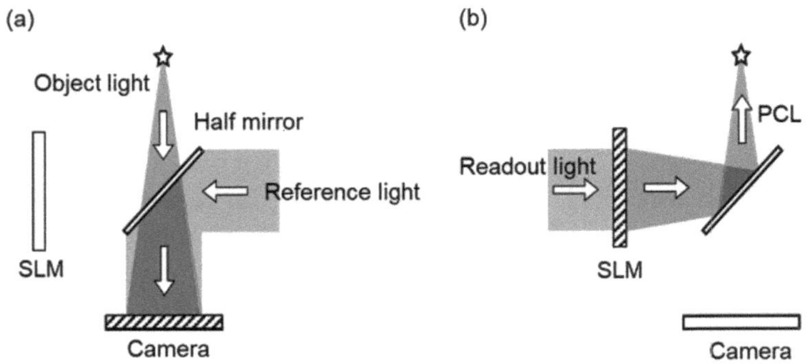

Fig. 9.3 Schematic of digital optical phase conjugation: Interference pattern (**a**) recording and (**b**) reading are performed simultaneously in separate locations. The spatial light modulator (SLM) is depicted as a transmission type for simplicity

recent liquid crystal SLMs are of the reflective type, called liquid crystal on silicon (LCOS), with high aperture ratios. The greatest challenge here is that the molecular reorientation time of the liquid crystals is generally slow (tens of hertz). However, the overdrive technique (Thalhammer et al. 2013) has been applied to solve this problem, and refresh rates exceeding 1 kHz have been achieved. The author measured a total delay of about 5.9 ms using a combination of high-speed camera (XIMEA MQ013MG-ON) and high-speed 1920 × 1152 LCOS SLM (Meadowlark optics HSP1920-500-1200). The digital mirror device (DMD) is another option that has been recently used in high-speed projectors. It comprises an array of micromirrors, each of which is switched on and off by tilting to ±12° mostly. The structure of the blazed grating necessitates off-axis illumination. A PCM using a 1920 × 1080 pixels DMD controlled by an FPGA has been reported, for which a response speed of 5.3 ms has been achieved (Wang et al. 2015).

For remote target tracking, a probe light is first emitted to illuminate a relatively large area. Part of the probe light reflected by the target is the object light. The object and reference light should interfere; therefore, a laser oscillator with a long coherence length is required. One of the methods to overcome this problem is to use a narrow bandwidth (few kHz) single-mode fiber laser. Another possible method is to divide the object light into two beams and spatially filter one of the beams to retain only the DC component. Because the object information is eliminated, the beam can be used as the reference light with zero path-length difference (Kumar et al. 2020).

References

Ahmad A, Alam MS, Vatsala, Varshney Y, Khan RH (2017) A state of the art review on wireless power transfer a step towards sustainable mobility. In: 14th IEEE India Council International Conference (INDICON)

Bonnefois AM, Conan J-M, Petit C, Lim CB, Michau V, Meimon S, Perrault P, Mendez F, Fleury B, Montri J, Védrenne N (2019) Adaptive optics pre-compensation for GEO feeder links: the FEEDELIO experiment. In: Proceedings of the SPIE 11180, International Conference on Space Optics—ICSO 2018, p 111802C

Brady A, Berlich R, Leonhard N, Kopf T, Böttner P, Eberhardt R, Reinlein C (2017) Experimental validation of phase-only pre-compensation over 494 m free-space propagation. Opt Lett 42: 2679–2682

Brignon A, Huignard J-P (2003) Phase conjugate laser optics. Wiley-Interscience, Hoboken, NJ

Bruesselbach H, Jones DC, Rockwell DA, Lind RC, Vogel G (1995) Real-time atmospheric compensation by stimulated Brillouin-scattering phase conjugation. J Opt Soc Am B 12: 1434–1447

Cui M, Yang C (2010) Implementation of a digital optical phase conjugation system and its application to study the robustness of turbidity suppression by phase conjugation. Opt Express 18:3444–3455

Fisher RA (1983) Optical phase conjugation. Academic Press, New York

Giuliano CR (1981) Applications of optical phase conjugation. Phys Today 34:27–35

Hayes CL (1977) High energy laser pointing and tracking system utilizing beam angle/focus dither method of operation, US patent 4,063,819

Hemmer PR, Katz DP, Donoghue J, Cronin-Golomb M, Shahriar MS, Kumar P (1995) Efficient low-intensity optical phase conjugation based on coherent population trapping in sodium. Opt Lett 20(9):982–984
Johnson K, Mao C, Moddel G, Handschy M, Arnett K (1990) High-speed, low-power optical phase conjugation using a hybrid amorphous silicon/ferroelectric-liquid-crystal device. Opt Lett 15(20):1114–1116
Kawakami K, Okamura H, Komurasaki K (2015) Simultaneous optical tracking of multiple targets in a field of view greater than 20°. Opt Express 23:9777–9783
Kawakami K, Okamura H, Komurasaki K (2017) Retrodirective tracking of a moving target using phase conjugate light generated in a Fabry-Perot Nd:YAG laser. J Appl Phys 121:093104
Kawashima N, Takeda K, Yabe K (2007) Application of the laser energy transmission technology to drive a small airplane. Chin Opt Lett 5:S109–S110
Kmetik V, Niino M (2002) Influence of the atmospheric effects on the lightwave energy transmission link performance. In: ISTS 2002-f-05
Kruzhilin YL (1978) Self-adjusting laser-target system for laser fusion. Sov J Quantum Electron 8(3):359–363
Kumar M, Quan X, Awatsuji Y, Cheng C, Hasebe M, Tamada Y, Matoba O (2020) Common-path multimodal three-dimensional fluorescence and phase imaging system. J Biomed Opt 25(3): 032010
Lebow PS, Ackerman JR (1989) Phase conjugation through Brillouin-enhanced four-wave mixing over an extended atmospheric path. Opt Lett 14:236–238
Lee S, Alexander JW, Jeganathan M (2000) Pointing and tracking subsystem design for optical communications link between the international space station and ground. Proc SPIE 3932:150–157
Markov V, Khizhnyak A, Zel'dovich B, Martinez T, Liu S (2003) Tracking system with PCM-laser for space control applications. In: Proceedings of the SPIE 5087, Laser Systems Technology
Markov V, Khiznyak A, Woll D, Liu S (2006) Field-testing of an active laser tracking system. In: Ryan S (ed) The advanced Maui optical and space surveillance technologies conference, Wailea, Maui, Hawaii. The Maui Economic Development Board, p E37
Murty SSR (1979) Laser beam propagation in atmospheric turbulence: Indian Academy of Sciences. Proc Sect C Eng Sci C2:179–195
Sakai J (1992) Phase conjugate optics (advanced science & technology). McGraw-Hill, New York
Schäfer CA, Matoba O, Kaya N (2007a) Optical retrodirective tracking system approach using an array of phase conjugators for communication and power transmission. Appl Opt 46:4633–4641
Schäfer CA, Matoba O, Kaya N (2007b) Tracking system by phase conjugation for laser energy transmission. In: Proceedings of the SPIE 6454, High energy/average power lasers and intense beam applications, p 64540A
Sergeyev AV, Roggemann MC (2011) Monitoring the statistics of turbulence: fried parameter estimation from the wavefront sensor measurements. Appl Opt 50:3519–3528
Thalhammer G, Bowman RW, Love GD, Padgett MJ, Ritsch-Marte M (2013) Speeding up liquid crystal SLMs using overdrive with phase change reduction. Opt Express 21:1779–1797
Tidwell TL, Gregory JT, Chalfant III CH, Orlando FJ, Leftwich M (2012) Rapid acquisition, pointing and tracking optical system for free space optical communications. US Patent 8,160,452 B1
Wang D, Zhou EH, Brake J, Ruan H, Jang M, Yang C (2015) Focusing through dynamic tissue with millisecond digital optical phase conjugation. Optica 2(8):728–735
Weichel H (1990) Laser beam propagation in the atmosphere: Tutorial texts in optical engineering: v. TT3.In: SPIE Optical Engineering Press
Xiong Y, Feng H (2003) Applications of adaptive optics in lunar laser ranging. In: Proceedings of the SPIE 4839, Adaptive Optical System Technologies II. https://doi.org/10.1117/12.456989
Zel'dovich B, et al (1995) Connection between the wave fronts of the reflected and exciting light in stimulated Mandel'shtem-Brillouin scattering. In: Landmark Papers on Photorefractive Nonlinear Optics, p 303–306

Chapter 10
Diode Laser-Sustained Plasma for Continuous-Wave Laser Propulsion

Makoto Matsui

Abstract This chapter delves into the promising method of laser-sustained plasma (LSP) for continuous-wave laser propulsion, with a particular emphasis on high-power diode lasers. LSP, a stationary thermal plasma generation approach, offers benefits like clean flows and the availability of active gases, as compared to arc discharge plasma and other electrodeless heating methods. Historically, the high costs and maintenance associated with CO_2 lasers for LSP generation hindered its industrial adoption. However, the emergence of high-power diode and diode-pumped fiber lasers, characterized by better maintainability, lifespan, and cost-effectiveness, paves the way for more accessible LSP generation. Experimental findings, including temperature measurements and fractional absorption and radiation loss, further illuminate the potential of this technique.

Keywords Laser-sustained plasma · Continuous-wave laser propulsion · High-power diode lasers · Electrodeless heating · CO_2 lasers · Diode-pumped fiber lasers · Temperature measurements · Emission spectroscopy · Laser-supported combustion wave

Introduction

Recently, there has been an increasing demand for the launch of nanosatellites weighing less than 1 kg. This is due to the government-led entry of private companies and research institutes, such as universities, into space development. Most of the current nanosatellite launches are piggyback launches, in which the nano-payload is carried on a large launch vehicle built for the original payload; however, a major challenge with the piggyback system is that the launch date, frequency, and orbit are completely dependent on the actual project. Therefore, a dedicated launch system with a low cost and a high launch frequency is required to meet the increasing

M. Matsui (✉)
Department of Mechanical Engineering, Shizuoka University, Shizuoka, Japan
e-mail: matsui.makoto@shizuoka.ac.jp

K. Mori et al. (eds.), *Beamed-mobility Engineering*, Springer Tracts in Electrical and Electronics Engineering, https://doi.org/10.1007/978-981-99-4618-1_10

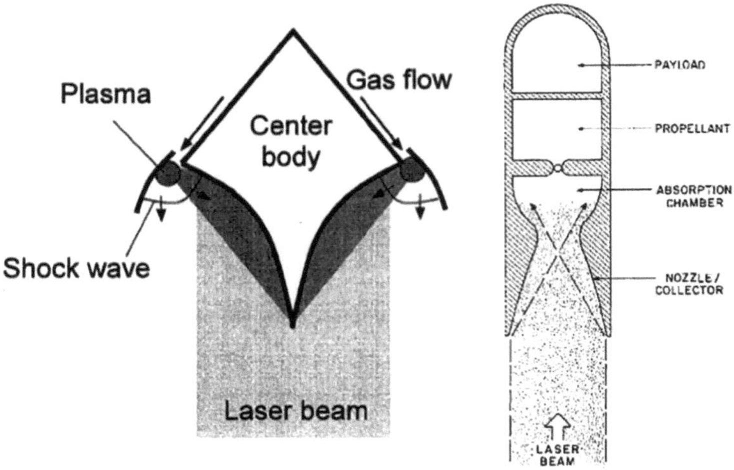

Fig. 10.1 Conceptual figures of RP-type (Myrabo 1982) (left) and CW-type (Legner and Douglas-Hamilton 1978) (right) laser propulsion

demand. An example of a nanosatellite launch system currently in operation is the Cab-3A in the USA. The Cab-3A is a system optimized for CubeSat launch. This system is characterized by launching a rocket from a supersonic aircraft. The launch cost is estimated to be approximately $50,000 per kg (Niederstrasser 2018). However, owing to the structural challenges of conventional chemical propulsion systems, including Cab-3A, launch costs cannot be dramatically decreased. Therefore, laser propulsion has been introduced as an alternative to chemical propulsion.

Laser propulsion is a space transportation system that generates thrust using a laser beam transmitted from the ground as the energy source. The concept of laser propulsion was first proposed by Kantrowitz (1972), and researchers have subsequently investigated various types of laser propulsion systems (Minovich 1972; Pirri and Weiss 1972; Pirri et al. 1974; Legner and Douglas-Hamilton 1978; Myrabo 1982; Phipps et al. 2006). Laser propulsion systems can be classified into repetitively pulsed (RP) and continuous-wave (CW) types, which use pulsed and continuous-wave laser beams, respectively. The conceptual figures for both types are shown in Fig. 10.1. For the RP type, an air-breathing method was proposed by Pirri et al. in 1972 to generate thrust using blast waves from a broken-down atmosphere (Pirri and Weiss 1972; Pirri et al. 1974); subsequently, similar methods have been studied (Ageev et al. 1980; Kare 1989; Myrabo and Messitt 1997; Lakatosh et al. 2018; Simons and Pirri 1977). In the 2000s, Myrabo et al. demonstrated a flight experiment using a 10-kW-class CO_2 laser and succeeded in launching a light craft weighing 50 g up to an altitude of 71 m (Myrabo 2001). Other methods using laser ablation (Pakhomov and Gregory 2000; Sasoh et al. 2005) or laser-driven in-tube accelerators have also been proposed (Sasoh et al. 2008; Watanabe et al. 2006; Onishi et al. 2006). The feasibility and cost of these methods have also been studied (Humble and Pierson 1995; Hong et al. 2000; Katsurayama et al. 2006).

In CW systems, Raizer et al. succeeded in generating stationary plasma called continuous optical discharge; later, this system was called laser-sustained plasma (LSP) (Raizer 1970). Legner et al. examined plasma as a heat source for propellant (Legner and Douglas-Hamilton 1978). Although several researchers have investigated hydrogen or argon LSP systems using a CO_2 laser, most studies have focused on the basic characteristics of the LSP system, and no flight test has been demonstrated as with the RP system (Jones 1980; Black et al. 1995; Kare 1995; Toyoda et al. 2000; Parkin et al. 2004; Parkin and Culick 2004; Komurasaki 2006). This was because a propellant tank was necessary for the CW system, whereas the air-breathing RP system used the atmosphere as the propellant, resulting in a much lower vehicle weight. The conventional CO_2 laser power was less than 10 kW, which is too low to demonstrate a CW flight test. Therefore, the CW system was not considered for launch from the ground but for the orbital transfer vehicle (Kantrowitz 1972).

However, with the recent development of the diode laser and fiber bundle technologies, the feasibility of a CW-type launch system is closer to reality. Through launch analysis, Kamei et al. shows that a 680 g payload can be launched by an existing 500 kW laser, and the launch cost becomes comparable to that of a conventional chemical rocket at 1000 times launch (Kamei et al. 2020). To realize a CW-type launch system, the characteristics of an LSP should be clarified.

LSP

Background of LSP

LSP was theoretically studied by Raizer in 1970 as a continuous optical discharge (Raizer 1970) and experimentally demonstrated by Generalov et al. in the same year using Xenon (Generalov et al. 1970). LSP is generated by irradiating a laser beam onto an initial plasma source via laser ablation using pulsed laser plasma or metal rods (Black et al. 1995; Toyoda et al. 2000). The laser absorption process is dominated by inverse bremsstrahlung radiation absorption (Raizer 1980). The LSP was maintained at a steady state in the direction of the flow velocity, where the energy was balanced with the laser-supported combustion wave propagating in the direction of laser incidence (Jackson and Nielsen 1974; Boni and Su 1974; Surzhikov 2000). A conceptual figure of an LSP is presented in Fig. 10.2.

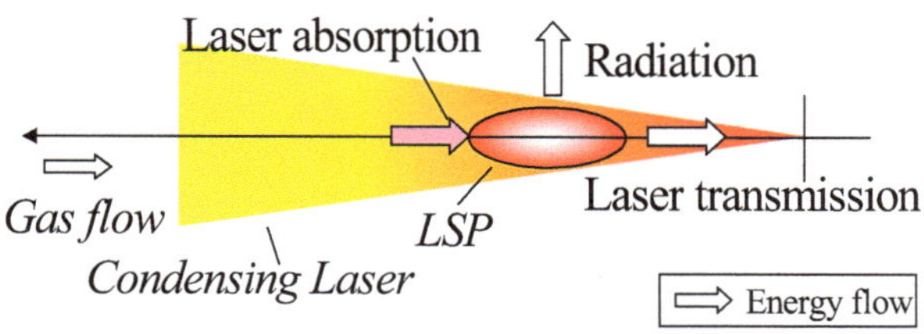

Fig. 10.2 Conceptual figure of LSP

Table 10.1 Specifications of high-power lasers

Laser source	CO_2 laser	Fiber laser	Diode laser
Maximum power	20 kW	500 kW	>1 MW
Energy conversion efficiency	10%	25%	50–80%
BPP[a] (5 kW)	6	2.5	100
Flour area (5 kW)	3 m^2	1 m^2	0.1 m^2
Wavelength	10.6 μm	1.07 μm	0.8–1 μm
Optics	Zn–Se	Glass	Glass

[a]*BPP* beam product parameter

LSP generation is a promising method for stationary thermal plasma generation (Matsui et al. 2009, 2012, 2013, 2014). Compared with arc discharge plasma, LSP has the advantages of electrodeless heating methods, such as clean flows and the availability of active gases. Compared with other electrodeless heating methods, such as inductively coupled plasma and microwave plasma, LSP can be stably generated at a pressure higher than the atmospheric pressure. Furthermore, it makes the high-temperature core region available for any purpose because the LSP can be easily moved by changing the focal lens position through the window, whereas arc discharge is sustained between the electrodes.

However, the LSP has hardly been applied to industries because kilowatt-class high-power lasers for LSP generation were formerly limited to CO_2 lasers, which have high initial and running costs, difficult maintenance due to the gas laser, low wall-plug efficiency (10%), large size, an invisible beam, and toxic nonclear Zn–Se optics because of their infrared wavelengths.

Recently, high-power diode and diode-pumped fiber lasers have been developed as alternatives to CO_2 lasers (Kan et al. 2008). Table 10.1 shows a comparison of the specifications of CO_2, fiber, and diode lasers (IPG Photonics n.d.). Compared with CO_2 lasers, the alternatives are much easier to maintain and last longer (>10,000 h). In addition, quartz-based optics are available in these lasers because of their near-infrared wavelength, resulting in low-cost optical systems and ease of alignment. In terms of the energy coefficient and installation cost, diode lasers are the optimum selection. Additionally, light sources well above the MW class can be realized by arraying them.

Principle of Laser Absorption Process

The absorption of the laser beam is expressed by the Beer–Lambert following Beer–Lambert law (Demtroder 1996).

$$dI = -kdx. \tag{10.1}$$

The integration of Eq. (10.1) over the absorption path x yields the following equation:

$$I(x) = I_0 e^{-kx}, \tag{10.2}$$

where I_0 and k denote the initial laser intensity and absorption coefficient, respectively. In LSP, the laser beam is mainly absorbed through the inverse bremsstrahlung radiation process. The absorption was caused by collisions between free electrons and heavy particles, which included neutral particles and ions. The absorption coefficient of the inverse bremsstrahlung radiation is expressed as the following equations (Kemp and Lewis 1988):

$$k_{\mathrm{en}} = \frac{k_B{}^2 T^2 A(T) \lambda^3}{hc} n_{\mathrm{e}} n_{\mathrm{n}} \left[1 - \exp\left(-\frac{hc}{\lambda k_{\mathrm{B}} T}\right)\right], \tag{10.3}$$

$$k_{\mathrm{ei}} = \sigma_{\mathrm{ei}} n_{\mathrm{e}} n_{\mathrm{i}} G_{\mathrm{B}} \left[\exp\left(\frac{hc}{\lambda k_{\mathrm{B}} T} - 1\right)\right], \tag{10.4}$$

$$\sigma_{\mathrm{ei}} = \frac{4}{3} \left(\frac{2\pi}{3 m_{\mathrm{e}} k_{\mathrm{B}} T}\right)^{\frac{1}{2}} \frac{z^2 e^6 \lambda^3}{hc^4 m_{\mathrm{e}}} \frac{1}{(4\pi\varepsilon_0)^3}. \tag{10.5}$$

Here, k_{en} and k_{ei} denote the absorption coefficients due to collision between electrons and neutral particles and between electrons and ions, respectively. The variables n_{e}, n_{n}, n_{i}, T, and λ denote the electron number, neutral particle, ion densities, temperature, and laser beam wavelength, respectively. The constants k_{B}, h, c, m_{e}, e, and ε_0 denote Planck's constant, mass of an electron, elementary charge, and electric constant, respectively. The factor $A(T)$ relating to argon was reported by John expressed as (John 1975)

$$A(T) = 5.74572 \times 10^{-38} \times T^{0.56646}. \tag{10.6}$$

The gaunt factor is calculated as (Demtroder 1996)

$$G_{\mathrm{B}} = 1.04 + 3.74 \times 10^{-5} \times T + 3.28 \times 10^{-10} \times T^2. \tag{10.7}$$

Assuming thermal equilibrium condition, the electron number density of the plasma n_{e} is expressed by the following Saha formula (Chen 2016):

$$n_{\mathrm{e}} \frac{n_{\mathrm{i}}}{n_{\mathrm{n}}} = 2 \frac{Z_{\mathrm{i}}}{Z_{\mathrm{n}}} \left(\frac{2\pi m_{\mathrm{e}} k_{\mathrm{B}} T}{h^2}\right)^2 \exp\left(-\frac{E_{\mathrm{j}}}{k_{\mathrm{B}} T}\right), \tag{10.8}$$

where Z is the distribution function expressed as

$$Z(T) = \sum\nolimits_{j=1}^{\infty} g_j \exp\left(-\frac{E_j}{k_B T}\right). \tag{10.9}$$

Here, E_j is the excited energy of state j, and g_j is the statistical weight of state j. Because Eq. (10.9) is an infinite series, and there are an infinite number of states, it is necessary to determine the upper limit of the state. When the Eq. (10.10) holds, further ranks can be neglected (Cambel 1963). In this case, a temperature range of up to approximately 30,000 K was considered.

$$\frac{\varepsilon_j}{k_B} >> T. \tag{10.10}$$

Assuming perfect gas on atoms, ions, and electrons, pressure p is expressed as follows:

$$p = (n_n + n_i + n_e) k_B T. \tag{10.11}$$

Considering the only single ionization, the relationship between electron number and ion densities becomes

$$n_e = n_i. \tag{10.12}$$

Therefore, the electron number density was calculated using Eqs. (10.8), (10.11), and (10.12). Finally, the absorption coefficients for pressure and temperature can be calculated using the above equations.

Figure 10.3 shows the typical absorption coefficient of argon. As the absorption coefficient strongly depends on the laser wavelength, the absorption coefficient of the near-infrared diode laser is two orders of magnitude lower than that of the infrared CO_2 laser at the same pressure and temperature. To compensate for the lower absorption coefficient, the pressure should be increased by more than one order of magnitude. This is because there are only a few reports on LSP generation using near-infrared lasers (Smith 2008; Bezel et al. 2015; Zimakov et al. 2016a, b; Matsui et al. 2019). In addition, owing to the lower beam quality, LSP generation using a diode laser is more challenging than that using other near-infrared lasers.

Experimental Apparatus

The experimental apparatus for the LSP generation system is illustrated in Fig. 10.4 (Nishimoto and Matsui 2016; Nishimoto 2016). A Xenon lump (USHIO Inc., UXL-500D) was used as the target. The lump was filled with 1 MPa Xenon, and a high-electron density was generated by the arc discharge plasma with an input power of less than 500 W. A continuous-wave diode laser bar module (HAMAMATSU Inc., LE0351) was used as the laser source. The maximum power and wavelength

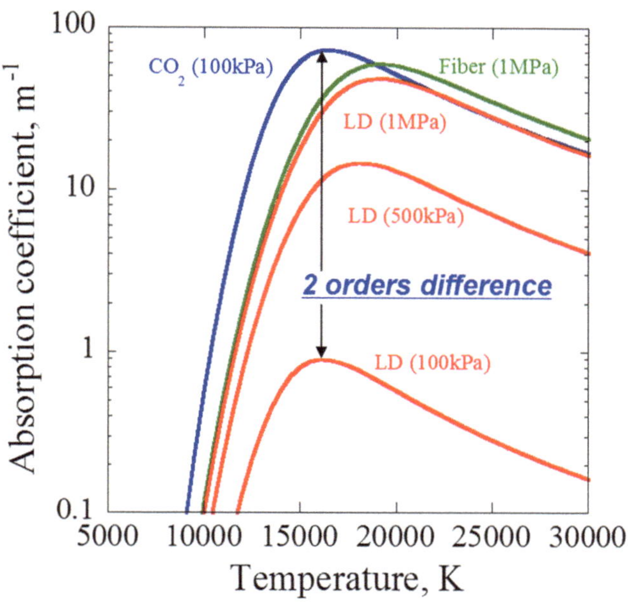

Fig. 10.3 Absorption coefficient for temperature and pressure

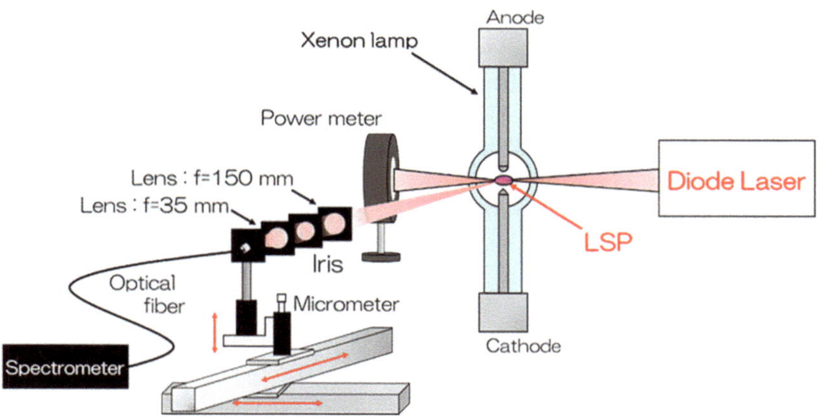

Fig. 10.4 LSP generation system

were 1.1 kW and 960 nm, respectively. The laser beam was focused on the center between the lump electrodes at a distance of 4 mm. The focal length was 100 mm. The laser was a stacked bar array diode laser, and the laser beam was directly collimated by fast- and slow-axis collimator lenses and then focused. The full width at half maximum at the focal point of the focal lens was 0.35 mm along the fast axis and 1.25 mm along the slow axis, as shown in Fig. 10.5. For safety reasons, the lump was covered with lump housing to trap the reflected laser beam. The laser

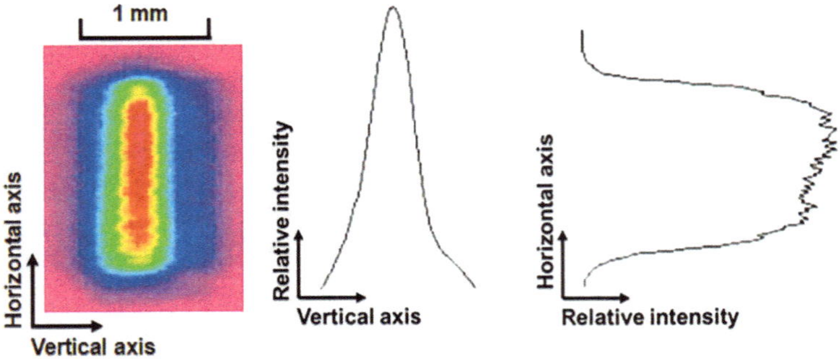

Fig. 10.5 Beam profile of the diode laser at focal point

beam was led through a quartz window to protect the diode laser from lump burst. The transmitted laser power and radiation power from the LSP were measured using a power meter (OPHIR Inc., 1000W-BB-34-V3).

The plasma film was recorded using a high-speed digital camera (Sony Corp., DSC-RX100M4) with a frame rate of 960 fps. The time variation of the plasma emission was measured using a high-speed photodetector (Thorlab Inc., DET10A/M) with a rising time of 1 ns. Data were recorded using an oscilloscope (Keystone, DSOX3104A) with a sampling frequency of 1 GHz.

Emission spectroscopy was used to estimate the LSP temperature. The emission was introduced to an optical fiber using a two-lens focusing system, whose focal lengths were 35 and 150 mm, and was detected by a spectrometer (Ocean optics Inc., HR4000). The detectable wavelength ranged from 200 to 1100 nm, and the wavelength resolution was 0.75 mm. The sensitivity was calibrated using a standard light source (Newport, 63,358), and the wavelength was calibrated using a mercury–argon lamp (Ocean optics, HG-1). The focusing system was mounted in a three-dimensional movable state. The origin was determined to be the strongest emission point, and the spatial resolution was 1 mm.

Results and Discussion

LSP Generation

First, arc discharge was generated in the Xenon lump at a minimum input power of 250 W to avoid exceeding the lump's pressure capacity with the laser beam injection. The laser beam was then focused on the arc discharge plasma, and finally, the arc discharge was turned off. Photographs of (a) arc discharge only, (b) laser beam injection to arc discharge, and (c) laser injection only are shown in Fig. 10.6. As seen in these images, the plasma was maintained without arc discharge. Therefore, the LSP can be successfully generated. The shape of the LSP was longer in the laser beam direction and slightly shorter in the arc discharge direction. For LSP

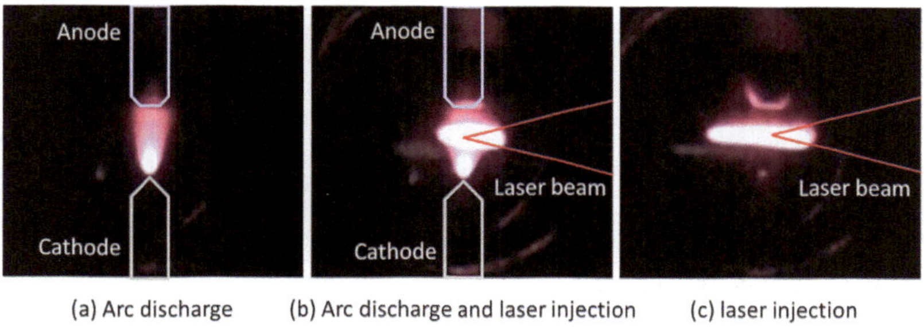

Fig. 10.6 Photos of (**a**) arc discharge, (**b**) laser beam injection to arc discharge, and (**c**) only laser injection

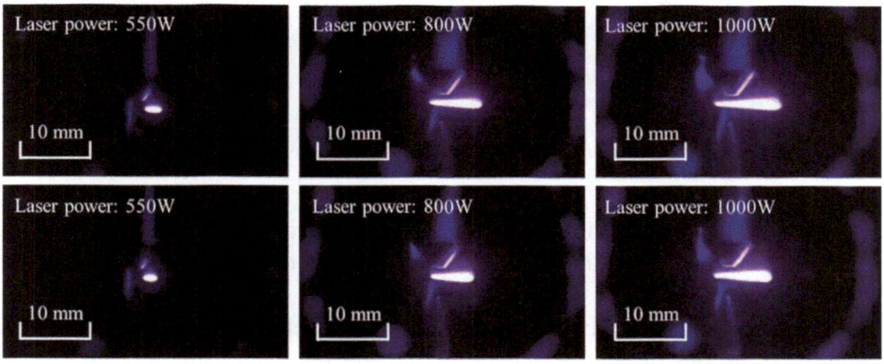

Fig. 10.7 Maximum (upper) and minimum LSP (lower)

generation, a laser-supported combustion (LSC) wave propagating in the direction of the inverse laser beam must balance the flow velocity. Convection flow is then induced in the direction of the laser beam. This was because the LSP was elongated in the direction of the laser beam.

The minimum laser power required to generate LSP was 526 ± 4 W. This value is one order of magnitude higher than that of a fiber laser (Bezel et al. 2015). Because the wavelength of the fiber laser, which is 1070 nm, is close to that of the diode laser of 960 nm, the absorption coefficient through the inverse bremsstrahlung radiation process is of the same order as that of the diode laser. In the case of fiber lasers, other absorption processes, such as bound-free or bound-bound line transitions, may be significant (Matsui et al. 2019).

Figure 10.7 shows the photos of the LSP at laser powers of 550, 800, and 1000 W. As shown in Fig. 10.2, the LSP changed in the direction of the laser beam, and the emission region of the LSP expanded and contracted periodically. The upper and lower images show the maximum and minimum emission regions, respectively. The emission region of the LSP increases with the increase in the laser power, and the maximum length of the emission region is approximately 9.7 mm.

Figure 10.7 shows the photos of the LSP at laser powers of 550, 800, and 1000 W. As shown in this figure, the LSP expanded and contracted periodically in the laser

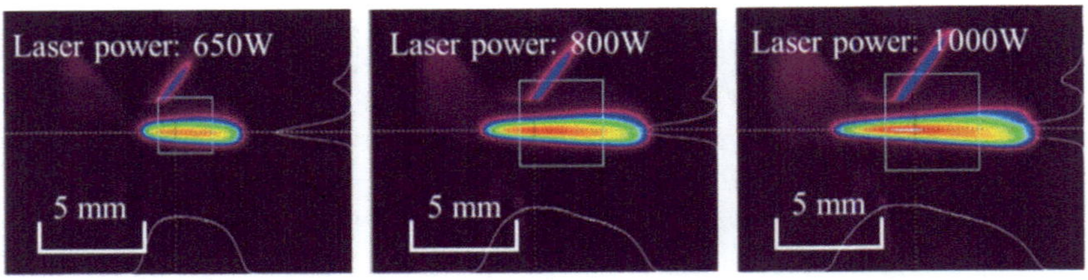

Fig. 10.8 Image in the direction of slow axis

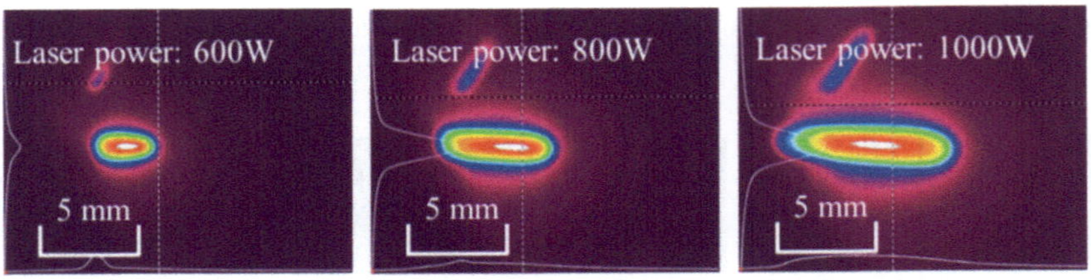

Fig. 10.9 Image in the direction of the slow (upper) and fast (lower) axes

beam direction. The emission region of the LSP increases with an increase in the laser power, and the maximum length of the emission region is approximately 9.7 mm.

For more details concerning the LSP emission intensity distributions, the LSP from the directions of the fast and slow axes was observed using a beam profiler (LBP2-VIS, New Port). Figures 10.8 and 10.9 show the image of the LSP in the direction of the slow and fast axes, respectively. From the images, the strongest emission intensity region in each axis is at the front and extends from the center to the back. This may be due to differences in thickness depths. The LSP is optically thick, and the emission around the plasma core is absorbed by the surrounding plasma. Therefore, from observations of the slow axis, the emission in the back region may be self-absorbed.

Fluctuation of LSP

The measurement video shows that the emission region of the LSP expanded and contracted in the direction of the laser beam. To clarify this fluctuation, fast Fourier transform (FFT) analysis was applied to the measured emission intensity using a high-speed photodetector. Figure 10.10 shows the power spectra of the emission intensity from the arc discharge and LSP at a laser power of 650 W. Here, the resolution frequency was 0.6 and 3 Hz for frequencies less than 500 Hz and between 500 and 5 kHz, respectively. The frequency peak of the arc discharge is 120 Hz, followed by its multiples. Thus, the fluctuation is thought to be caused by the power source frequency of 60 Hz. However, the power spectra of the LSP emission signal

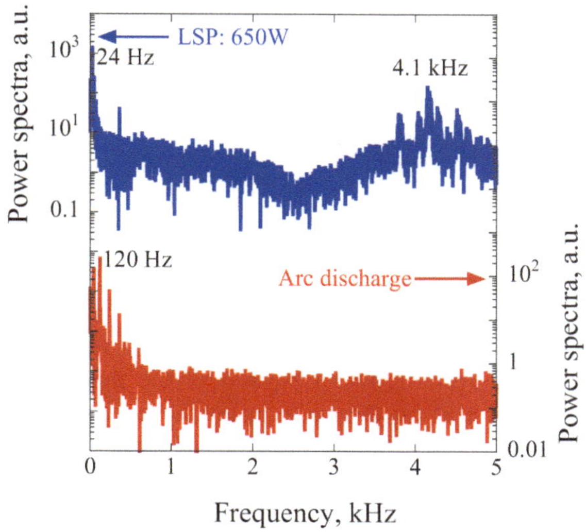

Fig. 10.10 Power spectra of arc discharge and LSP

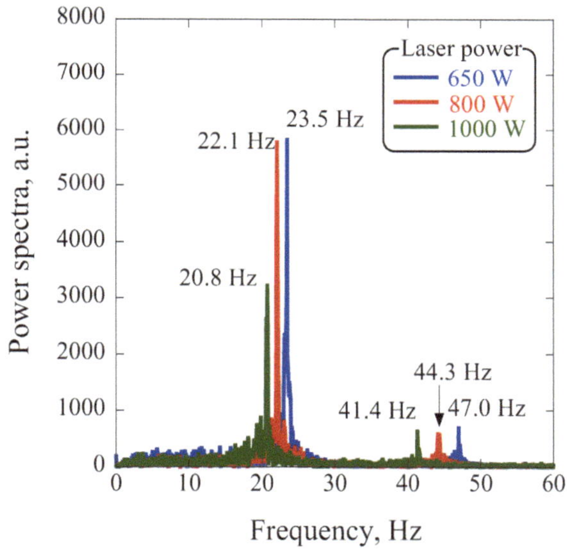

Fig. 10.11 Power spectra as a function of laser power

have no peaks originating from the power source. Instead, they exhibit strong peaks at a frequency of 24 Hz, followed by a frequency of approximately 4 kHz. The former low frequency, which corresponds to the fluctuation observed in the video, is caused by the convection flow described in the previous subsection.

The emission signals were measured by obtaining a large number of samples as a function of laser power. Figure 10.10 shows the measured power spectra with a frequency resolution of 0.07 Hz. The frequency peak decreased with the laser power. Because the velocity of the LSC wave is approximately proportional to the laser intensity, the result seems contradictory. However, as shown in Fig. 10.11, the LSP front moved far from the focal point. Then, the laser intensity at the LSP position may be inversely proportional to the laser power. Therefore, the convection velocity balanced with the LSC wave decreases with laser power, resulting in a decrease in the fluctuation frequency.

Temperature Measurement

The temperature of the LSP was estimated using emission spectroscopy (Nishimoto et al. 2016). Figure 10.12 shows typical the emission spectra of the LSP. The emission spectra were observed in the near-infrared and ultraviolet regions of Xe I, W I as an electrode material, and the whole area due to bremsstrahlung radiation. In this study, assuming that the LSP is in thermochemical equilibrium for a high pressure of over 1 MPa, the plasma temperature was estimated by the Boltzmann plot method.

The line–pair method can be employed to measure the temperature from the relative ratio of the emission coefficients of the atomic spectral lines. Assuming local thermal equilibrium, the atomic number density in each state obeys the Boltzmann distribution, expressed by the following equation:

$$\frac{n_\mathrm{j}}{n} = \frac{g_\mathrm{j}}{Z(T)} \exp\left[-\frac{E_\mathrm{j} - E_\mathrm{i}}{k_\mathrm{B}T} \right]. \tag{10.13}$$

In contrast, the emission coefficient of the spectral line is expressed by the following equation:

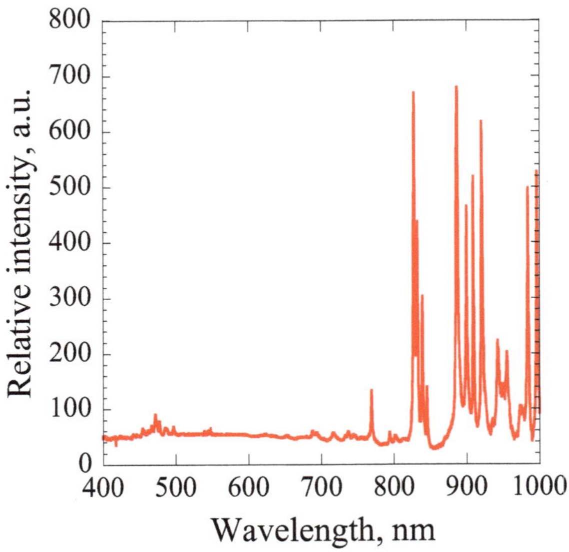

Fig. 10.12 Typical emission spectra

Table 10.2 Atomic data of xenon

λ, nm	A_{ij}, s^{-1}	$E_i - E_j$, eV	$g_i - g_j$
823.1634	2.86×10^7	8.315–9.821	5–5
828.0116	3.69×10^7	8.437–9.933	3–1
834.6822	4.2×10^7	9.570–11.055	3–5
840.919	3.06×10^6	8.3153–9.789	5–3

$$\varepsilon_{ji} = \frac{1}{4\pi} h\nu_{ji} A_{ji} N_j = \frac{1}{4\pi} \frac{hc_0}{\lambda_{ji}} A_{ji} N_j. \tag{10.14}$$

Substituting Eq. (10.13) into Eq. (10.14), we obtain

$$\varepsilon_{ji} = \frac{1}{4\pi} \frac{hc_0}{h_{ji}} \frac{g_j}{Z(T)} A_{ji} N \exp\left(-\frac{E_j}{k_B T} \right). \tag{10.15}$$

Rearrangement of the term in Eq. (10.15) yields the following equation:

$$\frac{\varepsilon_{ji} \lambda_{ji}}{A_{ji} g_j} = \exp\left(-\frac{E_j}{k_B T} \right) \frac{1}{4\pi} \frac{hc_0 N}{Z(T)} = K \exp\left(-\frac{E_j}{k_B T} \right), \tag{10.16}$$

where K is a common constant for all spectral lines of the same temperature and atom. By taking the logarithm of Eq. (10.16),

$$\ln \frac{\varepsilon_{ji} \lambda_{ji}}{A_{ji} g_j} = -\frac{E_j}{k_B T} + \ln K. \tag{10.17}$$

Plotting the left side of the above equation on the vertical axis and E_j on the horizontal axis produces a gradient that is proportional to $-1/k_B T$. The temperature was determined using this relationship.

Table 10.2 shows the atomic data of Xenon used in the Boltzmann plot. The spectra in the ultraviolet region were removed because the intensity was weak and, therefore, could not be separated owing to the low wavelength resolution of the spectrometer used in this experiment. Figure 10.11 shows an example of the Boltzmann plot. The value of each spectrum is obtained by conducting line fitting for the bremsstrahlung radiation and superpositioning of multi-peak fitting, assuming each Xe I spectrum as a Lorentz function. Figures 10.12 and 10.13 show the temperature and emission intensity distributions integrated for the entire wavelength region as a function of the laser power, respectively. The error bar represents the standard deviation of the five data points. As shown in the figures, the temperature distribution is smoother than that of the emission intensity. This is because the emission intensity is proportional to the fourth power of the temperature according to the Stefan–Boltzmann law, and it is reactive to temperature changes. The temperature increased slightly with an increase in the laser power, and the maximum temperature observed was in the range of 8800 to 9200 K (Figs. 10.14 and 10.15).

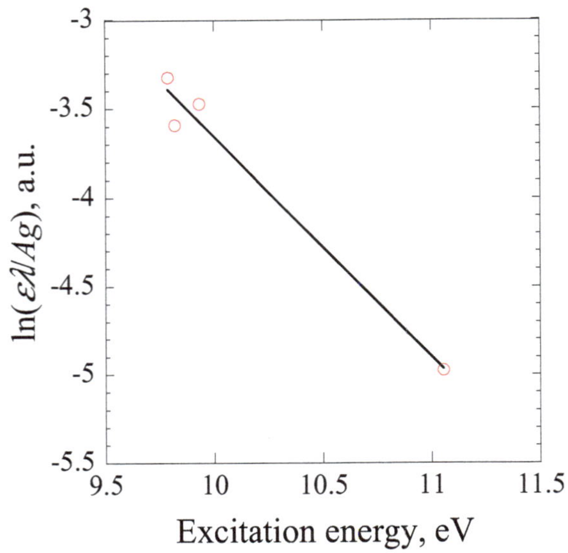

Fig. 10.13 Measured Xe I with liner fitting

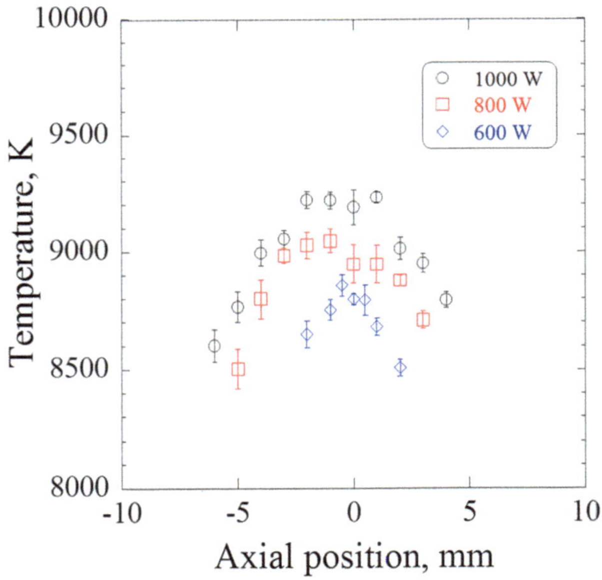

Fig. 10.14 Temperature distribution as a function of laser power

Fractional Absorption and Radiation Loss

To investigate the extent to which the laser beam was absorbed by the LSP, the transmitted beam intensity was measured using a power meter. Figure 10.16 shows the fractional absorption of the laser beam and net input power. The fractional absorption is 0.20 at a laser power of 538 W in the vicinity of the plasma generation

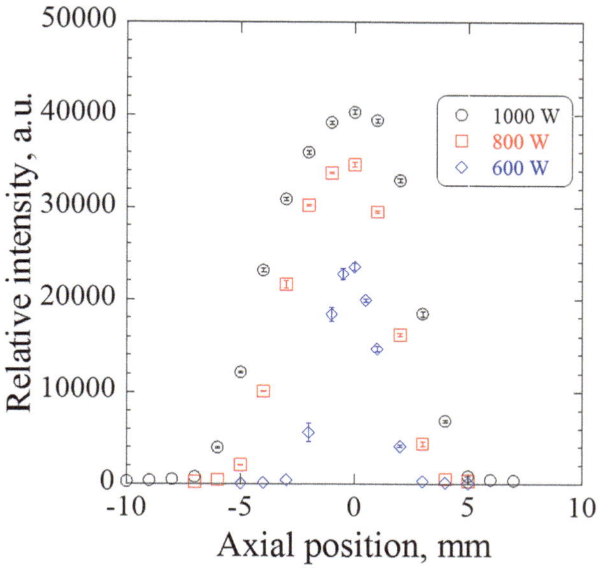

Fig. 10.15 Emission intensity distribution as a function of laser power

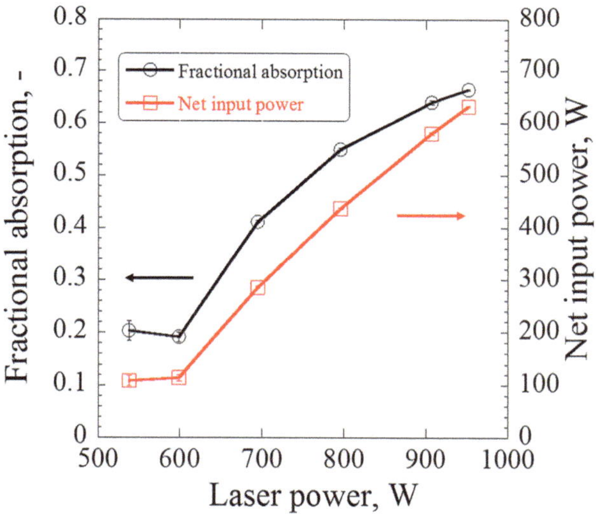

Fig. 10.16 Fractional absorption and net input power

threshold. Then, it reaches 0.66 at a laser power of 952 W. This is because the plasma length in the direction of the laser beam axis is directly proportional to the laser power, and the absorption length increases. The minimum net input power is 109 W, which is lower than the laser power of the generation threshold. Therefore, by decreasing the spot size and increasing the laser intensity, the LSP generation threshold is expected to decrease considerably owing to an increase in absorption.

To investigate the ratio of radiation loss to laser power, the radiation intensity of the LSP was measured using a power meter. Total radiation was obtained by integrating the measured radiation intensity with the solid angle. Figure 10.17

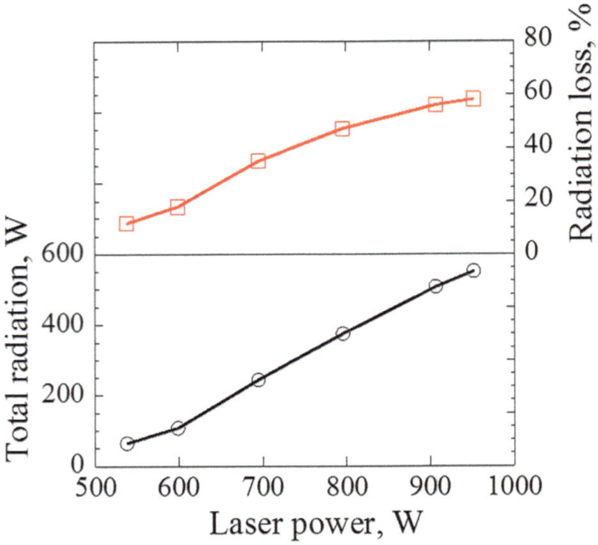

Fig. 10.17 Total radiation and radiation loss as a function of laser power

shows the total radiation and radiation loss. The total radiation and radiation loss are directly proportional to the laser power. At its maximum, the radiation loss reached 58%. Other losses are expected from the transmission of the laser beam and heat conduction.

Conclusion

The LSP was successfully generated using a 1 kW class diode laser, and the following results were obtained:

1. At a Xenon pressure of 1 MPa, the minimum laser power required to generate LSP was 526 ± 4 W.
2. The LSP expanded and contracted at a frequency of approximately 22 Hz along the direction of the laser beam.
3. The fluctuation was not caused by the power source but by the induced convection flow balanced with the laser-supported combustion wave.
4. The LSP temperature distribution was measured by emission spectroscopy, and it was found to have a gentle distribution in the direction of the laser incidence axis and increase slightly with an increase in the laser power. The maximum temperature measured was 9200 K.
5. Laser light absorption increased with an increase in the laser power, ranging from 0.20 to a maximum of 0.66.
6. The laser power required to produce the LSP was 538 W; however, the net laser absorption was 109 W.

Acknowledgments This research was partially supported by the Ministry of Education, Science, Sports, and Culture Grant-in-Aid for Young Scientists (A) 26709065. The authors gratefully acknowledge Mr. Koji Nishimoto and Takahiro Ono (past members of our laboratory) for their work in this research.

References

Ageev VP, Barchukov AI, Bunkin FV, Konov VI, Korobeinikov VP, Putjatin BV, Hudjakov VM (1980) Experimental and theoretical modeling of laser propulsion. Acta Astronaut 7:79–90

Bezel I, Delgado G, Derstine M, Gross K, Solarz R, Shchemelinin A, Shortt D (2015) High power laser-sustained plasma lightsources for KLA-Tencor broadband inspection tools. In: 2015 CLEO

Black J, Krier H, Glumb RJ (1995) Laser propulsion 10-kW thruster test program results. J Propuls Power 11:1307–1316

Boni AA, Su FY (1974) Propagation of laser supported deflagration wave. Phys Fluids 17:340–342

Cambel AB (1963) Plasma physics and magnetofluid mechanics. McGraw-Hill

Chen F (2016) Introduction to plasma physics and controlled fusion. Springer Cham Heidelberg

Demtroder W (1996) Laser spectroscopy, 2nd edn. Springer Verlag, Berlin

Generalov NA, Zimakov VP, Kozlov GI, Masyukov VA, Raizer YP (1970) Continuous optical discharge. Sov JETP Lett 11:302–304

Hong ZC, Hsu FC, Chem JS (2000) Vertical ascent to geosynchronous orbit with constrained thrust angle. J Spacecraft Rockets 37:64–70

Humble WE, Pierson BL (1995) Maximum-payload trajectories for a laser-propelled launch vehicle. J Guid Control Dyn 18:1259–1266

IPG Photonics (n.d.). https://www.ipgphotonics.com. Accessed 13 June 2019

Jackson JP, Nielsen PE (1974) Role of radiative transport in the propagation of laser supported combustion waves. AIAA J 12:1498–1501

John TL (1975) The free-free transitions of atomic, molecular negative ions in the infrared. Mon Not R Astron Soc 170:5–6

Jones LW (1980) Laser propulsion-1980. AIAA Paper

Kamei T, Matsui M, Mori K (2020) Feasibility study of a laser launch system for picosatellites and nano-satellites in low-earth orbits. Acta Astronaut 176:124–130

Kan H, Miyajima H, Zheng Y, Furuta S, Fujiwara H, Suzuki H, Torii K, Kageyama N, Uchiyama T, Oishi S, Watanabe A, Miyamoto M, Kanzaki T, Hiruma T (2008) Present status and future prospects of high power laser diode arrays. Laser Rev 36:268–272

Kantrowitz A (1972) Propulsion to orbit by ground-based lasers. Astronaut Aeronaut 10:74–76

Kare J (1989) Pulsed laser propulsion for low cost, high volume launch to orbit. In: IAF conference space power

Kare JT (1995) Laser-powered heat exchanger rocket for ground-to-orbit launch. J Propuls Power 11:535–543

Katsurayama H, Komurasaki K, Arakawa Y (2006) Feasibility for a single-stage-to-orbit launch to a geosynchronous transfer orbit by pulse laser propulsion. J JSAS 54:63–70

Kemp NH, Lewis PF (1988) Laser-heated thruster—interim report, NASA CR 1611665

Komurasaki K (2006) CW laser powered space propulsion system. Rev Laser Eng 34:414–418

Lakatosh BV, Abramenko DB, Ivanov VV, Medvedev VV, Krivtsum VM, Koshelev KN, Yakunin AM (2018) Propulsion of a flat tin target with pulsed CO_2 laser radiation: measurements using a ballistic pendulum. Laser Phys Lett 15:016003

Legner HH, Douglas-Hamilton DH (1978) CW laser propulsion. J Energy 2:85–94

Matsui M, Shinmi K, Ueno T, Komurasaki K, Arakawa Y (2009) Operation Characteristics of laser driven plasma wind tunnel. Trans JSASS Space Tech Jpn 7(ISTS26):31–39

Matsui M, Tanaka K, Nomura S, Komurasaki K, Yamagiwa Y, Arakawa Y (2012) Generation and diagnostics of atmospheric pressure CO_2 plasma by laser driven plasma wind tunnel. J Appl Phys 112:033301
Matsui M, Fukuji N, Nakano M, Komurasaki K, Arakawa Y, Goto T, Shirakata H (2013) Alumina reduction by laser sustained plasma for aluminum-based renewable energy cycling. J Renew Sustain Energy 5:039101
Matsui M, Yoneda S, Komurasaki K, Yamagiwa Y, Arakawa Y (2014) Atomic-oxygen-flow generation by laser-driven plasma wind tunnel as low-earth-orbit environment simulator. AIAA J 52(8):1806–1810
Matsui M, Ono T, Kamei T (2019) Generating conditions of argon laser-sustained plasma by disk, fiber and diode lasers. Vacuum 167:490–494
Minovich MA (1972) Reactorless nuclear propulsion–the laser rocket. AIAA Paper 719
Myrabo LN (1982) A concept for light-powered flight, AIAA Paper 82–1214
Myrabo LN (2001) World record flight of beam-riding rocket lightcraft: demonstration of "Disruptive" propulsion technology, AIAA Paper 01–3798
Myrabo LN, Messitt DG (1997) Ground and flight test of a laser propelled vehicle. AIAA J 1001:1–10
Niederstrasser C (2018) Small launch vehicles—a 2018 State of the Industry Survey. In: Proceedings of the small satellite conference, space access, SSC18-IX-01
Nishimoto K (2016) Generation of laser sustained plasma using 1kW class diode laser. Master thesis, Shizuoka University
Nishimoto K, Matsui M (2016) Generation of xenon laser sustained plasma using 1 kW class laser diode. Front Appl Plasma Tech 9(2):71–73
Nishimoto K, Ono T, Matsui M (2016) Temperature measurement of xenon laser sustained plasma by emission spectroscopy. J IAPS 24(2):79–83
Onishi N, Ogino Y, Sawada K, Ohtani T, Sasoh A (2006) Numerical simulation of laser-driven in-tube accelerator operation. AIP Conf Proc 830
Pakhomov AV, Gregory DA (2000) Ablative laser propulsion an old concept revisited. AIAA J 38: 725–727
Parkin KLG, Culick FEC (2004) Feasibility and performance of the microwave thermal rocket launcher. AIP Conf Proc 702:407–417
Parkin KLG, Didomenico LD, Culick FEC (2004) The microwave thermal thruster concept. AIP Conf Proc 702:418–429
Phipps CR, Luke JR, Helgeson W (2006) Performance test results for the lower-powered microthruster. AIP Conf Proc 830:224–234
Pirri AN, Weiss RF (1972) Laser propulsion. AIAA Paper 719
Pirri AN, Monsler MJ, Nebolsine PE (1974) Propulsion by absorption of laser radiation. AIAA J 12: 1254–1261
Raizer YP (1970) The feasibility of an optical plasmotron and its power requirements. Sov JETP Lett 11:120–123
Raizer YP (1980) Optical discharges. Sov Phys Usp 23(11):789–806
Sasoh A, Urabe N, Kim S, Jeung IS (2005) Impulse dependence on propellant condition in a laser-driven in-tube accelerator. Trans JSAS 160:63–70
Sasoh A, Jenug IS, Choi JY (2008) Access to space without energy and propellant on board. AIP Conf Proc 997
Simons GA, Pirri AN (1977) The fluid mechanics of pulsed laser propulsion. AIAA J 15:835–842
Smith DK (2008) Laser-driven light source. U.S. Patent 7435982
Surzhikov ST (2000) Numerical analysis of subsonic laser-supported combustion waves. Quantum Electro 30:416–420
Toyoda K, Komurasaki K, Arakawa Y (2000) Continuous-wave laser thruster experiment. Vacuum 59:63–72
Watanabe K, Mori K, Sasoh A (2006) Ambient pressure dependence of laser-induced impulse onto polyacetal. J Propuls Power 22:1148–1151

Zimakov VP, Kuznetsov VA, Solovyov NG, Shemyakin AN, Shilov AO, Yakimov MY (2016a) Interaction of near-IR laser radiation with plasma of a continuous optical discharge. Plasma Phys Rep 42(1):68–73

Zimakov VP, Kuznetsov VA, Solovyov NG, Shemyakin AN, Shilov AO, Yakimov MY (2016b) Interaction of near-IR laser radiation with plasma of a continuous optical discharge. Fizika Plasmy 42(1):74–80

MIX
Papier aus verantwortungsvollen Quellen
Paper from responsible sources
FSC® C105338

If you have any concerns about our products,
you can contact us on
ProductSafety@springernature.com

In case Publisher is established outside the EU,
the EU authorized representative is:
Springer Nature Customer Service Center GmbH
Europaplatz 3, 69115 Heidelberg, Germany

Printed by Libri Plureos GmbH
in Hamburg, Germany